COASTLINES OF THE
SOUTHERN NORTH SEA

This volume is part of a series of volumes on Coastlines of the World.
The papers included in the volume are to be presented at Coastal Zone '93

Volume Editors Roeland Hillen and Henk Jan Verhagen
Series Editor Orville T. Magoon

Published by the
American Society of Civil Engineers
345 East 47th Street
New York, New York 10017-2398

ABSTRACT

This proceeding, *Coastlines of the Southern North Sea*, contains papers presented at COASTAL ZONE 93, the eighth symposium on Coastal and Ocean Management held in New Orleans, Louisiana, July 19-23 1993. This volume is part the continuing series of volumes on Coastlines of the World. Some of the topics covered include environmental considerations, engineering and science; data gathering, and monitoring, legal, regulatory, and political aspects of coastal management planning, conservation and development and public information and citizen participation. This volume provides the professionals, decision-makers, and general public with a broad understanding of these subjects as they relate to the *Coastlines of the Southern North Sea*.

Library of Congress Cataloging-in-Publication Data

Coastlines of the southern North Sea / volume editor, Roeland Hillen and Henk Jan Verhagen
   p.cm.— (Coastlines of the world)
   Includes indexes.
   ISBN 0-87262-967-8
   1. Coasts—North Sea—Congresses. 2. Coastal zone management- -North Sea—Congresses. I. Hillen, Roeland. II. Verhagen, Henk Jan. III. American Society of Civil Engineers. IV. Series.
GB457.2.C63 1993
333.91'7'0916336—dc20   93-24858
                                         CIP

Cover photo courtesy of Rijkswaterstaat (NL), Survey Department

FOREWORD

Coastal Zone '93, is the eighth in a series of multidisciplinary biennial symposia on comprehensive coastal and ocean management. Professionals, citizens and decision makers met for five days in New Orleans, Louisiana, to exchange information and views on matters ranging from regional to international scope and interest. This year's theme was entitled "Healing The Coast," emphasized a recurrent focus on practical coastal problem solving.

Sponsors and affiliates included the American Shore and Beach Preservation Association, American Society of Civil Engineers (ASCE), Coastal Zone Foundation, Department of Commerce, National Oceanic and Atmospheric Administration, as well as many other organizations (see title page). The range of sponsorship hints at the diversity of those attending the Coastal Zone '93 Symposium. The presence of these diverse viewpoints will surely stimulate improved coastal and ocean management through the best of current knowledge and cooperation.

This volume of the Coastlines of the World series is included as part of the Coastal Zone '93 Conference. The purpose of this special regional volume is to focus on the coastline and coastal zone management of the *Coastlines of the Southern North Sea.*

Each volume of the Coastlines of the World series has one or more guest volume editors representing the particular geographical or topical area of interest.

All papers have been accepted for publication by the Volume Editors. All papers are eligible for discussion in the Journal of Waterway, Port, Coastal, and Ocean Engineering, ASCE.

A ninth conference is now being planned to maintain this dialogue and information exchange. Information is available by contacting the Coastal Zone Foundation, P.O. Box 279, Middletown, California 95461, U.S.A.

Orville T. Magoon
Coastlines of the World
Series Editor

PREFACE

The North Sea is a relatively small sea, with rough weather conditions, and surrounded by densely populated countries. The total volume of water in the North Sea is less than one hundredth of one percent of the salt water in the world. The sea itself, its bottom and the surrounding coasts have many functions, often with a conflicting character. Complicating the situation is the fact that the North Sea is bordered by seven countries, six of them bordering the southern part.

In historic times, many naval wars between these countries have been fought on the North Sea, starting in the Bronze Age, when the Kelts moved to Britain. Nowadays conflicts are solved in a different way: through common agreement on the use of the limited resources of the North Sea and the adjacent coastlines.

In this volume nearly sixty highlight aspects of the management of the coastlines of the southern North Sea. Experts in this field are mainly found in the governmental and semi-governmental organizations of the countries surrounding the sea. In this area we have no tradition of a regular flow of papers from government workers; papers are usually made by university scientist. Because of this fact, many aspects of management of the southern North Sea are until now only covered by governmental documents. Although these documents are available to the public, the accessibility is often a problem. Many of them are not in English, and only published as "grey-literature."

Therefore we were very happy to find national coordinators in all southern North Sea counties, who hunted for authors and (what is even a bigger problem) realized that all authors submitted their papers in time. So we are very indebted to Mr. Reginald Purnell from the Ministry of Agriculture, Fisheries and Food in the United Kingdom, to Mr. Per Roed Jakobsen (supported by Mr. Christian Laustrup) from the Danish Coastal Authority, to Mr. Bernard de Putter (supported by Mr. Peter de Wolf) from the Flemish Coastal Authority (Belgium), to Mr. Hans Kunz from the Ministry of the Environment of Lower Saxony (Germany) and to Mr. Alain Bryche of the Municipality of Dunkerque (France).

In order to realize one uniform style for all papers, a substantial load of word processing was required. This task was fulfilled in an excellent way by Ms. Thea Westerveld of the Tidal Waters Division of the Ministry of Transport and Public Works of the Netherlands.

Roeland Hillen
Henk Jan Verhagen
(Volume Editors)

CONTENTS

INTRODUCTION

GEOLOGY

COASTAL MORPHOLOGY

HYDRAULICS

WATER QUALITY

ECOLOGY

FUNCTIONAL USES OF THE COASTAL ZONE

OPERATIONAL ASPECTS OF COASTAL ZONE MANAGEMENT

CASE STUDIES IN COASTAL ZONE MANAGEMENT

The southern North Sea
Functional Uses and Institutional Arrangements

H.J. Verhagen[1], R. Hillen[2], B.G.M. van de Wetering[2]

Abstract

The southern North Sea is bordered by the United Kingdom, France, Belgium, The Netherlands, Germany and Denmark. The North Sea basin and its adjacent shorelines are intensively used. Management of the basin and the coastal zone is therefore essential. Because of the small scale of the area, the population density and the numerous interactions, international cooperation is vital.

In this paper an overview is given of both the morphologic-physical interactions and the administrative interactions between the governments involved.

A complicating factor in Coastal Zone Management around the southern North Sea is that CZM is highly connected to national spatial planning and to sea defence policy. In several of the North Sea countries this is not a subject of the national government, but is dealt with at regional level.

International agreements exist on most aspects of active use of the North Sea basin itself (navigation, oil and gas mining, fishery, etc) and its major shallow coastal areas, like the Wadden Sea. In the field of passive use (mainly pollution via rivers and the atmosphere) goals have been set through international agreements, but the majority of these goals have not yet been realized. International agreement on the recreation along the coastal strip does not yet exist. Also the standards for coastal protection against flooding by storm surges and against chronic erosion vary from each country to country.

General description of the southern North Sea area

The southern North Sea covers the continental shelf areas of Denmark, Germany, the Netherlands, Belgium and a part of the shelf areas of the United Kingdom and France (fig. 1).

The North Sea is a relatively enclosed and shallow basin with a total volume estimated at about 47,000 km^3, including the Skagerrak. The southern North Sea involves less than half of this volume. The water depth in the southern North Sea is

[1]   Int. Inst. for Hydraulic and Environmental Engineering, P.O. Box 3015, 2601 DA Delft, The Netherlands

[2]   Ministry of Transport, Public Works and Water Management, Tidal Waters Division, P.O. Box 20907, 2500 EX The Hague, The Netherlands

1

generally less than 50 metres. Over time, the sedimentation of sand and silt from the land has kept pace with the subsidence of the North Sea basin. In the northern part of the North Sea, water depths of over 500 meters occur; there the supply of erosion material was insufficient.

In the centre of the North Sea, roughly between 55° and 56° latitude lies the Dogger Bank, a shallow zone about half the size of the Netherlands. In this famous fishing area, water depths of less than 20 meters occur. During severe storms, the waves of the northern North Sea break here.

The bottom of the southern North Sea consists mainly of sandy sediments reworked by the forces of tides and waves. This is, amongst others, reflected by the undulating sands and elongated sandbanks like the Norfolk and Flemish Banks.

The water in the North Sea circulates according to a fixed pattern. Atlantic water enters through the English Channel in the south and along the Scottish and English east coast in the north. Outflowing water leaves the North Sea along the Danish and Norwegian coasts. On average, the water of the entire North Sea is refreshed every one to two years. Stratification of the water in the North Sea is restricted to the deeper parts (generally more than 30 meters deep) and to the summer period.

The various water masses in the southern North Sea are often separated by so-called fronts, transitions within a few kilometres distance. These transitions are measured in terms of salinity, temperature, nutrients and contaminants. During summer these fronts are more pronounced than during winter. Well known fronts in the southern North Sea are the Frisian Front, separating water from the Channel and the Atlantic Ocean, and the Danish Front, marking the boundary between 'continental' water and central North Sea water.

Several rivers from the European Continent and the United Kingdom flow into the southern North Sea. Total river inflow is estimated at 300 km³ per year. The most important rivers are the Scheldt (F,B,NL), Rhine (F,G,NL), Meuse (F,B,NL), Eems (G,NL), Weser (G), Elbe (G), Humber (UK) and Thames (UK). These rivers carry mud, nutrients and contaminants to the North Sea. Mixing of river and sea water occurs only gradually and over long distances. Water from the rivers Rhine and Elbe, for example, can be traced as far as northwest Denmark (De Ruyter et al, 1987).

Use of the basin and the coastal zone

The coastal areas surrounding the southern North Sea are densily populated. The highest population densities (> 1000 inhabitants per km²) are found along the coastlines of the Netherlands and Belgium. The number of inhabitants living in the entire catchment area of the southern North Sea is estimated at 150 million (NSTF, in prep). About 30 to 40 million people live in the lowlying areas fringing the southern North Sea.

For a description of the functional uses a distinction can be made between the offshore area, the adjacent shallow waters (like the Wadden Sea and the estuaries) and the surrounding land. This distinction in three categories is also recognized in (inter)national regulations, research and planning.

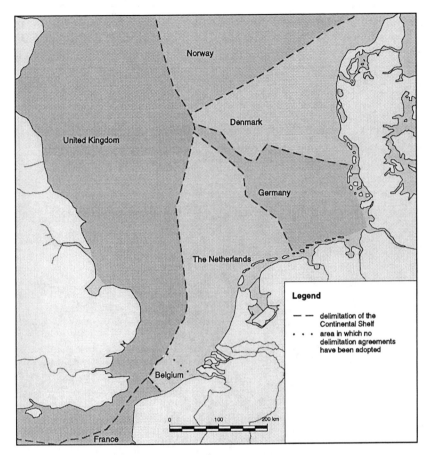

Figure 1:   The southern North Sea area (after ICONA, 1992).

Use of space in the southern North Sea
An impression of the present use of space of the southern North Sea is, among others, given in the North Sea Atlas (ICONA, 1992) and in the Quality Status Report of the North Sea Task Force (NSTF, in prep).
The southern North Sea is used intensively, mainly for shipping, dumping, fishing, offshore industry, pipelines and cables, military activities and extraction of raw materials (e.g. sand and gravel). As an example, an overview of the use of maritime space on the Netherlands Continental Shelf in 1990 is given in figure 2. Some of the functional uses will be discussed briefly in this chapter. More detailed information is given in other contributions to this volume.

The North Sea, and especially the southern part of it, is one of the world's busiest seas for shipping with over 400,000 shipping movements per year. There is

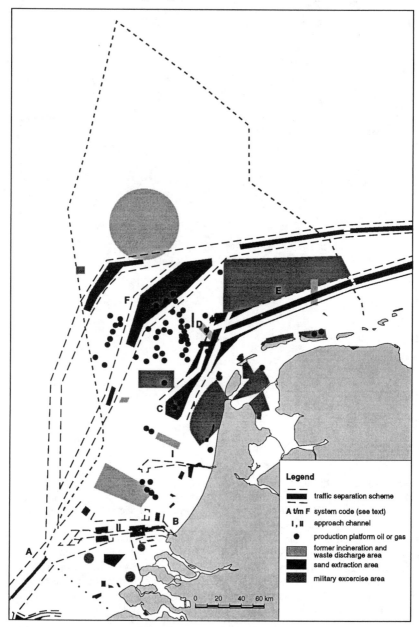

Figure 2:   The use of maritime space on the continental shelf of the Netherlands in 1990
              (after ICONA, 1992).

particularly heavy traffic through the Strait of Dover with approximately 150 ships per day sailing in each direction. In addition to that there are about 300 ferry movements per day on average. To ensure a smooth and safe movement of the shipping traffic, routing systems (i.e. separation schemes and deep water routes) are used since the late 1960s. The most important deep water channels lead from the Strait of Dover to the German Bight. Major harbours also have deep water approach routes, such as the Euro-channel to Rotterdam and the IJ-channel to Amsterdam. When required, approach routes are maintained at the desired depth through dredging.

The southern North Sea is rich in nutrients and contains a varied fish population. The rapid increase in catches over the past decades resulted in a threat to certain fish populations, for instance that of the herring in the 1970s (Popp Madsen & Richardson, 1993). To prevent the exhaustion of species, a fishing policy has been developed by the European Community. This policy includes total allowable catches (TAC) for certain species as well as quota-systems for the individual EC-Member Countries.

Exploration of the natural gas and oil reserves in the North Sea started in the 1960s. In the southern North Sea major gas fields are present on the continental shelves of the United Kingdom and the Netherlands. An overview of the production platforms and pipelines is given in figure 3. In the northern North Sea both gas and oil fields of importance are present on the Norwegian and UK shelves.
On the Dutch sector, 65 platforms produced about 19 billion cubic metres of gas and 2.5 million tonnes of oil in 1991. The total length of pipelines is about 1400 km. In the UK sector (entire North Sea), there are about 150 production platforms producing 84 million tonnes of oil and 45 billion cubic metres of natural gas. Total length of pipelines in the UK waters is over 5000 km (NSTF, in prep).

An extensive network of telecommunication cables exists between the coastal states bordering the southern North Sea. Since the beginning of this century, some 30 cable links were established, the greater majority of these connecting the United Kingdom to the European continent. Recently several older cables that are out of use have been removed (e.g. 13 on the Netherlands Continental Shelf; ICONA, 1992).

Other human uses of the southern North Sea include discharges of waste materials, both directly through dumping and indirectly via rivers or the atmosphere. Especially the rivers Elbe, Rhine, Thames and Weser, with their very densely populated and intensively used catchment areas, are major sources of contaminants and nutrients (Van de Kamer et al, 1993).
The area is also of great importance for recreational activities, especially bathing and sailing. Tourism and recreation continues to be a growth sector (OECD, in prep).

The intensive use of the southern North Sea results in various accidents and conflicts between functional uses. Examples are:
* in 1990, 10 major shipping accidents occured in the North Sea or adjacent waters. These included 5 cases of fire or explosion; 2 vessels foundered in bad weather, while structural failure, grounding and collision were each responsible for one

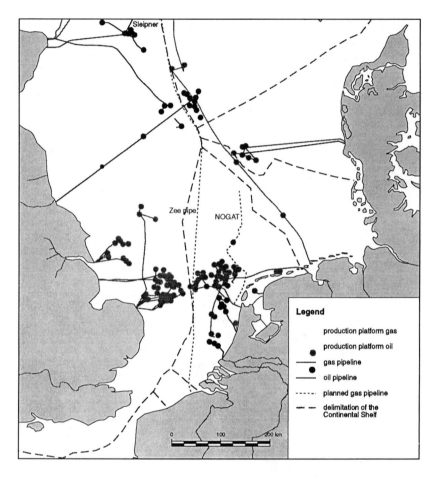

Figure 3:   Production platforms and pipelines in the Southern North Sea (after ICONA,1992)

loss. Two accidents involved major oil pollution (NSTF, in prep);
*   shipping and fishery frequently interfere with offshore industrial activities;
*   conflicts between fishery, extraction of raw materials and nature conservation are
    well known throughout the area. At present protection zones on the Dutch sector
    of the North Sea are being considered.
From the above it can be concluded that organisation and regulation per user function
is insufficient for the southern North Sea area. The need for comprehensive planning
and management of the area is evident. The appreciation of this fact is, amongst
others, illustrated by the international cooperation within the framework of the North
Sea Ministerial Conferences.

The shallow coastal waters at the fringes of the southern North Sea are mainly nature reserves with an important recreational function. They are also important for fishery, but mainly in an indirect way (spawning areas). These areas (e.g. the Wadden Sea, Eastern Scheldt and the Wash) are very sensitive to human action. In the past large parts of the shallow coastal waters were diked or reclaimed and subsequently used as agricultural area. Because of the overproduction of agricultural products at this moment the need for land reclamation decreases, while open water is valued higher in public opinion (Enemark, 1993).

Land use around the southern North Sea
The low lying areas surrounding the southern North Sea have been inhabited ever since several millennia ago. In the Middle Ages a considerable development started in this area. The main economic growth was in the 17th century. The inhabitants focused very much on international trade, resulting in great wealth in the area. In the 19th century industrialization started. Heavy industry requires good communication and a good infrastructure, especially for bulk goods. Railroads and river transport towards the seaports became very important. The population started to grow.

At this moment nearly all functions can be found in the land area around the North Sea. In fact the area is somewhat overused. Therefore in all countries relatively strict zoning regulations are in effect.
In the area a large number of harbours can be found, like Rotterdam, Hamburg, London, Antwerp, Amsterdam, Esbjerg, etc. Also many minor (or specialised) harbours can be found, like the ferry-harbours of Oostende and Harwich, the naval base of Den Helder and the fishing port of Hanstholm.

Concentrating on the coastal strip itself, outside the harbour areas, one can distinguish the following important functions:
* coastal defence (protection against flooding and/or erosion)
* nature reserve
* recreation (including housing)
* drinking water production (especially in the Netherlands)
Historically, coastal defence was the primary goal in the low lying countries, and all other uses were in fact prohibited. Because of this, in many areas nature was not disturbed. In less sensitive areas (from a point of view of flooding) much more human activities were allowed, resulting in more urbanised areas (e.g. in the Belgian coastal area). At this moment it is possible to maintain the required safety levels also when the coastal strip is used by other functions. This fact causes more stress on the coastal strip. Many conflicts arise between developers and nature conservationists. In order to solve these conflicts there are strict regulations of activities in nearly all areas. In general it is not allowed to develop coastal areas when this is not a part of a regional development plan. These plans have to be approved by authorities on various levels. Public opinion has a strong influence on this decision making process.
A large part of the dune area around the North Sea is at this moment appointed as nature reserve. It is the intention of most authorities to concentrate tourist development to a limited number of places.

Institutional arrangements

Since the 1960s, many bilateral delimitation agreements between the North Sea states were concluded. At present agreements on all continental shelf boundaries exists except for the boundary between the Netherlands and Belgium (fig. 1).

Within the continental shelf areas of the North Sea states several administrative zones are distinguished. Member States of the European Community are empowered to maintain an exclusive 12-miles zone in which fishing is reserved to vessels which traditionally fish in the waters concerned and operate from the ports of the zone in question. Other administrative zones are related to national or local boundaries (e.g. 1 km limit in the Netherlands) or to national legislation (e.g. mining acts). Under the MARPOL convention (marine pollution), boundaries of 25 and 50 miles from the coast apply (IMO, 1987).

During the second half of this century, population growth and economic development in coastal areas has increased rapidly. Many separate laws and regulations related to activities in coastal areas can be identified in the various coastal states of the southern North Sea area. Existing laws and institutional arrangements are generally oriented towards regulation of activities, especially through land use zoning. More recently also economic instruments (e.g. 'polluter-pays' principle, emission charges) are applied. The need for a good mixture of economic instruments and a tight regulatory framework is becoming evident.

Many coastal states aim at a more comprehensive planning and management in their coastal zones. However, at present there are no integrated coastal resource management policies in the countries bordering the southern North Sea. In many of these countries it is even difficult to identify which ministry is primarily responsible for the implementation of coastal zone management policies and plans (OECD, in prep).

In some countries of the southern North Sea area, coastal zone issues are handled by the national governments; in other countries federal states or provinces play a major role. This will be illustrated by two examples of coastal zone issues: coastal hazards (esp. flooding and erosion) and nature conservation in coastal areas.

In England the central government (through the Minister of Agriculture, Fisheries and Food) has policy responsibility for sea defence and flood protection. The Ministry pays grants to drainage authorities, such as the National Rivers Authority, to construct flood and sea defences. Likewise, grants are paid to maritime district councils to protect their coastlines against erosion.

In Belgium, the Ministry of the Flemish Community is responsible for coastal protection. It maintains a small governmental office for the co-ordination of coastal defence works, water infrastructure and maritime affairs. The actual projects are carried out by consultants.

In the Netherlands, the national government (through its Department of Public Works and Water Management) is concerned with coastal erosion issues. The maintenance of the sea defences is the responsibility of the local administration through the so-called Water Boards. Co-ordination of all activities related to coastal protection is on a provincial level through the so-called Provincial Consultative Bodies (Hillen & De Haan, 1993).

In Germany, coastal erosion and flooding affairs are the responsibility of the federal states. As far as the North Sea is concerned, four states are involved: Bremen, Hamburg, Niedersachsen and Schleswig-Holstein. The federal government provides a large part of the funds necessary for coastal defence works.

In Denmark, coastal protection issues are handled by the national government through the Danish Coast Authority. Again, this is a rather small administrative organisation merely co-ordinating coastal protection projects.

Ecological damage to the European coastal zones is severe. The coastlines (often referred to as the 'golden fringe of Europe') involve a large variety of habitats: sanddunes and beaches, brackish water areas, salt marshes, estuaries and cliffcoasts. The ecological importance of these areas is, amongst others, illustrated by the fact that over 50% of all plant species in western Europe occur in the coastal areas (Council of Europe, 1991).

From the ecological point of view, legislation and management are primarily directed to buying vulnerable coastal areas and establishing nature reserves, national parks and the like. This is realized through both private and governmental organisations. An outline of the relevant legislation in the various European countries is given by the Council of Europe (1991).

In the United Kingdom, the National Trust (NGO) owns and manages over 800 km of coastline (situation 1990). Moreover, about 30% of the coasts forms part of the 'Heritage Coasts', a joint initiative of the Countryside Commission and local authorities.

In Belgium, the national government sponsors NGO's and private initiatives to buy and manage nature reserves in the coastal zone.

The Netherlands policy of nature conservation in the coastal area is outlined in the Nature Policy Plan of the central government. The most important point of action is to bring the entire coastal dune area and large parts of the intertidal zones under the Nature Conservancy Act before 1998.

The German North Sea coasts are well protected. The federal states of Schleswig-Holstein, Niedersachsen and Hamburg, responsible for nature conservation on their territories, have all established national parks in the Wadden Sea.

In Denmark, the Act on Nature Conservation offers possibilities to effectively protect biotopes and contains guidelines for the protection of special zones.

The above may only serve to illustrate the variety of legislation and management in the countries bordering the southern North Sea. In some countries the national government plays a key-role (e.g. the Netherlands). In Germany the federal states have established their own legislation and management. And in countries like Belgium the government merely coordinates activities in the coastal zone.

In 1992, ministers of the North Sea countries and a representative of the Commision of the European Community signed a declaration on coordinated extension of coastal state jurisdiction by the establishment of exclusive economic zones (EEZ). The existence of EEZs in the North Sea is expected to result in a better and more effective enforcement of international rules of environmental protection.

International cooperation

International cooperation in the North Sea was a fact long before the United Nations Convention on the Law of the Sea (adopted in Montego Bay, 10 December 1982) placed a strong emphasis on such cooperation. In fact, international cooperation covered at that moment the most important aspects of management of human uses of the area.

Cooperation started in 1969 on the topic of oil pollution, shortly after the disaster with the Torrey Canyon, with the signature of the Agreement for Cooperation in dealing with Pollution of the North Sea by Oil, known as the Bonn Agreement. Strictly speaking, the Bonn Agreement provides the mechanism for assistance and its scope was limited, at least in the first version, to pollution from oil spills.

The Conventions of Oslo (pollution from dumping and incineration, signed in 1972) and Paris (pollution from land-based sources, signed in 1974) provide since the early seventies a more comprehensive framework with the aim of protecting the marine environment of the northeastern Atlantic (incl. the North Sea).
As often is the case, a concrete example reminded the countries concerned that the unlimited dumping of (industrial) waste into the sea could lead to an unacceptable situation. This example was provided by a Dutch ship, the "Stella Maris" which, having sailed from the port of Rotterdam on 16 July 1971 to dump chlorinated waste in the North Sea, was obliged to return to port on 25 July (without carrying out her mission) because of the combined weight of public opinion and of the Governments of Norway, Iceland, Ireland, the Netherlands and the United Kingdom. Within eight months the Oslo Convention was agreed upon and signed (Anonymous, 1984).
At that time it was felt necessary to draw up a similar document, dealing this time not with the prevention of pollution by dumping, but with the prevention of marine pollution by discharges of dangerous substances from land-based sources, water-courses or pipelines. Negotiations on this topic resulted in the completion of the Paris Convention which was opened for signature in June 1974.

The ideas, implemented in the Oslo Convention were also used, with only minor changes, on a global scale in the completion of the so called London Dumping Convention, which was opened for signature in December 1972.

Very clearly, the early seventies was a fruitful period for international cooperation on the prevention of marine pollution. In this period also the Convention for the Prevention of Pollution by ships, known as MARPOL 73/78, was signed (London, 2 November 1973).
All North Sea countries are Contracting Party to the above conventions that all deal with the prevention of pollution, be it on different aspects. Other main human influences (or uses) of the North Sea are shipping, offshore industry and fisheries.

Matters regarding shipping are dealt with within the framework of the International Maritime Organisation (IMO). IMO is a global framework within the United Nations, aimed at promoting cooperation between states on technical matters regarding shipping. One of the main result of the IMO-framework in the North Sea area is the establishment of traffic separation schemes.

Three other (more regional) fora are competent on shipping matters in the North Sea:
* the Regional Conference of Ministers on Marine Safety. Within this forum, the Memorandum of Understanding on Port State Control (adopted in Paris, 1982) has a major place;
* The regional Meeting on Safety of Navigation in the North Sea (SONNOS, first meeting in 1981, all North Sea states are participating since 1987), with as a main task the coordination with regard to routing and safety of shipping;
* The European Community and the activities developed within that framework.

In the framework of the Common Fisheries Policy (CFP) of the EC, the member states have transferred their competence in the field of fisheries to the EC (EC 170/83, 25 January 1983, the Establishing a Community System for the Conservation and Management of Fishery Resources). This implies that non North Sea coastal states (but members of the EC) participate in the negotiations related to the North Sea.

Norway, as non-EC-member but important fishery country, and the EC have concluded an agreement on fisheries (Brussels, 27 June 1980). Under this agreement vessels from other EC-member states are allowed to fish in Norwegian waters and Norwegian vessels are allowed to fish in the exclusive EC-fisheries zone.

In contrast to shipping and fishery, oil and gas exploitation in the North Sea takes place within the sovereign power of the coastal state. This implies that the regulation of the exploitation of the continental shelf takes place to a very large extent within national legislation. The two most important categories of restrictions on the sovereign power of a coastal state are environmental protection and safety aspects. Environmental aspects of the operation of offshore installations fall under the scope of the Paris Commission (mainly regarding the use of drilling mud and chemicals and the discharge of contaminated cuttings and production water) and the MARPOL Convention (mainly operational discharges from machine rooms). Safety aspects of the offshore industry are discussed within the framework of the IMO and, at the regional level within SONNOS.

A special problem that has environmental and safety aspects is the removal of abandoned offshore installations. Environmental aspects are discussed within the Oslo Commission and the London Dumping Convention while safety aspects are discussed within the IMO.

From the above it becomes clear that the traditional attitude of environmental management (deal with each separate problem at a time, on the assumption that if problems on a case by case basis are well solved, the total is well managed), undoubtedly resulted for the North Sea in a situation where none of the many competent international organizations is exclusively competent to deal in an comprehensive way with the management of the North Sea. There is even overlap in competence between different organizations (IJlstra, 1988).

An initiative in Germany to improve this situation resulted in the first International Conference on the Protection of the North Sea (INSC-I, Bremen, 1984). Although the aim of the initiative (to draw up one "umbrella" convention for the protection of the North Sea) was not reached, it was the first occasion for an active and direct involvement of the political level. Final decisions were taken by competent ministers. Since then, the North Sea Conference evolved into a forum where far-reaching decisions on the prevention of pollution are taken (for instance on a 50% reduction of the input of hazardous substances between 1985 and 1995 or on the phasing out of marine incineration by 1995 as agreed at INSC-II in London, 1987), but also decisions with regard to nature conservation (Memorandum of Understanding on Small Cetaceans, INSC-III, The Hague, 1990). A regular Fourth Conference will be held in 1995 in Denmark. In December 1993, environmental and agricultural ministers of North Sea countries will jointly meet at an Intermediate Conference, especially to discuss problems related to eutrophication of the North Sea.

A forum for cooperation on the scientific level, installed by INSC-II, is the North Sea Task Force. A main task of this cooperation is the elaboration of a "Quality Status Report of the North Sea", to be published in December 1993. This report should be the basis of further political decisions to be taken by North Sea ministers in 1993 and 1995.

References

Anonymous (1984), The Oslo and Paris Commissions, The first Decade.

Council of Europe (1991) Naturopa, special volume on European coasts; *Centre Naturopa*, Strasbourg, France.

De Ruyter, W.P.M., L. Postma & J.M. de Kok (1987) Transport Atlas of the southern North Sea. *Rijkswaterstaat & Delft Hydraulics*; The Hague/Delft, the Netherlands.

Doody, E.J.P. (1991) Sand Dune Inventory of Europe; *UK Joint Nature Conservation Committee & European Union for Coastal Conservation*, Peterborough, UK.

Enemark, J., The protection of the Wadden Sea and the International cooperation. *This volume.*

Hillen, R. & Tj. De Haan (1993) Development and implementation of the coastal defence policy for the Netherlands. *This volume.*

ICONA [Interdepartmental Co-ordinating Committee for North Sea Affairs] (1992) North Sea Atlas for Netherlands Policy and Management. *Stadsuitgeverij*; Amsterdam, the Netherlands.

IJlstra, T. (1988). Regional cooperation in the North Sea: an inquiry. *International Journal of Estuarine and Coastal Law*, 3, 181-207.

IMO [International Maritime Organisation] (1987) Final Act on the Conference on Marine Pollution, 1973. *Reprint of the 1977 edition*; IMO, London, UK.

NSTF [North Sea Task Force] (in preparation) Quality Status Report of the North Sea; *NSTF secretariat*, London, UK.

OECD [Organisation for Economic Co-operation and Development] (in preparation) Coastal Zone Management: Integrated Policies; *OECD secretariat*, Paris, France.

Popp Madsen, K. & K. Richardson (1993) Fisheries in the southern North Sea. *This volume.*

Van de Kamer, J.P.G., K.J. Wulffraat, A. Cramer & M.J.P.H. Waltmans (1993) Water Quality Management in the southern North Sea. *This Volume.*

Geology of the Southern North Sea Basin

D. Cameron[1], D. van Doorn[2], C. Laban[2], H.J. Streif[3]

Abstract

The southern North Sea Basin has had a long and complex geological history. Between Late Carboniferous and the end of Triassic times, the basin was largely confined between two ancient east-west trending Palaeozoic upland areas, the London-Brabant Massif and the Mid North Sea High/Ringkøbing-Fyn High. Crustal stretching, followed by thermal subsidence, then enabled up to 3500 metres of Jurassic, Cretaceous, Tertiary and Quaternary sediments to accumulate, with the depocentre striking NNW from the Netherlands towards the central North Sea. Late Miocene, Pliocene and Early Pleistocene sedimentation were dominated by the build-out of major delta systems across the basin, principally from its eastern seaboard. The subsequent history of the basin has included three episodes of regional glaciation, punctuated by strongly tidal, marine environments similar to those of the present day.

Introduction

Since the search for hydrocarbons beneath the southern North Sea began in the 1960s, more than 2000 wells have been drilled to a maximum depth of 5000m, and more than a million kilometres of digitally recorded seismic-reflection data have been acquired by the oil exploration companies. Much geological information has been published in many hundreds of papers, and the structural setting and stratigraphy of the Late Carboniferous to Tertiary sediments have been particularly well established in the area north of 52°30' in which the hydrocarbons occur.

Reconnaissance surveys by the British Geological Survey (formerly the Institute of Geological Sciences) and Rijks Geologische Dienst (Geological Survey of the Netherlands) have led to the publication of maps at a 1:250,000 scale of the solid (pre-Quaternary) and Quaternary geology and of the sea-bed sediments of the UK and Dutch sectors. These maps have been based on interpretation of more than 50,000 kilometres of high-resolution, seismic-reflection data, calibrated by several thousand shallow cores and more than 100 boreholes drilled up to 274m beneath the sea bed, and on the published hydrocarbon-exploration data. Participation by the geological

---

[1] British Geological Survey, Murchison House, West Mains Road, Edinburgh EH9 3LA, Scotland, U.K.

[2] Dutch Geological Survey, P.O. Box 157, 2000 AD Haarlem, The Netherlands

[3] Niedersächsisches Landesamt für Bodenforschung, Postfach 510153, D 3000 Hannover 51, Germany

surveys of Belgium, Denmark, Germany, the Netherlands and the UK in the southern
North Sea Quaternary Project, funded by the Commission of European Communities
since 1989, has provided much new information on the late Neogene and Quaternary
evolution of the whole of the offshore area.

Figure 1:   Regional structure of the Southern North Sea Basin.

Pre-Permian basin development

Basinal marine and volcaniclastic Lower Palaeozoic sediments, mildly
metamorphosed during the Caledonian orogeny, form the basement to much of the
southern North Sea; crystalline Pre-Cambrian rocks are probably also present at depth
beneath the London-Brabant Massif. Except over the crest of the London-Brabant
Massif, where they approach to less than 1 km below the sea bed, the Lower
Palaeozoic rocks and their patchy cover of mainly alluvial Devonian sediments are
now buried beneath many kilometres of younger Palaeozoic, Mesozoic and Cenozoic
sediments. The thickest Devonian deposits occur towards the east of the Mid North
Sea High, where they contain evidence for temporary Middle Devonian northward

ingression of an embayment of the Proto-Tethys Ocean along a lineament adopted later by the Central Graben of the North Sea (Ziegler, 1990).

The London-Brabant Massif and Mid North Sea High/Ringkøbing Fyn High (fig. 1) became established as stable upland areas early in Carboniferous times. Up to 4000m of deep-water and deltaic sediments were deposited in rapidly subsiding grabens and half-grabens during an early Carboniferous phase of crustal extension (Leeder, 1987), whereas contemporary horsts accumulated condensed sequences including platform carbonates. As crustal extension diminished, relatively uniform deltaic sedimentary facies spread across central areas of the southern North Sea by late Namurian times, and up to 3000m of combined Namurian and Westphalian sediments have been proved by drilling. Although Dinantian and Namurian organic shales and minor coal seams are present, the Lower Westphalian coal measures have been the principal source rocks for the many gas fields of the southern North Sea. These were deposited on a low-lying, paralic delta plain (Guion and Fielding, 1988). Later in the Westphalian, regional uplift caused the poorly-drained deltaic facies to be superseded by better-drained, alluvial-plain sedimentation.

Early in Westphalian times, the southern North Sea had become a foreland area to a chain of Variscan mountains that extended from south-west England through France into eastern Europe (Glennie, 1990a). Its Carboniferous rocks were gently folded, faulted and peneplaned, and up to 1500m of sediments were eroded from parts of the current offshore area (Cameron et al., 1992).

Permo-Triassic basin development

Early in the Permian, much of the southern North Sea began to subside gently once more, as the area north of the London-Brabant Massif now lay within an east-west trending Variscan foreland, post-orogenic collapse basin (Ziegler, 1990). This basin, which extended between eastern England and Poland until late Triassic times (Glennie, 1990a), was bounded to the north by the Mid North Sea High and Ringkøbing-Fyn High. These highs accumulated condensed sequences of Permian and Triassic sediments, whereas the London-Brabant Massif continued as a stable upland area. Basal Permian volcanics have been recorded in the Horn Graben of the German and Danish sectors, at the Ringkøbing-Fyn High, and locally elsewhere (Glennie, 1990b).

During much of the Permian, the southern North Sea Basin lay in a rain shadow north of the Variscan mountains (Glennie, 1990b). Lower Permian sediments, including aeolian and fluvial sands, are generally less than 400m thick and were deposited in an arid desert environment. The aeolian sands occur for up to 150km north of the London-Brabant Massif, and provide the principal reservoir for the many gas fields in the current offshore area. Contemporary sabkha deposits accumulated around the margins of a desert lake that extended from just off the Yorkshire coast as far east as Germany. The lacustrine sediments include silty mudstones and evaporites, and these are up to 1500m thick beneath the German Bight of the North Sea.

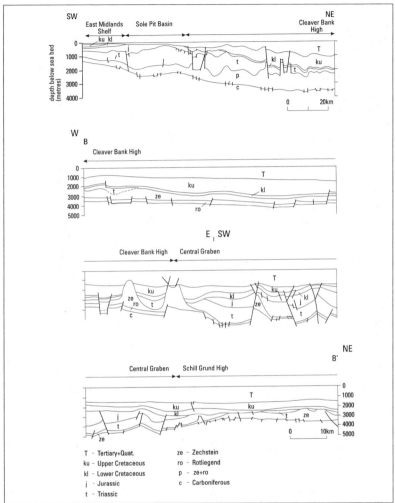

Figure 2:   Schematic profiles across the central UK and Dutch sectors of the southern
             North Sea.

Five short-lived but widespread transgressions of the Boreal Ocean entered the basin
by a temporary seaway late in the Permian (Taylor, 1990); two further transgressions
have been recorded in the east of the basin (Best, 1989). Marine shales and
carbonates were deposited following the first, second and third transgressions, while
evaporites accumulated during periods when constriction of the seaway led to
increased salinity in the basin. These evaporites include local salt-pan deposits of the
first cycle and basinwide halites, polyhalites and subsidiary salts of the second, third
and fourth cycles. The evaporites of the second cycle are especially thick, and have

been deforming intermittently by halokinesis since mid-Triassic times, and especially during regional episodes of rifting and basin inversion; Triassic rifting across the Central Graben initiated the salt diapirs illustrated in Figure 2 (Section B - B').

Triassic sediments of the southern North Sea are locally several thousand metres thick. Although dominated by reddish-brown mudstones, there are sandstones forming important gas reservoirs in the Lower Triassic, and beds of dolomite, anhydrite, sandstone and widespread halite horizons in the Middle and Upper Triassic. Early Triassic sediments were deposited in playa-lake, floodplain and fluvial environments (Fisher and Mudge, 1990) at a time of major rifting of the Horn Graben and Central Graben of the North Sea (Ziegler, 1990). Later Triassic sediments also include coastal sabkha deposits and evaporites that record four basinwide incursions of marine influences associated with transgression of the Tethys Ocean into east European parts of the basin. A more permanent connection with the Tethys Ocean caused fully marine conditions to spread across the southern North Sea during Rhaetic, Late Triassic times. These marine conditions have continued intermittently during the past 200 million years.

Jurassic to mid-Miocene basin development

It was during late Triassic and early Jurassic crustal extension beneath its central axis that the North Sea first adopted its present NNW-SSE alignment. Early Jurassic extension led to particularly rapid subsidence of the Central Graben and the Sole Pit Basin. Predominantly argillaceous marine Lower Jurassic sediments are up to 2000m thick in the former area, where they include the organic-rich, gas- and oil-prone, Posidonia Shale.

Subsidence was temporarily interrupted by Middle Jurassic domal uplift and deep erosion along the central axis of the North Sea; perhaps as much as 2000m of sediments were eroded from the the Cleaver Bank High at this time (Glennie, 1990a) and from the Schill Grund High. Middle Jurassic sediments, best preserved from later erosion in the western UK sector and in the contemporary rift zone of the Central Graben, include a variety of shallow-marine, brackish-water, and fluviodeltaic sediments.

Renewed late Middle Jurassic to early Cretaceous extension of the crust beneath the Central Graben by about 18km (Clark-Lowes *et al.*, 1987) and associated strike-slip movements along the faults bounding the Broad Fourteens Basin (Glennie, 1990a) then enabled up to 2000m of Upper Jurassic and Lower Cretaceous sediments to accumulate in these basins (Herngreen and Wong, 1987). Local sediment thicknesses and facies were strongly influenced by the contemporary halokinesis of the deeply buried Upper Permian evaporites. The Upper Jurassic sediments are mainly of shallow marine, paralic and continental facies in the southern Dutch sector; they comprise paralic overlain by marine argillaceous sediments in the Central Graben, and consist of relatively thin organic-rich, marine shales and minor limestones in the UK sector (Brown, 1990). The Lower Cretaceous sediments are marine, calcareous mudstones and minor sandstones.

Crustal extension ended in mid-Cretaceous times, but the regional thermal anomaly that it generated has been decaying ever since, and this has been the principal mechanism by which the southern North Sea Basin has continued to subside. The influence of fault-bounded basins as the principal depocentres diminished as the whole of the southern North Sea, including the London-Brabant Massif, Mid North Sea High and Ringkøbing-Fyn High, subsided in response to regional downwarping. At various intervals, notably towards the beginning and at the end of the Cretaceous, the fault-bounded basins and adjacent areas were locally subjected to significant uplift and partial erosion of their previously deposited sediments (basin inversion).

A uniform blanket of white, coccolith-rich chalk accumulated in warm, oxygenated waters during Late Cretaceous times, and is up to 1500m thick. This chalk deposition was interrupted around contemporary basin-inversion structures by local episodes of uplift and erosion. An unconformity separates the Chalk Group from overlying Palaeogene sediments across much of the southern North Sea (fig. 2), and this records widespread temporary emergence of the area above sea level around the Cretaceous/Palaeogene boundary. Following submergence and renewed subsidence, up to 1000m of mainly argillaceous, marine Palaeogene sediments were deposited. Their depocentre lay approximately above the Central Graben and its southward continuation into the Netherlands, whereas the London-Brabant Massif and Ringkøbing-Fyn High accumulated condensed Palaeogene sections.

Mid-Miocene to mid-Pleistocene basin development

Tectonic uplift caused most of the southern North Sea to emerge above sea level once more during mid-Miocene times, in response to regional stresses generated by continental convergence of the Alpine orogeny (Glennie, 1990a). Relaxation of this stress system then enabled subsidence to resume, and between the late Miocene (about 10 million years ago) and late mid-Pleistocene times (0.4 Ma) the southern North Sea Basin became the site of one of the world's major delta systems. The associated fluvio-deltaic plain eventually covered an area of at least $150,000km^2$, and the delta-related sediments are between 500m and 1500m thick along the central axis of the North Sea.

Delta systems built out at first from the eastern seaboard of the North Sea, fed by former Baltic rivers that drained from the Fennoscandian Shield (Bijlsma, 1981) and by westward-flowing north German rivers that drained from central Europe (Lüttig, 1974; Lüttig and Meyer, 1974; Gibbard, 1988). Build-out of marginal deltas associated with the Rhine, Meuse, Scheldt, and relatively minor British rivers then occurred from the early Pleistocene, about 1.8 Ma (Zagwijn, 1979), and helped to deflect the principal direction of delta growth from the west or south-west to the north-west. These rivers were discharging into the southern embayment of an epicontinental sea, as there was little if any connection of the North Sea through the English Channel into the Atlantic Ocean from Miocene until comparatively recent times (Cameron et al., 1992).

Miocene deltaic deposits, largely confined to the German and Danish sectors, were deposited as water depth in the east of the North Sea Basin increased rapidly by between 300m and 500m (Gramann and Kockel, 1988), an amount that can only have been caused by tectonic subsidence of the German Bight and Ringkøbing-Fyn High, as it was far in excess of contemporary sea level fluctuation (Haq *et al.*, 1988). Pliocene deltaic deposits are thickest in the central Dutch sector, and were deposited as maximum water depth in the basin decreased to less than 150m (Cameron *et al.*, in press). Deltaic deposits in the UK sector are largely Pleistocene in age, and their deposition resulted in complete filling in of the southern North Sea Basin, and expansion of its fluvio-deltaic plain to at least as far north as 56° by mid-Pleistocene times.

Regional seismic interpretation, calibrated by the results of boreholes, has shown that the deltaic deposits comprise an upward-coarsening megasequence at any location in the southern North Sea (Cameron *et al.*, 1987). Argillaceous, often intensely bioturbated prodelta sediments are overlain successively by muddy delta-front sands and clays, by delta-top facies including shallow subtidal, intertidal and non-marine sands and clays, and by clean, locally gravelly, fine- and medium-grained fluvial sands. These facies record the approach and overwhelming of that location by the advancing deltas. The fluvial sediments are thickest in the east of the basin, where they have the greatest, Pliocene to mid-Pleistocene age range.

The regional seismic interpretation has also revealed that deltaic sedimentation was punctuated many times by intervals of regional stratigraphic hiatus or of basinwide change in sedimentary environment. Such intervals generate sequence boundaries on the seismic profiles, and they may be the product of fluctuation in relative sea level within the basin, change in subsidence rates, or of sediment supply to the basin. Whatever their cause, they suggest that progressive infill of the southern North Sea Basin was accompanied by cyclic fluctuation of its coastlines, perhaps by many tens of kilometres, in response to one or more of these factors.

Mid-Pleistocene to Holocene basin development

There is abundant evidence from the palaeobotanical record of the Netherlands for cyclic fluctuation between warm temperate (interglacial) and cold climatic conditions in the North Sea Basin and its hinterland from early Pleistocene times (Zagwijn, 1989). Climatic severity notably increased during the cold (glacial) climatic stages of the Middle Pleistocene, leading to glaciation in the basin's peripheral sediment source areas (Boulton, 1992), but it was not until the Elsterian Stage that began about 0.4 Ma that there was the first of three widespread invasions of ice across the basin itself. Following each of these glaciations, climatic amelioration led to the development of strongly tidal, marine environments similar to those of the present day.

Elsterian ice cover was accompanied by the erosion of a swarm of anastomosing, but mainly north-south or north-north-west trending glacial palaeovalleys cut into the Pleistocene and pre-Pleistocene strata (Long *et al.*, 1988). The largest valleys, up to 12km wide and 450m deep, are boat-shaped in longitudinal section with an uneven thalweg. In the southern North Sea, they are mainly concentrated between 53° and 54°N, but contemporary valleys are also widespread beneath the northern Netherlands and north Germany. The mechanism by which such valleys formed is still a matter of controversy, but most authors favour a subglacial meltwater (Boulton and Hindmarsh, 1987) or jökulhlaup origin (Wingfield, 1990). The valleys were partially filled by diamictons, outwash gravels and sands, thick lacustrine clays, and fluvial sands before the climatic amelioration of the Holsteinian Stage. If there were contemporary glacial or periglacial deposits south of 53°, then they have been completely removed by subsequent erosion.

Early in the Holsteinian interglacial period, rising sea level led to the re-establishment of a shallow sea, the shorelines of which lay partly landward of those of the present North Sea (Cameron et al., 1992). Those glacial palaeovalleys which had not already been filled by Elsterian deposits contain an additional component of marine Holsteinian sediments, typically clays or fine-grained sands. These accumulated in a quiet, shallow-water environment, but open-marine conditions were eventually established during deposition of up to 25m of sparsely shelly sands in the Dutch sector (Laban et al., 1984). Farther south, Holsteinian sediments were deposited along the Franco-Belgian border during transgression from the south-west (Sommé, 1979; Paepe and Baeteman, 1979); there was probably no marine connection between the North Sea and the English Channel through the Dover Strait at this time (Zagwijn, 1979; Hinsch, 1985).

Following climatic deterioration and marine regression, thick ice sheets spread across Denmark, northern Germany, and the northern Netherlands during the Saalian glaciation, but the ice cover extended no more than 100km north-west of the present Danish and Dutch coastlines (Foged, 1987; Joon et al., 1990). Spreads of till continue offshore, and ice-pushed ridges and subglacial valleys formed within the limits of the ice sheet (Fig. 3). Proglacial lacustrine clays, outwash sands, and wind-blown sands were deposited in a periglacial environment beyond the ice margin.

Following deglaciation, rising sea level led to the establishment of a shallow sea once more during the Eemian Stage. At the sea's maximum extent, when north European winter temperatures were significantly warmer than those of today (Gerasimov and Velichko, 1982), its shorelines extended partly beyond those of the present North Sea. Eemian marine sands locally have a much higher shell content than those deposited by the Holsteinian or Holocene transgressions (Cameron et al., 1989), and the molluscan fauna has a strong lusitanian affinity (Spaink, 1958).

Regional climate began to deteriorate late in the Eemian Stage and early in the succeeding Weichselian Stage. When sea level had fallen by about 40m, a brackish-water lagoon occupied much of the southern Bight of the North Sea, between East Anglia and the Netherlands (Cameron et al., 1989). This became a freshwater lagoon as sea level continued to fall; there is no record of the sedimentary environment during the early stages of climatic deterioration elsewhere in the North Sea.

Regional climate continued to deteriorate until, at the height of the late Weichselian glaciation, sea level fell to at least 110m below that of the present day (Jansen et al., 1979). In the west, a lobe of grounded ice extended from the English coast into the

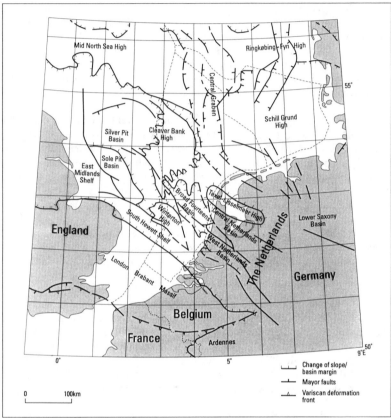

Figure 3:   Limits of the Elsterian (E), Saalian (S), and Weichselian (W) ice sheets in the
                 North Sea area.

central Dutch sector of the North Sea (Fig. 3). This ice sheet deposited a blanket of
pebbly clay-rich till that merges northeastwards with a proglacial, water-laid
diamicton and glaciolacustrine clays (Jeffery in: Cameron *et al.*, 1992). A system of
glacial palaeovalleys was eroded during ice decay; their orientations are
approximately normal to the inferred ice margin (fig. 3), and their dimensions are
notably smaller than those of the Elsterian glaciation, implying thinner ice cover
(Jeffery, *op cit.*). A second lobe of grounded Scandinavian ice extended into the
North Sea from the coast of Jutland, Denmark. Its limits are poorly defined offshore
but there was no connection across the North Sea between the UK and Scandinavian
ice sheets at this time (Cameron *et al.*, 1987; Long *et al.*, 1988).

Aeolian and fluvioglacial sands were deposited over large areas of the North Sea
beyond the late Weichselian ice margin. Cryoturbation and frost-wedge structures
indicate that periglacial conditions prevailed across these newly emergent areas. The

Elbe river system, with its tributaries the Weser and Eider rivers, flowed northwestwards from the German coast across the present sea floor in a 30-40 km wide valley system (Figge, 1980), and the Rhine and Meuse rivers deposited extensive, and locally gravelly sands offshore of their modern estuaries (Cameron *et al.*, 1989).

As the late Weichselian ice sheet began to decay, rising sea level enabled glaciomarine muds to be deposited in relict palaeovalleys and in present deep-water, northwestern areas of the southern North Sea. The earliest brackish-water incursion into remaining deep-water areas may have occurred as early as 10 ka BP (Eisma *et al.*, 1981), when sea level was about 65m below present (Jelgersma, 1979). With continuing sea level rise, tidal sand ridges formed west of the Dogger Bank, which became isolated by tidal erosion of the east-west trending Outer Silver Pit to form a temporary island. Tidal flat sedimentation became more widespread between 9 ka and 8 ka BP; the southern North Sea was connected with the English Channel through the Dover Strait about 8.3 ka BP, and fully marine conditions became established widely after 7 ka BP (Eisma *et al.*, 1981). Deposits of the early Holocene transgression indicate a unidirectional landward migration of the shoreline during this period. Such deposits are thicker than 10m only in the tidal sand ridges, in the Elbe palaeovalley, and in the southeastern German Bight.

Detailed records of more recent sea level change are complicated by the effects of isostatic uplift following the last glaciation, and of continuing tectonic subsidence along the central axis of the North Sea. Such records are largely determined by radiocarbon dating of peat layers that lie directly beneath transgressive marine beds at the Dogger Bank and at various sites between there and the modern coastline. Streif (1985; 1990) has deduced the history of sea level change in the German Bight from a study of the complex, interdigitating succession of channel, tidal flat and brackish-water deposits, intercalated peat layers and beach sands that occurs along the north German coastline. These deposits are up to 35m thick, completely masking the relief of the underlying Pleistocene land surface. Tidal-flat sedimentation extended to the Frisian Islands at 7.5 ka BP when sea level was 25m below present (Streif, 1985; 1990). The shoreline has migrated landwards by between 10 and 20 km since then, but sedimentary facies indicate that there were periods of reduced sea level rise between 4.8 and 4.2 ka BP and between 3.3 and 2.3 ka BP, and sea level lowered temporarily at about 2.7 ka BP and 2.0 ka BP.

Acknowledgement

This paper is published with the permission of the Directors of the British Geological Survey (N.E.R.C.), Rijks Geologische Dienst, and Niedersächsisches Landesamt für Bodenforschung. The authors wish to thank Dr. F. Kockel (BGR) for his particularly helpful comments on an early draft of the manuscript.

References

Best, G. 1989. Die Grenze Zechstein/Buntsandstein in Nordwest-Deutschland und in der südlichen deutsche Nordsee nach Bohrlochmessungen. *Zeit. deut. Geol. Ges.* 140:73-85.

Bijlsma, S. 1981. Fluvial sedimentation from the Fennoscandian area into the North-West European Basin during the Late Cenozoic. *Geol. Mijnb.* 60:337-345.

Boulton, G.S. 1992. Quaternary. *In*: P.McL.D. Duff and A.J. Smith (editors). *Geology of England and Wales*. London. Geological Society. pp. 413-444.

Boulton, G.S. and Hindmarsh, R.C.A. 1987. Sediment deformation beneath glaciers: rheology and geological consequences. *J. Geophys. Res.* 92:9059-9082.

Brown, S. 1990. Jurassic. *In*: K.W. Glennie (editor). *Introduction to the petroleum geology of the North Sea* (3rd edition). Oxford. Blackwell Scientific Publications. pp. 219-254.

Cameron, T.D.J., Bulat, J. and Mesdag, C.S. 1993. A high-resolution seismic profile through a late Cenozoic delta complex in the southern North Sea. *Mar. Pet. Geol.* in press.

Cameron, T.D.J., Crosby. A., Balson, P.S., Jeffery, D.H., Lott, G.K., Bulat, J. and Harrison, D.J. 1992. United Kingdom Offshore Regional Report: *The Geology of the southern North Sea*. London. HMSO for the British Geological Survey. 150 pp.

Cameron, T.D.J., Schüttenhelm, R.T.E. and Laban, C. 1989. Middle and Upper Pleistocene and Holocene stratigraphy in the southern North Sea between 52° and 54°N, 2° to 4°E. *In*: J.P. Henriet and G. De Moor (editors). *The Quaternary and Tertiary Geology of the southern Bight, North Sea*. Brussels. Belgian Geological Survey. pp. 119-135.

Cameron, T.D.J., Stoker, M.S. and Long, D. 1987. The history of Quaternary sedimentation in the UK sector of the North Sea. *J. Geol. Soc. Lond.* 144:43-58.

Clark-Lowes, D.D., Kuzemko, N.C.J. and Scott, D.A. 1987. Structure and petroleum prospectivity of the Dutch Central Graben and neighbouring platform areas. *In*: J. Brooks and K.W. Glennie (editors). *Petroleum Geology of North West Europe*. London. Graham and Trotman. pp. 337-356.

Eisma, D., Mook, W.G. and Laban, C. 1981. An early Holocene tidal flat in the southern Bight. *Spec. Publ. Int. Ass. Sed.* 5:229-237.

Figge, K. 1980. Das Elbe-Urstromtal im Bereich der Deutschen Bucht (Nordsee). *Eiszeitalter Gegenw.* 30:203-211.

Fisher, M.J. and Mudge, D.C. 1990. Triassic. *In*: K.W. Glennie (editor). *Introduction to the petroleum geology of the North Sea* (3rd edition). Oxford. Blackwell Scientific Publications. pp. 191-218.

Foged, N. 1987. The need for Quaternary geological knowledge in geotechnical engineering. *Boreas*. 16:419-424.

Gerasimov, I.P. and Velichko, A.A. 1982. Paleogeography of Europe during the Last One Hundred Thousand Years. Moscow. Nauka. 154pp. (in Russian with English summary).

Gibbard, P.L. 1988. The history of the great northwest European rivers during the past three million years. *Phil. Trans. R. Soc. Lond.* 318B:559-602.

Glennie, K.W. 1990a. Outline of North Sea history and structural framework. *In*: K.W. Glennie (editor). *Introduction to the petroleum geology of the North Sea* (3rd edition). Oxford. Blackwell Scientific Publications. pp. 34-77.

Glennie, K.W. 1990b. Lower Permian - Rotliegend. *In*: K.W. Glennie (editor). *Introduction to the petroleum geology of the North Sea* (3rd edition). Oxford. Blackwell Scientific Publications. pp. 120-152.

Gramann, F. and Kockel, F. 1988. Palaeogeographical, lithological, palaeoecological and palaeoclimatic development of the Northwest European Tertiary Basin. *In*: R. Vinken (compiler). *The Northwest European Tertiary Basin*. Geol. Jahr. A100:428-441.

Guion, P.D. and Fielding, C.R. 1988. Westphalian A and B sedimentation in the Pennine Basin, UK. *In*: B.M. Besly and G. Kelling (editors). *Sedimentation in a synorogenic basin complex: the Upper Carboniferous of Northwest Europe*. London. Blackie & Son. pp. 153-177.

Haq, B.U., Hardenbol, J. and Vail, P.R. 1988. Mesozoic and Cenozoic chronostratigraphy and cycles of sea level change. *In*: C.K. Wilgus, B.J. Hastings, H. Posamentier, J.C. Van Wagoner, C.A. Ross and C.G.St.C Kendall (editors). *Sea level changes: an integrated approach*. Soc. Econ. Paleontol. Mineral., Spec. Publ. 42:71-108.

Herngreen, G.F.W. and Wong, Th.E. 1987. Revision of the 'Late Jurassic' stratigraphy of the Dutch Central North Sea Graben. *Geol. Mijnb*. 68:73-105.

Hinsch, W. 1985. Die Molluskenfauna des Eems-Interglazials von Offenbüttel-Schnittlohe (Nord-Ostsee Kanal, Westholstein). *Geol. Jahr*. A86:49-62.

Jansen, J.H.F., Van Weering, T.C.E. and Eisma, D. 1979. Late Quaternary sedimentation in the North Sea. *In*: E. Oele, R.T.E. Schüttenhelm and A.J. Wiggers (editors). *The Quaternary history of the North Sea*. Uppsala. Acta Universitatis Upsaliensis. pp. 175-187.

Jelgersma, S. 1979. Sea level changes in the North Sea Basin. *In*: E. Oele, R.T.E. Schüttenhelm and A.J. Wiggers (editors). *The Quaternary history of the North Sea*. Uppsala. Acta Universitatis Upsaliensis. pp. 233-248.

Joon, B., Laban, C. and van der Meer, J.J.M. 1990. The Saalian glaciation in the Dutch part of the North Sea. *Geol. Mijnb*. 69:151-158.

Laban, C., Cameron, T.D.J. and Schüttenhelm, R.T.E. 1984. Geologie van het Kwartair in de zuidelijke bocht van de Noordzee. *Meded. Werk. Tert. Kwart. Geol.* 21:139-154.

Leeder, M.R. 1987. Tectonic and palaeogeographic model for Lower Carboniferous Europe. *In*: J. Miller, A.E. Adams and V.P. Wright (editors). *European Dinantian Environments*. Chichester. John Wiley & Sons. pp. 1-20.

Long, D., Laban, C., Streif, H., Cameron, T.D.J. and Schüttenhelm, R.T.E. 1988. The sedimentary record of climatic variation in the southern North Sea. *Phil. Trans. R. Soc. Lond.* 318B:523-537.

Lüttig, G. 1974. Geological history of the River Weser (Northwest Germany). *In*: P. Macar (editor). *L'évolution Quaternaire des bassins fluviaux de la Mer du Nord méridionale*. Cent. Soc. Géol. de Belg. Liège. pp. 21-34.

Lüttig, G. and Meyer, K.-D. 1974. Geological history of the River Elbe, mainly of its lower course. *In*: P. Macar (editor). *L'évolution Quaternaire des bassins fluviaux de la Mer du Nord méridionale.* Cent. Soc. Géol. de Belg. Liège. pp. 1-19.

Paepe, R. & Baeteman, C. 1979. The Belgian coastal plain during the Quaternary. *In:* E. Oele, R.T.E. Schüttenhelm and A.J. Wiggers (editors). *The Quaternary history of the North Sea.* Uppsala. Acta Universitatis Upsaliensis. pp. 143-146.

Sommé, J. 1979. Quaternary coastlines in northern France. In: E. Oele, R.T.E. Schüttenhelm and A.J. Wiggers (editors). The Quaternary history of the North Sea. Uppsala. Acta Universitatis Upsaliensis. pp. 147-158.

Spaink, G. 1958. De Nederlandse Eemlagen. I. *Algemeen overzicht* nr. 29. KNNV, Wet. Med.

Streif, H. 1985. southern North Sea during the Ice Ages - inundations and ice-cap movements. *In: German research: reports of the DFG 1985.* Weinheim. VCH-Verlagsgesellschaft. pp. 29-31.

Streif, H. 1990. Das ostfriesische Küstengebiet - Nordsee, Inseln, Watten, und Marschen. *Sammlung Geol. Führer.* 57:2. Berlin/ Stuttgart. Borntraeger. 376 pp.

Taylor, J.C.M. 1990. Upper Permian - Zechstein. *In:* K.W. Glennie (editor). *Introduction to the petroleum geology of the North Sea* (3rd edition). Oxford. Blackwell Scientific Publications. pp. 153-190.

Wingfield, R.T.R. 1990. The origin of major incisions within the Pleistocene deposits of the North Sea. *Mar. Geol.* 91:31-52.

Zagwijn, W.H. 1979. Early and Middle Pleistocene coastlines in the southern North Sea basin. *In:* E. Oele, R.T.E. Schüttenhelm and A.J. Wiggers (editors). *The Quaternary history of the North Sea.* Uppsala. Acta Universitatis Upsaliensis. pp. 31-42.

Zagwijn, W.H. 1989. The Netherlands during the Tertiary and the Quaternary: a case history of Coastal Lowland evolution. *Geol. Mijnb.* 68:107-120.

Ziegler, P.A. 1990. *Geological Atlas of Western and Central Europe* (2nd edition). Amsterdam.778 Shell Internationale Petroleum Maatschappij BV.

The Shaping of the French-Belgian North Sea Coast throughout Recent Geology and History

R. Houthuys[1], G. De Moor[2], J. Sommé[3]

Abstract

The present-day French-Belgian North Sea coastal barrier and coastal plain were shaped after the last Ice Age, when especially between 10,000 and 5,000 years ago the global sea level underwent a rapid rise and the gently sloping northern part of the Flemish plain and the incised river mouths were flooded. The present paper brings an integrated overview of the French-Belgian coastal plain's recent geological and historical evolution, such as it is documented by the sedimentary record as well as by historical sources. The geologic approach provides indirect and environmental evidence for the coastal barrier dynamics. Also, some implications on present-day coastal management are given.

Introduction

The French-Belgian North Sea coast is a 120 km long, almost rectilinear sand beach barrier stretching from the Cap Blanc Nez chalk cliffs (France) to the mouth of the Westerschelde (Belgium/the Netherlands). East of Calais, the shoreline consists of a gently sloping sandy beach backed by coastal dunes. The morphology of this part of the southern North Sea coast is described in more detail in this volume by De Putter *et al.* (this volume).

These sandy beaches have been built up through thousands of years by the action of wind, waves and tides on a supply of loose sand grains at the seaward side of a complex coastal barrier, perhaps an island barrier, whose location changed with time. At the same time, in the more sheltered area landward of the coastal barrier, an intertidal flat developed by deposition of mainly fine-grained material and by intermittent peat formation.

The French-Belgian coastal plain extends some 10 to 20 km landward from the beach barrier. Most of the plain's present-day elevation is between mean and high tide sea level. As such, the plain would be inundated by the sea twice a day, should it not be protected by a continuous system of beaches, dunes and dikes.

---

[1] Project Engineer, Eurosense, Nerviërslaan 54, B-1780 Wemmel (Belgium)

[2] Professor, Laboratory of Physical Geography, Ghent University, Krijgslaan 281, B-9000 Gent (Belgium)

[3] Professor, Science and Technology University of Lille, Geomorphology and Quaternary Laboratory and UPR 7559 CNRS, F-59655 Villeneuve d'Ascq Cédex (France)

The first geologic studies of the coastal plain date back to the 19th century (Belpaire, 1855; Rutot, 1897). Later pioneer work included the definition of nowadays well-known stratigraphic units such as the Calais and the Dunkerque Formations (Dubois, 1924; Briquet, 1930; Halet, 1931). In most of the coastal plain the Calais deposits and their cover of Dunkerquian sediments are separated by the Upper Peat complex (fig. 2), also called surface peat, at least where this latter unit is not eroded by tidal gullies. Modern geological and morphological research in the Belgian coastal plain was initiated by Tavernier & Ameryckx (1947), and by Tavernier & Moorman (1954), while Stockmans et al. (1948) introduced pollen analysis as a dating tool and Verhulst (1959) started historical research on the Dunkerque transgressions and on the impact of man. Their results are summarized by Tavernier et al. (1970). An overview of more recent research can be found in De Moor & Pissart (1992a), in Maréchal (1992) and in Sommé (1988). In the Belgian coastal plain recent geological

work has especially been carried out by Baeteman (1981, 1991) in the western part, by Mostaert (1985) and by De Moor et al. (1992b) in the eastern part, while De Ceunynck (1985) studied the dunes and Thoen (1978) and Augustijn (1992) resumed the historical approach.

The present description of the Holocene coastal evolution is mainly based on published data relating to the coastal plain's sedimentary record and historical sources, without presenting new evidence. This paper for the first time synthesizes data from both the French and the Belgian parts of the coastal plain.

The present-day coastal barrier and coastal plain, whose position and shape were conditioned by the pre-existing topography, have been shaped in a very recent and short time, at least geologically speaking, known as the Holocene (table 1). Since the end of the last Ice Age, especially between 10,000 and 5,000 years ago, the global sea level underwent a rapid rise. As a result of this, the gently sloping northern part of the coastal plain

Geological time	standard classification of marine deposists
Subatlantic (from -900 on)	Dunkerque III (from 12th century on)
	Dunkerque II (+250 to +800)
	Dunkerque I (-500 to -200)
Subboreal (-3000 to -900)	Dunkerque O (?) (-1500 to -1000)
	Calais IV (?) (-2700 to -1800)
Atlantic (-6000 to -3000)	Calais III (-3300 to -2700)
	Calais II (-4300 to -3300)
	Calais I (-6000 to -4300)
Boreal (-7000 to -6000)	
Preboreal (-8000 to -7000)	

Table 1:   Subdivision of the Holocene (-8000 to present) and transgressive periods. Time is given in years AD (positive numbers) and BC (negative numbers). Based on Van Staalduinen (1979) for the Rotterdam area.

and the incised river mouths were flooded. The local rate of the Holocene sea level rise is illustrated in fig. 1.

The sea level rise still goes on today. Over the period 1927-1990, an increase of mean sea level of approximately 1.5 mm/year was measured in Oostende. In the same period, the tidal range increased by approximately 0.9 mm/year[4].

The causes of the sea level rise are complex and will not be discussed in detail in this paper. The main components of the experienced rise are generally agreed to be the melting of the Weichselian ice caps and the corresponding increase of seawater volume, the thermal expansion of sea water due to higher temperatures, and, locally in north-western Europe, a relative subsidence of the land surface to compensate for the glacio-isostatic rebound of Scandinavia after the end of the glaciation when the pressure of the ice cap was taken away. Other reasons for a local relative sea level rise include compaction of unconsolidated sediments such as clay and peat, and the continuation of restricted tectonic movements such as those related to the North Sea - Rhine graben system. Land movements during the Holocene due to tectonics and glacio-isostatics appear to be negligible in the Belgian part of the coastal plain (De Moor & De Breuck, 1973).

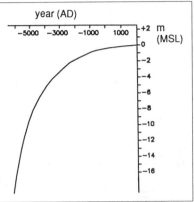

Figure 1: The rise of mean sea level, as documented by sediments in the coastal plain (after Köhn, 1989).

Though the rate of sea-level rise decreases with time (fig. 1), many scientists nowadays believe that rates may increase again as a result of the global warming, attributed to the greenhouse effect. Apart from the interesting scientific aspects, the study of the origin and development of the coastal barrier and plain is relevant to predict long-term responses of the coastal system to a possible further rise in sea level.

Another very important factor in the evolution of the coastal barrier and the coastal plain is the frequency and intensity of storms. All major breachings and inundations documented since the Middle Ages, are related to storm events. This holds true for the episodes of severe erosion in our time as well (De Wolf et al., this volume).

---

[4]

Source: Hydrographic Service, Oostende, Belgium.

The formation of a coastal barrier and the development of a coastal plain

The French-Belgian coastal barrier is essentially a long, almost rectilinear sandbody whose geomorphological units are the dunes, the beach, and the nearshore (the latter being sometimes called the underwater part of the beach). In cross section, this sandbody is typically a few hundreds of metres to a maximum of about 2 km wide, while its surface may reach heights of 15 to 20 m (fig. 2). At irregular distances the coastal barrier was interrupted by tidal inlets and river mouths, most of which are nowadays transformed to harbour access or drainage channels.

The present-day position of the coastal barrier is the result of a dynamic equilibrium between coast-building and coast-destructing forces, acting on loose grains of sand. These are supplied by coastal currents, winds and waves. Position and shape of the coastal barrier has changed significantly throughout geological and historical time. The main factor triggering these changes appears to be the changes in sea level, i.e. both the Holocene sea level rise and the changes in tidal range. Sea level strongly determines the impact of the forces acting directly on the coast : waves, wave driven currents, tidal currents, and winds. Most of these factors have a dual action, i.e. they can both be constructive and destructive forces, depending on their magnitude, frequency, direction, etc.

It is not the purpose of this paper to review the action of these natural forces, although their impact on coastal barrier formation is still a major topic in present-day coastal research.

The variability in position of the coastal barrier is illustrated in fig. 4. It is believed that, after the last Ice Age, no coastal barrier was present at the French-Belgian coast until the rate of sea level rise slowed down during the Atlantic. The reasons for sediment accretion at the southeast side of the southern North Sea, including the formation of a coastal barrier, an inshore sediment accumulation, and the development of offshore sand banks, are complex. According to Eisma (1980), the following conditions explain the trend of accretion during the Holocene :
-       the presence of large amounts of unconsolidated material, predominantly sand, deposited on the emerged shelf surface during the Ice Age low stand;
-       the continuing supply of sand and mud by the rivers;
-       the easterly, net coastward direction of coastal transport.

During the Holocene sea level rise, the coastal barrier formed at some distance seaward within the newly inundated area. The flat area between the coastal barrier and the elevated inland was regularly flooded by tide and storm water through tidal inlets, river mouths and occasional overwash events. These tidal flats, sheltered by the coastal barrier, were subject to differential sedimentation, sand being deposited in tidal channels and on tidal flats, and clayey sediments on higher marshes. The different facies of the coastal plain Holocene sediments testify of the changes in landscape that occurred throughout time. Fig. 2 is a schematic overview of the different sedimentary facies and their geographical relationship, representative of the coastal plain in the IJzer river area.

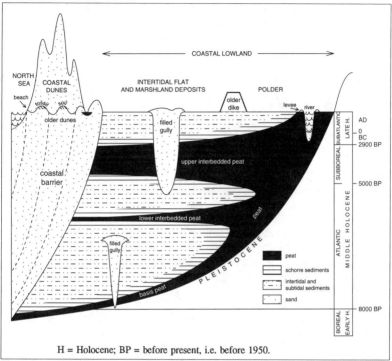

H = Holocene; BP = before present, i.e. before 1950.

Figure 2:   Simplified cross-profile representative of the coastal plain sedimentary system near Nieuwpoort and the IJzer river. Shown with strong vertical exaggeration. Modified after Köhn, 1989.

Peat layers developed in predominantly freshwater bogs, which possibly reflect a rise in ground water level induced by the rising sea level. The intervals between the peat layers consist of a complex succession of compact clays, units of interlayered sand and mud, and metres thick masses of either subhorizontally bedded or cross-bedded sand. These sediments reflect environments of coastal lagoons, salt marshes, intertidal flats, shorefaces, and tidal channels, respectively.

The succession of the Dunkerque III deposits (table 1) is probably not due to fluctuations in sea level but rather reflects fluctuations in sediment availability and supply and in hydrodynamic characteristics. Important factors in this are the geometry of the coastal barrier and the occurrence and intensity of storms, but also the growing interference of man. The more recent inundations of the coastal plain often have a differing geographical extent, due to the network of dikes gradually being established.

Geographical distribution and thickness of the Holocene sediments

The Holocene deposits in the French-Belgian coastal plain consist of two successive units, named Calais and Dunkerque deposits, which are described below. The total thickness of the Holocene sediment body may locally exceed 50 m, more particularly in the coastal dune belt. In most of the coastal plain however, the top surface of the Holocene is so flat that the relief of the Holocene base provides a good image of its thickness.

Fig. 3 is a contour line map of the Holocene deposits' base surface in the coastal plain. It shows essentially a surface dipping towards the present-day coastline and cut by the paleo-thalwegs of Aa (north of Watten), IJzer (north of Diksmuide) and Zwin (near Cadzand). Note the striking difference in thickness of the deposits between the eastern and western parts.

In the coastal plain east of Diksmuide, the Holocene base is mostly above −5 m and has a very gentle slope towards the coast. In most places, the Holocene deposits cover a basal peat that overlies Pleistocene sediments. Near the coast, the base surface slope increases and thicknesses exceed 20 m.

West of Diksmuide, the slope of the Holocene base is steeper. Thicknesses are mostly between 10 and 20 m, while in the very coastal zone the surface is incised to depths of over 25 m below mean sea level. In this western part, the contour lines suggest a more pronounced embayment during the Flandrian sea level rise than that formed by the present-day coastline.

Here also the Holocene base fossilizes an erosional relief developed in Pleistocene and locally Tertiary sediments, and west of Calais even in Cretaceous sedimentary rocks. The fossilized relief of the westernmost area shows some control by tectonics and lithology of the substratum (Sommé, 1988b). Elsewhere in the coastal plain however, there is no clear evidence for a lithological or tectonic control. The Pleistocene substratum shows no significant lithological differentiation. Neither is there any important change in substratum lithology at the transition from the thick western to the thin eastern part.

In the coastal plain few outcrops of Pleistocene sediments exist. South of Calais, the elevated remains of marine Pleistocene spits pierce the Holocene tidal flat deposits at Coulogne and Les Attaques (Sommé, 1977, 1988a, b). Near Bergues and Gistel are outcrops of the Tertiary substratum. North of Gistel, at Oudenburg, a low Tardiglacial (late Pleistocene) sand ridge was high enough to escape immersion in Holocene time.

Evolution during the Atlantic (Calais Transgression)

Calais deposits are found in most of the French-Belgian coastal plain, especially where the base of the Holocene is deeper than −5 m with respect to mean sea level: the large embayment-like area between Calais and Oostende, and some deeper seaward parts of the area east of Oostende (fig. 3). As they are overlain by the younger Dunkerque deposits, they crop out only in areas such as De Moeren at the French-Belgian border, where peat has been dug during the Middle Ages.

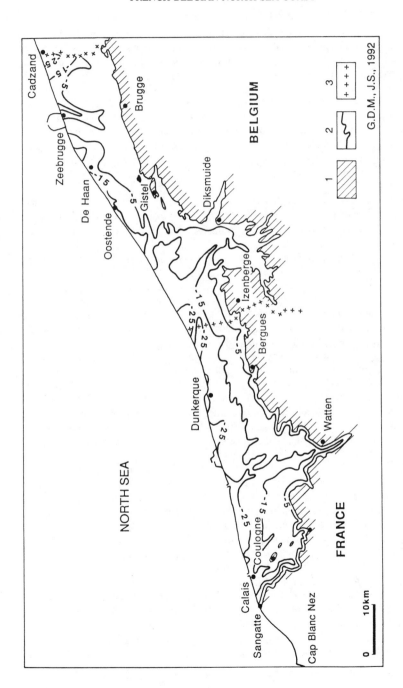

Figure 3:   Base of the Holocene deposists in the French-Belgian coastal plain.
Key: 1. Outcrop of pre-Holocene; 2. Depth contour lines of the base of the Holocene (in metres below mean sea level); 3. State boundary.

The depth and facies of dated sediments and their position in relation to the present-day coastline allow to retrace the successive locations of the coastline during the postglacial sea level rise. They show that already during the earlier stages of the Flandrian transgression the sea entered the deeper parts of the Early Holocene French-Belgian coastal zone, especially the large Aa embayment west of Dunkerque and a few smaller channels northwest of Diksmuide (Baeteman, 1981; fig. 3).

At about 8,000 years BP, at the end of the Boreal, the sea already reached Watten in the Aa estuary, 20 km from the present-day coastline, where marine sediments have been found at a depth of -17 m (Sommé et al., 1992).

A more generalized marine inundation of the Atlantic marine embayment took place around 7,000 years BP reaching a depth of -10 m. Around 6,600 years BP a larger part of the IJzer area was flooded, the sea level reaching -7 m.

A temporary outcrop in that area, described by De Moor et al. (unpublished), showed the base of the Calais deposits over a postglacial podsolic soil developed in the top of the Pleistocene sediments at a level of about -7 m. The Calais deposits probably did not develop above 0 m.

West of Dunkerque, the Calais deposits are mainly composed of tidal flat and beach sediments, deposited in the Early and Middle Atlantic. The sediments are predominantly sandy ("*sables pissards*"), though the upper part may consist of more silty or clayey intertidal deposits. In the inner reaches of the coastal plain, the deposits are significantly more silty and clayey and contain peaty horizons which possibly indicate transgressive and regressive oscillations (Sommé, 1979, 1988a, 1992; Van der Woude & Roeleveld, 1985).

More to the east the Calais deposits consist mostly of fine sands, with clayey and peaty intercalations becoming more abundant upward.

The coarser facies of the Early Atlantic Calais deposits suggests that initially no coastal barrier was present, and this was probably so up to 6,800 years BP (Baeteman, 1978). Afterwards finer-grained sediments indicate a more or less sheltered intertidal flat environment, where tidal channels supplied sediment. In this sheltered environment, lagoonal deposits, brackish reed swamps and even moors could develop when conditions were favourable.

In Atlantic times, the coastal barrier, locally covered with dunes, was less rectilinear than the present coast. East of Nieuwpoort, it had a much more seaward position than the later Dunkerque shores, while at the French-Belgian border it had a more landward position, such as indicated by the Ghyvelde Atlantic dune ridge (fig. 4). Near Calais, a remarkable remainder of the Atlantic coastal barrier is formed by the "Banc des Pierrettes", a shingle spit complex (fig. 4). This sediment body contains a lot of flint and chalk pebbles, derived by erosion from the chalk cliffs of Cap Blanc Nez. The complex spit shows a permanent vertical growth at a constant position (Dubois, 1924; Sommé, 1977, 1979).

During the Late Atlantic a marked decrease of marine influence is observed in the inner and central parts of the coastal plain, and finally peat growth became dominant at the brink of the Subboreal.

Evolution during the Subboreal

In the French-Belgian coastal plain the Subboreal was a period of decreased marine influence, in which tidal flats silted up and were increasingly sheltered by coastal shoals and barriers. A vegetation cover spreaded over most of the coastal plain. Due to the low and wet conditions, peat formation in coastal moors extended over all the lower grounds behind the coastal barrier system. Towards the end of the Subboreal, the peat grew into a raised bog, the top layers of which consisted of *Sphagnum*. Locally it formed dome-shaped bodies that would not be flooded during the later Dunkerque inundations. Peat was also present and developed to relative high levels in all of the Westerschelde area (fig. 4).

The Subboreal peat of the inner reaches of the coastal plain is mostly a single layer that today, even after important compaction due to sediment load and dewatering, locally still reaches thicknesses of 4 to 6 m. In the outer reaches, especially in the western part of the French-Belgian coastal plain, the peat often splits up into a number of peat beds, separated by sandy, silty or clayey layers. They indicate minor transgressions, due to changes in coastal barrier morphology, storm events, or short transgressive episodes.

The Subboreal peat is generally called surface peat. Where it has not been taken away by tidal channel erosion or exploitation by man, it is at the origin of many stability problems of roads and constructions in the coastal plain.

Evolution during the Subatlantic (Dunkerque Inundations)

Successive inundations occurred during the Subatlantic. The three main transgressive phases are referred to as Dunkerque I, II, and III (table 1). The latter is often subdivided in Dunkerque IIIa (12th century) and IIIb (post-12th century inundations). Dunkerque II and III are well documented by both geological and historical sources. The extent of these inundations is shown in fig. 4. In Belgium, their extents were mapped by Tavernier *et al.* (1970) using pedo-morphological characteristics and by Verhulst (1959) using historical data.

The two main regressive stages are known as the Roman and Carolingian regressions. During the regressions the emerging peat surfaces were occupied and the upsilted marshlands reclaimed. At the onset of Dunkerque IIIa, a few defensive dikes were built. Since the first Dunkerque IIIb inundations however, dikes have not only been built to defend but as well to gain land. Drainage after land reclamation has provoked differential setting and relief inversions. The latter refer to sand-filled tidal channels that after dewatering are now slightly higher than the surrounding clay and peat flats. Relief inversions are particularly clear in the area left unaffected by the Dunkerque III inundations.

At the beginning of the Dunkerque inundations, the coastline was more or less that from the Atlantic, remains of which are found in the Pierrettes spit and the Ghyvelde dunes. In the Hem-Aa estuary system, an opening towards the sea continued to exist. The coastal dunes, as relatively safe (high) and interesting (between sea and marshland) locations, are the site of early human settlements (Thoen, 1978; Termote, 1992). Remnants of Neolithic habitation have been found in the older dunes.

During the Dunkerque I transgression, most of the French part of the coastal plain was inundated, with the exception of the very southern edge of the Calais marshes near Ardres, where locally lacustrine marls were deposited. Near Calais, a new system of sandy barrier ridges covered by low stabilized dunes marked a change in the general orientation of the coastline in this area (from W-E to WSW-ENE). This change is thought to reflect erosion in the cliff region west of Sangatte

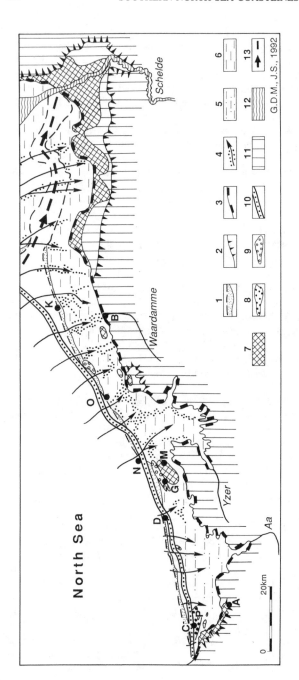

Key: 1. Present-day coastline with dune belt; 2. Not inundated parts of the coastal plain; 3. Limit of Dunkerque II tidel deposits; 4. Dunkerque II beach barrier breachings and intertidal channels with sandy infill; 5. Intertidal sand flat deposits; 6. Intertidal mud flat deposits; 7. Outcrop of surface peat; 8. Gravelly pre-Dunkerque spits; 9. Older dunes (Mid-Holocene and Early Subatlantic); 10. Beach barrier, breached at the Dunkerque II transgression; 11. Outcrop of Pleistocene deposits; 12. Pre-Dunkerque III Schelde; 13. Westerschelde Estuary, developed during the Dunkerque III transgression.

A = Ardres; B = Brugge; C = Calais; D = Dunkerque; G = Ghyvelde; K = Knokke; M = De Moeren; N = Nieuwpoort; O = Oostende; P = Pierrettes spit.

Figure 4:   The French-Belgian coastal plain and the Schelde Estuary at the Dunkerque II transgression.

(Sommé, 1979). In Belgium, the existing shore was strongly attacked and several breachings occurred during Dunkerque I (Tavernier *et al.*, 1970). The landward extension however was rather restricted and a large part of the Belgian coastal plain remained a peat surface.

During the Roman regression, widespread human occupation appeared at the southern margins of the coastal plain. Also in the coastal dunes, remains have been found of Roman occupation.

The Dunkerque II transgression was the largest in extent (fig. 4). At the southern borders of the coastal plain, Dunkerque II tidal flat deposits may directly cover the Tertiary substratum. Particularly in the area south of Dunkerque, Dunkerque II sands overlie Weichselian loess or cover sands (Paepe, 1960; Sommé, 1977).

In some areas bogs grew sufficiently high to escape flooding: fringes of the plain near Ardres and southeast of Oostende, and a wide bog named De Moeren. Also some existing dunes escaped flooding. Numerous breachings of the beach barrier occurred. East of Oostende, the coastline underwent a remarkable landward retreat. A wide gap in the coastal barrier must have existed at the Aa mouth. In this area large sand flats developed; elsewhere, the sedimentation was more clayey (fig. 4).

The origin of the recent dune belt dates back to the late Dunkerque II and the subsequent Carolingian regression. From that time, the coastal plain itself showed a scarce occupation, first for open-range sheepherding and afterwards, using dikes for the first time to protect against spring tide flooding, for growing crops. In the Belgian part of the coastal plain, areas with outcropping Dunkerque II deposits are known as "Oudland".

At the Dunkerque IIIa transgression, the extent of the inundations of the coastal plain was restricted by man-built dikes. Wide open inlets developed at the Aa and IJzer river mouths, and a tidal inlet east of Knokke, named Zwin, was considerably widened. This inlet reached its maximum development in the 12th century. Brugge was connected to the Zwin inlet by the Damme canal and this link to the open sea was fundamental to the city's famous commercial and economic bloom. The inundated areas have gradually been reclaimed, mainly during the 12th century. Areas where Dunkerque IIIa deposits crop out are named "Middelland".

In Belgium, the Dunkerque IIIb transgression primarily affected the area northeast of Brugge, especially in the present-day Westerschelde area, where very rapidly since the 12th century an important complex of tidal channels developed. These would eventually, due to regressive erosion, capture and divert the river Schelde north of Antwerpen. At the same time, the Zwin inlet was subject to rapid upsilting. Areas where Dunkerque IIIb deposits are outcropping are known as "Nieuwland".

During the Dunkerque III transgressive period, which is actually still going on, the main coastline changes can be summarized as follows.

West of Calais, an erosional tendency predominates near the Sangatte cliff. Between Calais and Dunkerque, the sea front of the estuarine zone is marked by a series of low sandy barrier ridges that successively take in more seaward positions. These tend to close the river mouths by easterly accretion. In the Gravelines area (Aa estuary), the shoreline has advanced about 2 km since the end of the Middle Ages. East of Dunkerque, the coast has retreated, but not over substantive distances. Here, older and younger dunes form an elevated single ridge subject to erosion. Also east of Nieuwpoort and Oostende, dune development went on. The major trend in coastline development from Dunkerque to Oostende was rectification through local advances and retreats of the coast. Near Knokke the coastal retreat was more important, related to the development of the Westerschelde estuary. Well-coordinated attempts to consolidate the coastline at its existing position characterize the 19th but especially the 20th century. They essentially included the construction of sea walls and groynes, and, more recently, the application of beach scraping and beach nourishments (De Wolf et al., this volume). The coastal barrier nowadays is still the sea-defence for the coastal lowlands, and this fact along with the outstanding economic and touristic significance of the sea front area itself, justifies the French and Belgian governments' efforts for the rational management and better understanding of this dynamic environment.

Conclusion

Information on the Holocene development of the French and Belgian part of the North Sea coastal plain and coastal barrier has been collected and presented in an integrated way in this paper. The reconstruction of their Holocene evolution is mainly based on the sedimentary record present in the coastal plain. The available evidence clearly shows the triggering function of the Holocene sea level rise, especially for the Atlantic Calais transgressions. The most recent coastal barrier breachings were more localized and were, in addition to sea level and tidal-range changes, more heavily related to the occurrence of storm events and the morphological balance of the coastal system.

The geologic approach provides indirect and environmental evidence for the coastal barrier dynamics. A coastal barrier has been present ever since the rate of sea level rise slowed down during the Atlantic. Important changes have occurred in its position, the main trend being a seaward shift in the embayment-like French North Sea coast and a landward movement in Belgium, especially when approaching the present-day Westerschelde estuary.

Some implications on present-day coastal management arise from the geological experience. Even though in the more recent evolution of the coastal barrier sea level rise has not been a dominant factor, it is clear that even a slight change in sea level would occasion significant and rapid morphological adaptations. Equally intense impacts are to be expected if the natural processes of sediment supply and transport are artificially disturbed. The fact that no new deposits are allowed to form in the reclaimed coastal lowlands is another cause of long-term inequilibrium in the natural coastal evolution.

References

Augustijn, B., 1992. Zeespiegelrijzing, transgressiefasen en stormvloeden in maritiem Vlaanderen tot het einde van de XVIde eeuw. Brussel, Algemeen Rijksarchief, Vol. I (p. 1-389), Vol. 2 (p. 390-731).

Baeteman, C., 1981. De Holocene ontwikkeling van de westelijke kustvlakte (België), Brussel, VUB, PhD. Thesis.

Baeteman, C., 1989. Quaternary sea level investigations from Belgium, Brussel, BGD, Professional Paper 241.

Baeteman, C., 1991. Chronology of coastal plain development during the Holocene in west Belgium, *Quaternaire*, 2, p. 116-125.

Belpaire, A., 1855. Etude sur la formation de la plaine maritime depuis Boulogne jusqu'au Danemark. Antwerpen.

Briquet, A., 1930. Le littoral du Nord de la France et son évolution morphologique. Orléans, 439 p.

De Ceunynck, R., 1985. The evolution of the coastal dunes in the western Belgian coastal plain. *Eiszeit Gegenw.*, 35, p. 33-41.

De Moor, G., & De Breuck, W., 1973. Sedimentology and stratigraphy of Pleistocene deposits in the Belgian coastal plain. *Natuurwetensch. Tijdschr.*, 55, p. 3-96

De Moor, G. & Pissart, A., 1992a. Het reliëf. In: Geografie van België. Nationaal Comité voor Geografie, Brussel, *Gemeentekrediet*, p. 129-215.

De Moor, G., Mostaert, P., Libeer, L., Moerdijk, Fl. & Van Den Abeele, V., 1992b. Geomorfologische kaart van België: kaartblad Oostende. Nationaal Centrum voor Geomorfologisch Onderzoek, 1 map.

De Putter, B., Lahousse, B., Clabaut, P., Chamley, H. & van der Valk, L., 1993. Coastal morphology and sedimentology of the Belgian, French and Dutch coast. *(This volume)*.

De Wolf, P., Fransaer, D., Van Sieleghem, J. & Houthuys, R., 1993. Morphological trends of the Belgian coast shown by 10 years of remote-sensing based surveying. *(This volume)*.

Dubois, G., 1924. Recherches sur les terrains quaternaires du Nord de la France. *Mém. Soc. Géol. Nord*, VIII, 356 p.

Halet, F., 1931. Contribution à l'étude du Quaternaire de la plaine maritime Belge. *Bull. Soc. Belge Géolog.*, 41, p. 141~166.

Köhn, W., 1989. The Holocene transgression of the North Sea as exemplified by the southern Jade bay and the Belgian coastline. *Essener Geogr. Arbeiten*, 17, p. 109-152.

Maréchal, R., 1992. De geologische structuur. In: Geofie van België. Nationaal Comité voor Geografie, Brussel, *Gemeentekrediet*, p. 37-86.

Mostaert, F., 1985. Bijdrage tot de kennis van de Kwartairgeologie van de oostelijke kustvlakte op basis van sedimentologisch en lithostratigrafisch onderzoek. Gent, R.U.G., PhD. Thesis, 588 p.

Paepe, R., 1960. La plaine maritime entre Dunkerque et la frontière belge. *Bull. Soc. belge Et. Géogr.*, XXIX, p. 47-66.

Rutot, A., 1897. Les origines du Quaternaire de la Belgique. *Bull. Soc. Belge Géologie*, 11, Mém., 140 p.

Sommé, J., 1977. Les plaines du Nord de la France et leur bordure. Etude géomorphologique. Champion, Paris and Lille, 2 vol., 810 p.

Sommé, J., 1979. Quaternary coastlines in northern France. *In* The Quaternary History of the North Sea, *Acta Universitatis Uppsaliensis*, Uppsala, p. 147-158.

Sommé, J., 1988a. La plaine maritime française de la mer du Nord : évolution holocène et héritage pléistocène. *In : Campagnes et littoraux d'Europe*, Mélanges P. Flatrès, *Hommes et Terres du Nord*, Lille, p. 273-281.

Sommé, J., 1988b. Géomorphologie de la zone terminale du Tunnel sous la Manche dans le Nord de la France. *Hommes et Terres du Nord*, 3, p. 155-161.

Sommé, J., Munaut, A.V., Emontspohl, F., Limondin, N., Lefèvre, D., Cunat, N., Mouthon, J. & Gilot, E., 1992. Weichselien ancien et Holocène marin à Watten (Plaine maritime, Nord, France). *Quaternaire* (in press).

Stockmans, F., & Van Hoorne, R., 1954. Etude botanique du gisement de tourbe de la région de Pervijze (Plaine maritime belge). *Mém. Inst. Roy. Sc. Nat. Belg.*, 130, 144 p.

Tavernier, R., 1947. L'évolution de la plaine maritime belge , *Bull. Soc. Géol. Belgique*, 56, p. 332-343.

Tavernier, R., Ameryckx, J., Snacken, F. & Farasyn, D., 1970. Kust, duinen, polders. Brussel, Nationaal Comité Geografie, Atlas van België, Blad 17, verklarende tekst, 32 p.

Tavernier, R. & Moorman, F., 1954. Les changements du niveau de la mer dans la plaine maritime flamande pendant l'Holocène. *Geologie en Mijnbouw*, 16, p. 201-206.

Thoen, H., 1978. De Belgische kustvlakte in de Romeinse tijd. Bijdrage tot de geschiedenis van de landelijke bewoningsgeschiedenis. Brussel, *Verhandel. Kon. Academie Wetensch., Lett. & Sch. Kunst. België, Klasse Letteren*, n° 88.

Van der Woude, J.D. & Roeleveld, W., 1985. Paleoecological evolution of an interior coastal zone: the case of northern France coastal plain. *Bull. Ass. Fr. Et. Quatern.*, 1, p. 31-39.

van Staalduinen, C.J., 1979. Toelichtingen bij de Geologische Kaart van Nederland 1:50000, blad Rotterdam West (37 W). Haarlem.

Verhulst, A., 1959. Historische geografie van de Vlaamse kustvlakte tot 1200. *In: Bijdragen tot de geschiedenis der Nederlanden,* 14, 's-Gravenhage, 27 p.

Morphology of the Danish North Sea Coast

P.R. Jakobsen[1], H.Toxvig Madsen[2]

Abstract

The Danish North Sea coast consists of several elements:
- *the headland coast of northern Jutland, with long uninterrupted sandy beaches;*
- *the vulnerable, sandy western coast, with several former coastlines;*
- *the Wadden Sea coast, dominated by tidal action and accretionary landforms.*

Historical coastline movements are illustrated by map compositions and by analysis of results of the West Coast monitoring system. A morphological description of five areas is given: the Skaw Spit, the Wadden Sea, the Torsminde - Hvide Sande Barrier coast, Skallingen and the Thyborøn Barrier Beach.

Today tourism and recreation set the agenda for many coastal activities, hence it is fortunate that the technical trend in coastal protection clearly is toward beach nourishment.

Geological Background

The genesis of the Danish North Sea coast has by and large its origin in the post Quaternary, except along its northern part, where headlands of chalk formations are found, between which long uninterrupted sandy beaches are suspended in gentle forms.

The ongoing battle between land and sea may best be understood by considering three postglacial extremes:
- The Ice-Sea, which left the northern Jutland as an archipelago, the evidence of which is found today in fossil coastlines in elevations of 10 - 60 m above present sea level (fig. 1 and 2).
- The Continental Period, during which the present coastline everywhere was sheltered (fig. 1).
- The Littorina or Stone Age Sea, which transgressed considerable parts of the present coastal landscape (fig. 1). The coasts of this period were contemporary to the stone age civilization.

To a large extent the present coastline in general thus represents a fragile balance between possible extremes.

This coastal evolution allows us to classify the Jutland coasts in three major categories:

---

[1] Director, Danish coast authority, Højbovej 1, DK7620 Lemvig, Denmark

[2] Head of West coast branch, Danish Coast Authority

-      The Northern Jutland headland coasts;
-      The central West Coast, which is dominated by barrier beaches, constitutes
       the most vulnerable part of our coastal system;
-      The Wadden Sea coast to the south which is part of the Frisian Wadden Sea
       coasts, extending from Den Helder in Holland to Blåvands Huk in Denmark.

The total length of the coasts is approximately 450 km, of which the headland coast
is 200 km, the central West Coast is 140 km and the Wadden Sea coast is 110 km.

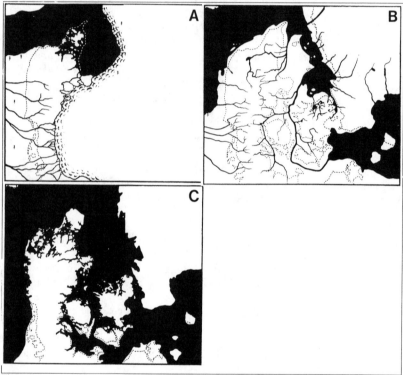

Figure 1:   The sea around Denmark in the postglacial period. A: The Ice sea; B: The sea
               during the Continental Period; C: The Littorina sea (A. Schou).

Land and Sea

The post Littorina relative rise of the northerly part of Denmark, which finds its
explanation in the glacial rebound, is presented in figure 2, which primarily has been
prepared on the basis of levelling of fossil beach ridges.

The "tilting line" theory, which is illustrated by the figure, was prevailing till the
middle of this century. However, contemporary trends show, that the North Sea coast
is now completely south of the "tilting line", which then runs north west/south east
at the Skaw, figure 3 and 4.

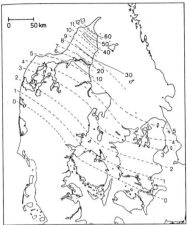

Figure 2:   The level of the Ice sea
             sediments (solid lines), and
             the level of the Littorina sea
             (dotted lines) (Mertz, 1924).

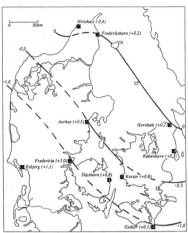

Figure 3:   An indication of the
             waterlevel rise in mm/year
             in relation to local land
             level based on 100 years of
             water level registration at 10
             different localities (Duun-
             Christensen, 1990).

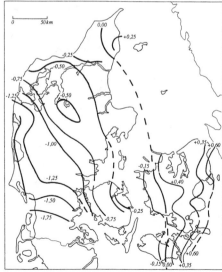

Figure 4:   The vertical land movements in
             mm/year (Remmer, 1991).

The relative sea level rise of 1 - 2 mm/year along the West Coast results in an increase in erosion level of about 0.2 m/year. This shall be viewed in relation to the fact that the erosion level of unprotected beaches along the West Coast, till the middle of the seventies, was about 0.5 m/year. So, roughly speaking the secular relative movements of land and sea may explain about half of the coastal erosion.

However, the situation during the latest 15 years has been different, since the coast has suffered an increasing number of erosion events, notably high water level combined with waves.

According to one example out of many it can be seen from the statistics in figure 5, that in popular terms the number of severe high water levels along the coast has been doubled over the latest 20 years.

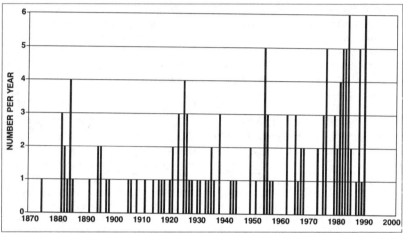

Figure 5:   Annual number of high water situations above 2.4 m DNN at Esbjerg.

We have also calculated, that the erosion level along the West Coast has raised from about 0.5 m/year to 1.8 m/year for unprotected beaches over the latest 15 years.

If the background erosion level of about 0.2 m/year caused by the secular rise in water level is deducted, it may be postulated, that the recent weather conditions have raised the erosion level from 0.3 m/year to 1.6 m/year i.e. a factor of about 5.3.

The reflections on orders of magnitude may be one of the tools for the coastal engineer assisting him in bringing some order into the daily data flow. This flow often confronts him with conflicting evidences and may leave him with a feeling of confronting a chaotic nature.

Morphological description of five areas

The coastal engineer has often chosen his profession because he is fascinated by the morphological scenarios, the understanding of which are the prerequisites for his understanding of the present day's problems.

In the following a number of interesting examples will be presented and references given to more detailed works (fig. 6).

The Skaw Spit (6,000 years)

The Stone Age man could not enjoy the huge Skaw Spit formation because it is younger than him and just about 6,000 years old.

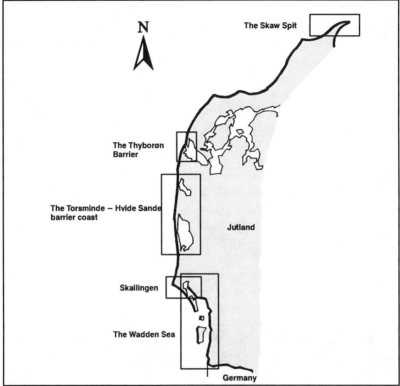

Figure 6:   Areas given a detailed morphological description.

Despite its young age, it is an impressive spit formation of active growth towards north east, with a rate of about 7 m/year. Its base at the Littorina coast is about 40 km wide, and its length perpendicular to the base about 33 km.

Today the consequences of the continued development are especially related to the town of the Skaw and the intense recreational development in this area, where two seas meet, and the light reflects from both. Its successive growth is illustrated on figures 7 and 8.

The Wadden Sea (5,000 years)
It is assumed, that the Danish part of the Wadden Sea has evolved during a long transgressional phase, which may have started acting on the land area left after the Littorina period (fig. 1).
Gradually offshore barrier islands have formed, been rolled over to their east and in modern time found a state of dynamic equilibrium at the present position (fig. 9).

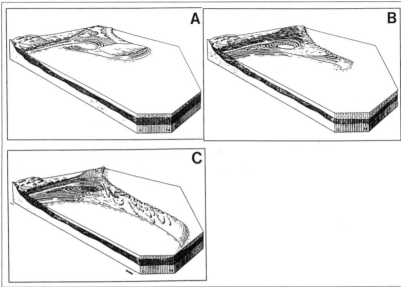

Figure 7:   The formation of the Skaw Spit by growth of the beach ridge system (Skou, 1945).

Figure 8:   The system of beach ridges at the tip of the Skaw Spit.

Dynamic equilibrium may also be a correct statement for the entire Wadden Sea as such, since it appears, that the relative sea level rise is compensated by sedimentation in the area. The large islands of Fanø and Rømø are, however, subject to continued growth towards the west. A couple of hundred years ago only the central nucleus of Rømø existed, around 1800 AD a brim of about 1 km was added and in the latest

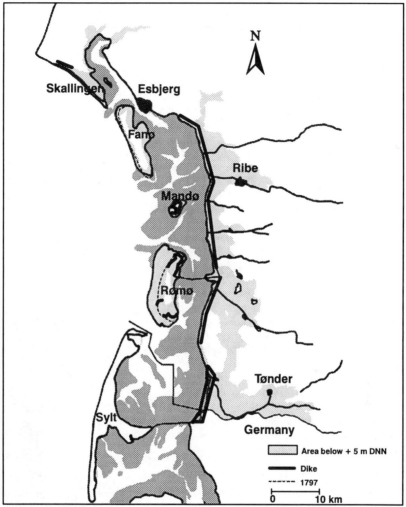

Figure 9: The Wadden Sea area.

century Juvre Sand and Sea Sand came about and have thereby added another 1.5 km to the width of the island.

If we relate this development to the low demographic pressure on the Wadden Sea islands and the peninsula Skallingen it is easy to comprehend, that major schemes for combatting classical coastal erosion have not been required in the front line. This may be in contrast to the much less favourable conditions in our neighbouring areas to the south.

However, large dike schemes for the protection against inundation of lowlands, notably valuable farmland have been implemented. A total of about 80 km of dikes fronting the sea today protects more than 40,000 ha land below +5 m DNN.

This is a classical situation for coasts flanking the Wadden Sea, but the Danish situation differs in the sense, that the degree of development of the lowlands has been relatively low except for the two urbanized areas of Ribe and Tønder. Dike reinforcement schemes in Ribe and a forward/second dikeline in Tønder were implemented during the years 1978 - 1981.

Thereby a satisfactory safety level for about 60% of the total lowland within the region has been obtained by improvement of the dikes over 25% of the total length. Comparative benefit/cost calculations have indicated that ratios for the remaining areas are well below Ribe and Tønder. Nevertheless the "roaring 80s" have justified the improvement of remaining dike sections as well.

The Torsminde - Hvide Sande Barrier Coast (1,000 years)
This part of the coast is together with the Thyborøn Barriers the most vulnerable part of the West Coast because of the narrow and low barriers between the sea and the inlets.

As can be seen on figure 10, the inlets were part of the Littorina Sea. When the water level decreased, land spits were formed from the material in the littoral drift. These spits finally closed the inlets.

In the days with no coastal protection the barriers moved east concurrently with the coastline retreat. Severe storms often resulted in breaches in the barriers. Also the entrances to the inlets moved according to the outer impacts.

Not until about 1930 the entrances were stabilized with the building of sluices and a few years later small fishing harbours were established.

To avoid inundation of the former Littorina plans it has been necessary to build dikes or reinforce the dune system on the spits between the sea and the inlets. In addition smaller dikes have been built along parts of the inlets.

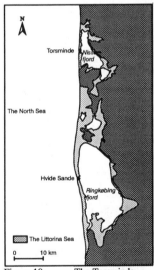

Figure 10:   The Torsminde - Hvide Sande barrier coast.

Skallingen (500 years)
It is generally assumed, that the relative sea level rise in the Wadden Sea area decreased following the Littorina period and that it has been compara-tively small

during the latest 2,000 years. Trends over the past 100 years have been 1-1.5 mm/year.

Skallingen, the northernmost peninsula in the Wadden Sea (fig. 11) is in principle a barrier island, which has emerged from the sea and partly stabilized during the past 500 years only. It is still in a transgressional phase, but comprehensive studies carried out during recent years have shown, that its main characteristics will be preserved during the process.

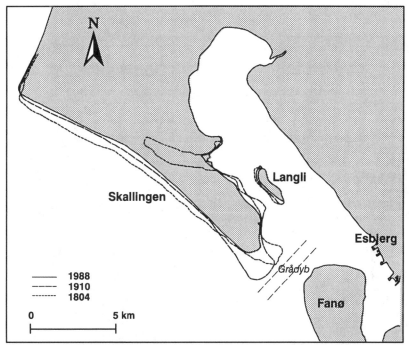

Figure 11:   Skallingen (Geographical Institute of the University of Copenhagen).

The development of Skallingen has over the latest 100 years possibly also been related to the tidal inlet of Grådyb. Grådyb gives access to the Port of Esbjerg through a dredged channel crossing the tidal bar.

The annual amount of maintenance dredging is about 1.5 Mm^3/year. During the latest 25 years, the spit of Skallingen has receded considerably towards north west (in all about 1.5 km). This trend has apparently slowed down, while a related period of growth of the tidal flat on its eastern side has gained increasing importance.

We are thus in this area exposed to a morphological complex - barrier island, tidal flat, tidal bar - in dynamic change, in a situation with considerable impact by man in terms of a dredged access channel.

The ongoing studies have clarified a number of the issues discussed, but have also shown, that it is very difficult to interrelate the causing factors.

It may be postulated, that the major effects of the dredged channel is, that we are deprived of a dynamic development of the morphological complex, which otherwise would have occurred.

The Thyborøn Barrier Beach (130 years)

No coastal case in Denmark has caused more public debate and been devoted more attention politically and economically than the Thyborøn Barriers.

It is known, that the Vikings invaded England from positions around the Lime Fiord and it is known, that during the beginning of our millennium the western part of the fiord was rich on herring, so there must have been an inlet already at that time.

During the Middle Age, read also the "Little Ice Age", the inlet closed, and the barrier was known for centuries as the "Amber Track".

During the 19th century the coastal climate increased in severity, and a breach occurred in 1825 and was followed by a permanent major breach at Thyborøn in 1862 (fig. 12). The consequence was dramatic. The now separated two barriers swung inwards towards the east, and about 5 minor fishing villages were lost during the subsequent period up to 1875, where stabilization work commenced with the building of about 80 large groynes.

Over the years the coastal development was followed with much concern because the coast continued to erode and the coastal profile steepened.

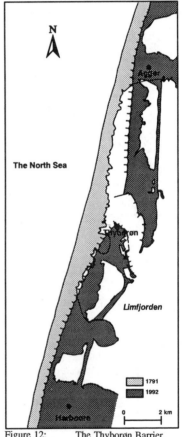

Figure 12:   The Thyborøn Barrier.

Several commissions tried to analyze the problems and the earlier ones erroneously believed in the so-called "theory of catastrophe" which predicted, that further natural development would lead to collapse
of the coastal system.

Dr. Per Bruun showed in his doctor thesis "Coast Stability" (Bruun, 1954), that the theory was wrong and that the future coastal evolution would be controlled coastal retreat, while the coastal profile would attain a new dynamic equilibrium. This theory was further substantiated by the works of Lundgreen and Sørensen and revised legislation in 1970 stamped the new technical insight.

It is the task of the Coast Authority to monitor the development and to give advise on future action. The situation is today well under control, however, it remains the most demanding coastal problem on the Danish North Sea coast and a complete review, new research and update has commenced 1992 and will continue through 1994.

References
Bruun, P., 1954. Coast Stability. Ph.D. thesis.
Duun-Christensen, J.T., 1990. *Journal of Coastal Research*, Special Issue No. 9.
Soerensen, T., 1960. The Development of Coast Profiles in a Receding Coast
    Protected by Groynes. *Coastal engineering Conference*, 7, p. 836 - 846
Mertz E.L., 1924. Overview of the Late- and Postglacial Level Changes in Danmark
    (in Danish).
Remmer, O., 1991. Unpublished results.
Schou, A., 1945. The Maritime Foreshore. Ph.D. thesis (in Danish).

The Morphology of the Anglian Coast

J. Pethick[1], D. Leggett[2]

Abstract

The morphology of the Anglian coast consists of three Integrated Scale Coastal Evolution (ISCE) units. A bay, marked by the 10m bathymetric contour, extends from Flamborough in the north to Cromer in Norfolk and is a response to north easterly extreme wave events while a series of offshore Banks extending from Suffolk to north Norfolk is a response to south easterly extreme wave events. The estuaries of the coast form the third ISCE system in which both local and external sediment sources play a part in the post-glacial morphological evolution. The management of the conventional, high water mark coast should recognise the importance of these integrated scale units if long term solutions are to be provided to coastal problems.

Coastal scale

The problem of the scale at which landform analysis and synthesis should take place has been examined by numerous geomorphologists: (e.g. Carter, 1988; Schumm & Lichty, 1965). All conclude that the spatial and temporal scales, at which landforms respond to environmental forces, are linked and that as the spatial scale of an investigation increases, so too the process elements of the landform are also modified. Stive *et al.* (1990) have developed an important model for the Dutch coast in which they identify three scales of coastal evolution. A small scale coastal evolution (SSCE) exhibits a morphodynamic length of 100m and a time scale from individual storm events to seasonal changes. A medium scale evolution component (MSCE) which has a morphodynamic scale of 1km, a time scale measured in years and a large scale coastal evolution component (LSCE) with a morphodynamic length of 10km and a time scale measured in decades.

In the present paper the MSCE and LSCE models have been incorporated into a broader integrated view of the coast: the Integrated Scale Coastal Evolution model (ISCE). This integration not only extends the time frame to include the response of the coast to extreme events with return intervals between 1:10 and 1:250 years, but also extends the spatial scale both longshore and, crucially, into the offshore zone where these extreme, high magnitude, low frequency (HMLF) waves will

[1] Institute of Estuarine and Coastal Studies, University of Hull, Hull, UK.

[2] National Rivers Authority, Anglian Region, Aqua House, Orton Goldhay, Peterborough, PE2 0ZR U.K.

break. The result is the development of a large scale coastal morphology, lying some distance from the high water mark shoreline and responding only to occasional catastrophic events but, nevertheless, a component of the coast which must be considered by coastal managers as integrating the smaller coastal units with which they are principally concerned.

Description

It is not our intention to describe the Anglian coast in detail. Readers are instead referred to the many texts which exist, such as for example that of Steers (1927, 1949, 1972) or the detailed accounts in many papers, some of which are referred to below. It is, however, important for our purposes that the reader unacquainted with the Anglian coastline be given some indication of the variation in morphology and process over its length. The account given here divides the coast into four regions for convenience, rather than intending any morphodynamic integrity within the sub-divisions.

The Holderness to Lincolnshire Coast

South of the massive chalk headland of Flamborough, the Holderness coast comprises a 45km long cliff, averaging 15m in height, cut into Devensian glacial tills. These tills suffer mass failures which cause the cliff to retreat by an average of 2m a year -a rate which has certainly been proceeding at least throughout recorded history. The debris from each cliff failure is removed by wave action and transported alongshore with a net southerly drift. A spit, Spurn Point, marks the southern limit of Holderness before the Humber estuary interrupts the coastal development.

The Humber is a macro-tidal estuary, tidal for approximately 100km inland. It is one of the more important sediment sinks on this coast, accreting at the rate of over $1Mm^3$ a year for at least the past 150 years (Thunder, de Boer & Wilkinson, 1973; Pethick, 1993). South of the Humber the coast consists of a multi-barred sandy foreshore merging into the accreting dune and salt marsh coast of north Lincolnshire (Robinson, 1970).

The Wash and its adjoining coast

South of Mablethorpe (fig. 1) the coastal zone consists of a late Holocene reclaimed marshland protected by a sea embankment and more recently an artificially nourished beach. This beach is suffering erosion and foreshore steepening (NRA, 1991). This section of the coast is terminated by the complex spit of Gibraltar Point (King, 1964) before the shore turns towards the west into the massive, Jurassic floored, embayment of the Wash. The Wash itself is progressively infilled with Holocene marine sediment at the rate of over $5Mm^3$ per year (Evans & Collins, 1987) a process accelerated by reclamation of inter-tidal areas for agricultural land. The source of this sediment is the subject of some controversy (Dugdale, Plater & Albanakis, 1987); opinion is, however, unanimous that the main source is from the North Sea but whether this is derived from the erosion of adjoining coastal regions or from a more diffuse source is presently under debate.

Figure 1:   General location map

The southern termination of the Wash embayment is marked by the chalk cliffs of Hunstanton - one of the four chalk outcrops on the Anglian coast. To the east the coast of North Norfolk runs east-west and is characterised by a number of alternating units of sand dune, salt marsh, open beach, barrier islands and spits, including the classic coastal landforms of Scolt head Island and Blakeney Point and also less well known, but geomorphically important, features such as the Stiffkey Meals, a system of low water sand bars and dunes. This complex coast has been described best by Steers (1972) but the extensive literature includes, for example, studies by Hardy (1961); Pethick (1980); Funnell & Pearson (1989); Shih-Chiao & Evans (1992).

The Norfolk/Suffolk coast

The north Norfolk coast terminates at Weybourne where the Anglian glacial tills form a 15m to 20m high cliff. This till lies on a chalk abrasion platform which outcrops at low water and forms the third of the four chalk outcrops on the Anglian coast. These cliffs are eroding at between 1 and 2m per year (Cambers 1975). Longshore sediment transport directions here are complex with a major divide occurring at Cromer. To the west, longshore transport has a slight westerly net drift while to the east the movement is southerly (Shih-Chiao & Evans, 1992) although Clayton, McCave & Vincent (1983) suggested that an offshore move-ment of material takes place at Cromer, feeding an offshore sediment bank known as the Norfolk Banks. The eroding glacial till cliffs of Norfolk continue south to Happisburgh where a narrow sand dune ridge, artificially maintained, marks the edge of a coastal zone marshland stret-ching inland to the Norfolk Broads. This dune and marsh coast continues south to Lowestoft, broken only by the now urba-nised spit at Great Yarmouth, which has developed southwards across the mouth of the River Yare. At Lowestoft the southerly sediment movement is inter-rupted by another offshore movement along a transport pathway known as a Ness (McCave, 1987). South of Lowestoft the cliffed coast reappears, cut
in glacial till of Anglian age, which is eroding at the rate of approximately 1m a year. At this point on the coast the net southerly drift of sediment derived from the erosion is again interrupted by offshore movements at Nesses of which several examples have been noted in the literature: Winterton, Lowestoft, Benacre Ness and Thorpeness (fig. 2). The sediment moving offshore from the coast is held in an offshore bank - which extends from Thorpeness in a north easterly direction for approximately 50km (fig. 2).

The southern estuaries

South of Thorpeness the coastline forms a deep and complex embayment formed by series of estuaries. They include the Deben, Orwell, Stour, Blackwater, Crouch, Thames and Medway (fig. 1). The geology of this area is London Clay and Pliocene Crag with only a thin veneer of glacial till in some places. These estuaries have received relatively little attention in the literature but there is some evidence to suggest that they are no longer acting as net sediment sinks, although it is clear from their depositional history that they must have done so until comparatively recently. A discussion of the sediment balance of these southern estuaries is given later in this paper.

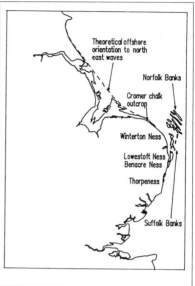

Figure 2:   Nearshore topography of the Anglian coast (10m bathymetric contour)

Integrated scale coastal evolution

The ISCE hypothesis proposed above suggests that, on a coast such as that of the Anglian region, with few geological controls, coastal evolution takes place at a scale which is determined by high magnitude events occurring with low frequency (HMLF). Such events will set up morphological responses further seaward than the conventionally defined upper or high water shoreline. The morphological response to such extreme events is not expected to be different in kind from that at any other scale, for example the shoreline profile will be expected to develop a slope and length appropriate to the dissipation of wave energy while the orientation of the shoreline will be such that long shore transport will be reduced to a minimum, a limit set perhaps by sediment inputs from rivers or coastal erosion (Komar, 1976).

The intention here is to identify the maximum scale of coastal unit which can be expected to respond in a unified manner to such extreme wave events. In order to do so we begin by examining two coastal areas whose morphodynamic scale is considerably larger than the ~10km proposed under the Stive et al. (1990) model for large scale coastal evolution. The first example we use is that of the Holderness coast in the extreme north of the Anglian region: a 45 km cliff coastline stretching south from Flamborough Head. Our second example is the Norfolk/Suffolk coast stretching south from Cromer in the north to Thorpeness in the south. Both coasts have been briefly described above.

The Holderness coast

Work by Pethick, Leggett & Lorriman (1980) on the sediment budget of this coastline shows that the annual sediment input into the shoreline from mass failures of the glacial till cliffs is approximately 1Mm³ of fine grained sediment (<100µm) and 250K m³ of coarse sediment (>100µm). Examination of the mechanism by which this debris is transported away from the coast by wave action reveals that, as may be expected, the fine fraction is carried as suspended load and diffuses seawards while the coarse fraction is transported as longshore bed-load and is dependent on wave approach angle and wave energy. A statistical analysis of the sediment transport movements over a year, measured at 3 hourly intervals, showed no significant net transport in either direction for wave return periods of less than 8 months. When the return intervals for > 8 months were included, however, it was found that a net southerly drift was imparted to the sediment transport. It appears that the density of wave energy – approach angle for this part of the North Sea is asymmetric since the longest fetch is to the north east. Thus, as longer return intervals are examined, the probability increases that these larger waves will come from the north east. More significant in the context of the present paper is the fact that the analysis showed that when waves up to and including a return interval of 15 months were examined the cumulative net transport to the south was 300,000m³ - the volume of sand-sized material input into the shore system by the mass failure mechanism during that period.

It appears from this that work that the Holderness coastline has become adjusted to the prevailing wave climate in such a way as to maintain a net sediment transport which is exactly that input to the system from independent, mass failure, sources. Such a conclusion is not unexpected (see for example fig. 10-11 in [Komar, 1976]); what is interesting is that the whole of this 45km coast has adjusted as a single unit, so that the whole coast has oriented itself to the prevailing wave approach angle in such a way as to provide exactly that longshore energy gradient needed to transport the input of sediment from individual mass failures. We suggest that this is an example of an integrated scale coastal evolution - one in which each smaller unit contributes sediment and energy in the longshore direction but the whole coast reacts in a co-ordinated manner within a longer time frame.

A significant corollary to this conclusion is the fact that wave return intervals greater than 15 months did not increase the volume of southerly transport. Instead the shoreline of the entire coast develops a series of offshore bars during extreme wave events which are oriented in a south east -north west direction. These bars develop a length of over 1km during extreme events (>15 month wave return intervals) and are shore attached at their northern end. Their orientation is normal to the extreme wave approach angle so that sediment transport along their length is minimised at these times. Thus two forms of shore morphology have evolved along this Holderness coast. The first of these is one which is adjusted to low magnitude, high frequency (LMHF) events and which maintains a morphological equilibrium despite the continuous input of large volumes of sediment from cliff failures. The second morphological shore is one which responds to HMLF events and therefore, as envisaged under our hypothesis, is formed to seaward of the LMHF event shore. It can be argued, of course, that this series of offshore bars does not constitute a shore morphology in the normal sense of the term especially since they are an ephemeral feature. Nevertheless, we consider that this offshore morphology is an essential part of the ISCE model, allowing the entire coast to develop a co-ordinated response to wave and sediment inputs so as to maintain a morphodynamic equilibrium.

The Norfolk-Suffolk coast

The main morphological features of this section of the coast, running from Cromer in the north to Thorpeness in the south, have been described above. The upper shore consists of two eroding cliff sections (Cromer to Happisburgh and Lowestoft to Thorpeness) interrupted by a section of low backshore fronted by a narrow dune ridge. Offshore a marked and permanent bar is present - the Banks (fig. 2). The wave climate of this section of the coast differs in one important respect from that of the Holderness coast. Although the LMHF waves maintain the same north easterly asymmetry, the HMLF waves ap-proach from the south east rather than the north east (fig. 3).

We suggest that this distinction results in a composite shoreline in which the inner or landwards shore is a response to the north easterly LMHF wave events but the outer Banks are a response to the HMLF waves approaching from the south east. Such an hypothesis would explain the controversy which has been rehearsed in the literature over the years (Carr, 1981; Lees, 1981; McCave, 1987; Robinson, 1966) as to the direction of sediment transport on the Banks. Sediment transport on these offshore banks may only be expected to occur during extreme wave events when the net drift would be north eastwards. At other times, sediment movement along the landward or upper shoreline, inshore of the Banks, is in a southerly direction. Sediment is brought south by LMHF waves events to the Nesses along this shore where it is carried seawards to the Banks and subsequently moved back north along the Banks during extreme wave conditions.

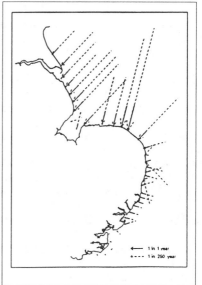

Figure 3:   Wave approach angles on the Anglian coast

The protection afforded to the inner coast by the offshore Banks was noted by Carr, (1979) who calculated that waves with H_s greater than 2.2m would break on the Banks and not propagate landwards to the inner shore, thus he envisaged, as we do here, the existence of two 'shorelines': an upper low magnitude shore marked by the glacial cliff and a net southerly sediment transport and an outer, high magnitude shore, the Suffolk Banks, exhibiting a net northerly sediment transport. The Norfolk-Suffolk coast may therefore be likened to a barrier island coast in which the outer barrier responds to extreme wave events, while protecting the inner lagoon from all but the smallest waves. In the case of the Anglian coast the tidal range is too great to allow true barrier island formation in the sense of a well developed supra-tidal beach (Hayes, 1975), but the wave induced morphodynamics on barrier island and submerged bank are, nevertheless, similar.

The integrated Anglian coast

Integration of coastal process and response scales is essential if effective strategies are to be developed for managing the coast. The interaction of time and space on the coastline is highly complex and we have argued that the ISCE model, developed above, helps to conceptualise this system. The adoption of an ISCE scale for the analysis of the Anglian coast allows the identification of a basic morphological   unity. An inner inter-tidal shoreline, developed in response to LMHF waves, is protected by an outer sub-tidal shore which is affected by HMLF waves. The difference between the Holderness coast and the Norfolk/Suffolk coast, identified above, lies in the direction of the extreme wave events. In the case of the Holderness coast both LMHF waves and HMLF waves approach from the north east so that the two morphological responses run almost parallel to each other and the sequence offshore bars meet the upper shore in a series of *en echelon* ridges. In the Norfolk/Suffolk case, however, the approach angles for LMHF and HMLF waves differ by almost 90° so that the inner and offshore morphologies form an acute angle to each other with apex at their common origin at Thorpeness, north of which they lie at an increasing distance from each other.

Between these two coastal sections, however, lies the embayment of the Wash and its adjacent coastal areas of Lincolnshire and North Norfolk. In order to identify a complete Anglian coastal scale it is necessary to consider the links between these areas. It is relatively simple to identify in the Wash and its adjacent coasts a series of medium to large size coastal units. There has, however, been no attempt to examine the integration of these units into a single interacting coastal model. The analysis presented above for the Holderness and Suffolk coasts suggests that such a scale of interaction should be looked for in the offshore zone rather than attempt to link together high water landforms which represent a response to low magnitude wave events.

This section of the Anglian coast, however, differs from those already described in that it has two significant geological controls. Once these are recognised the integration of the medium to large scale coastal units becomes more obvious. Two chalk outcrops occur on the North Norfolk coast - one at Hunstanton where it forms a 20m eroding cliff face, the other between Cromer and Weybourne, where it forms the low tide platform lying below the glacial till of the Anglian glaciation. We suggest that this latter outcrop at Cromer, despite the fact that it does not form a major cliff and indeed remains largely unrecognised, forms a major natural headland on the Anglian coast, equal in importance to the massive headlands of Flamborough or the North Foreland. This chalk outcrop acts as the southern headland of a bay whose northern headland is Flamborough Head and within which the orientation of the offshore morphology is a direct response to north easterly, high magnitude wave events constrained at either end by the geological structures. The 10m contour acts as a convenient cartographic approximation to this offshore coastline; fig 2 shows that it forms a bay running across the Wash re-entrant and merging ultimately with the Holderness coast. Towards the southern end of this outer 'bay' it merges with the upper shore morphology of the north Norfolk coast - forming the barrier islands and spits of Blakeney and Scolt. The second chalk headland at Hunstanton lies within the Wash embayment where it is protected from HMLF wave attack by the offshore morphology although its cliff is eroded by LMHF waves. Along the Weybourne to Cromer section, however, erosion of the headland by both LMHF and HMLF wave events has resulted in deep water close onshore so that here the offshore and onshore morphologies are superimposed. It is interesting to note that south of the Cromer headland the coast resumes its orientation to north easterly wave approach angles as far south as Happisburgh suggesting that it is only the presence of the Cromer chalk outcrop which forces the east-west orientation of the north Norfolk offshore morphology.

The estuaries

It is not possible in the context of this paper to provide an account of the morphology of the estuaries of the Anglian coast. The Humber, Wash and the southern estuaries complex can only be briefly examined here in relationship to the morphological development of the open coast described above and to the general problem of sediment supply. Several authors have attempted to provide a macro-sediment budget for this coast for example Eisma (1981); Eisma & Kalf (1987); McCave (1987); Pethick (1992); Shih-Chiao & Evans (1992). Some of the problems facing such an evaluation may be summarised using the Humber estuary as an example. Here Pethick, (1993b) has shown from yearly bathymetric surveys made over the past 80 years that fine sediment accretion (<100µm) in the Humber inter-tidal zone is taking place at an average rate of 0.84Mm³ per year while coarse sediment accretion in the estuary sub-tidal channels amounts to 0.6Mm³ per year – although much of this latter amount may be recycled dredged material from navigation channels. These amounts may be compared to the sediment production from the erosion of the Holderness coast, which, as described above, amount to 1.0Mm³ of fines and 0.25Mm³ of coarse sediment. No direct connection between this sediment source and sink is implied here but nevertheless it is clear that an approximate balance is achieved between the two.

That such a balance exists may at first sight appear obvious enough until the interaction between these two coastal units is examined. The demand for sediment within an estuary such as the Humber during the Holocene is made of two main components: the initial response of the estuarine morphology to the post-glacial rise in sea level and, once such an equilibrium morphology has been achieved, the continuing demand for sediment required for its maintenance under a slow tectonic subsidence. The provision of sediment from the open coast to the estuarine sinks is, as we discuss above, a long term response to wave climate. There appears to be no feedback from sink to source which could control the amount of sediment being released from such erosion. This being so it is difficult to see why a balance should exist between source and sink as demonstrated by the Humber estuary budget outlined above. The only explanation can be that the demands of the sink are always greater than the supply from the source so that the apparent balance is in fact the manifestation of an under supply and may suggest that the Humber is still demanding large amounts of sediment in order to achieve its initial morphological response to the post glacial sea level.

If this hypothesis for the Humber is now applied to the Wash it will be seen that the under supply here is even more marked since the only adjacent source is the Norfolk coast - apart from the Holderness coast whose supply is entirely utilised, directly or indirectly, in the Humber sink. The Norfolk cliff erosion is estimated to supply some 1.3Mm³ of sediment to the North Sea system (McCave, 1987). Even if this were to be supplied in its entirety to the Wash it would not satisfy the sink there whose sedimentation rate is presently estimated at between 2.6 and 5.4Mm³ per year (Dugdale *et al.*, 1987).

The situation in the southern estuaries is more difficult to assess due to a lack of any data, even the type of general estimates provided above for the Wash and Humber. Frostick & McCave (1979) in their study of the Deben suggested that most of the accretion here was merely internal redistribution caused by winter storms. Pethick (1993a) used data from two years of continuous monitoring of an estuarine cross section in the Blackwater estuary to provide a first approximation of a sediment budget for the whole estuary . This showed that the Blackwater is suffering net loss of sediment but that this loss amounts to only 1000m³ per year - a negligible amount which suggests that the estuary can be regarded as in long term sedimentary balance but with minor adjustments occurring to its morphology. Kirby (1990), in his study of the Medway estuary in the extreme south of the Anglian coast, suggested that his results would extrapolate to give a net sediment deficit of approximately 2.0Mm³ per year for the whole estuary. If this figure is accepted then it indicates a major reorganisation of the estuarine geomorphology, which Kirby suggest began some 200 years ago.

All of these preliminary findings suggest that the southern estuaries are not demanding more sediment than can be supplied by the available sources. Thus these estuaries can be regarded as quite distinct from the northern estuaries of Wash and Humber. The reasons for this distinction may be connected to the supply of sediment in the southern North Sea. Eisma (1981) and Eisma & Kalf (1987) showed that the anti-clockwise flow of suspended sediment in the North Sea led to the development of a zone of high sediment concentration within a tidal gyre off the Anglian coast (fig. 4). Further north, the southerly movement of suspended sediment prevents any major suspended concentrations or related deposition so that any material deposited in these areas is, according to Eisma, derived from local sources. Thus, it could be suggested that the southern estuaries, lying within the high suspended sediment zone of the southern North Sea, have not needed to rely on local sources to supply the material needed to develop their post-glacial morphology. Accordingly, these estuaries have already attained some degree of equilibrium, while the northern estuaries, relying to a far greater extent on the local sources of suspended material have not yet reached any such equilibrium and thus continue to demand more sediment than can be supplied to them - leading to a sediment budget in which demand is always greater than supply.

Discussion and conclusions

The analysis presented above suggests that the Anglian coast is composed of three superimposed ISCE units. The first of these is a bay, lying between the headlands of Flamborough and Cromer, within which the orientation of the offshore morphology is shown to be a response to extreme north easterly wave approach angles but is constrained by the two headlands. In the northern part of this bay the Holderness coast exhibits a complex dual morphology in which the inner shore is a response to sediment transport of cliff erosion debris, while a series of outer sand bars reduce the sediment transport during extreme wave events to a minimum. The middle of this bay is marked by the 10m bathymetric contour which again merges with the upper shoreline towards the southern headland at Cromer (fig. 2).

The second integrated coastal scale model lies south of Cromer where the offshore morphology - the

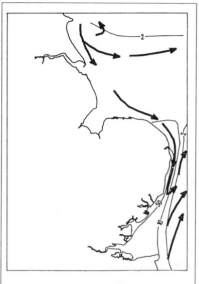

Figure 4:   Suspended sediment concentration (parts per million) and transport directions on the Anglian coast (from Eisma, 1987)

Norfolk/Suffolk Banks - are a response to extreme wave events which here exhibit a south easterly approach angle - quite distinct from that of the first ISCE unit in the north of the Anglian coast. It is suggested that the outer Banks are analogous to barrier islands chains forming on coasts with a lower tidal range. The apex of the acute angle formed by these inner and outer coastal morphologies lies at Thorpeness and marks the southern termination of the ISCE unit. Towards the north of this coastal section, that is north east of Cromer, these Banks come under the increasing influence of the north easterly wave approach angle of the Flamborough to Cromer ISCE unit. Here the Norfolk Banks lie at an increasing distance offshore and break down into a series of disconnected sand waves whose orientation is parallel to the Holderness and Lincolnshire coasts and mark a transition zone between the two ISCE units.

South of Thorpeness lie the complex estuarine embayment of the southern estuaries and these form part of the last of the three ISCE units - the estuarine unit – which also includes the composite estuary of the Wash and the single estuary of the Humber. These estuaries are shown to occupy both different time and spatial scales from the integrated open coast model. Their long term adjustment to post-glacial sea level change involves sediment supply from a number of different sources, some external to the North Sea, some more local. It is suggested here that the northern estuaries of Humber and the Wash have relied on local sources for their post-glacial development - sources such as the eroding cliffs of Norfolk and Holderness. This supply has been inadequate to allow rapid development of a morphodynamic equilibrium so that the demand for sediment from both these estuarine areas is in excess of the supply. The suspended sediment transport system of the North Sea bypasses these northern estuaries, sweeping sediment south from its source in the North Atlantic to the coast of Norfolk and Suffolk it where it moves across the North Sea towards the Wadden Sea. The tidal gyre set up along the Norfolk/Suffolk coast by this movement traps suspended sediment and this has allowed extremely high sediment concentrations to develop here. The Holocene demand for such sediment from the developing southern estuary system has therefore been adequately met by this North Sea source so that their present demand is minimal and indeed there is some evidence that they are now marginal net exporters of sediment. The estuaries of the Anglian coast therefore may be seen as superimposed on the integrated coastal scale model described above for the open coast - rather than forming part of it. The only interaction between open coast and estuaries is a one way system - with no feed-back between sink and source and even this interaction is confined to the northern estuaries.

The Anglian coast has been described in this paper as one in which the conventional, high water mark, shoreline is only one of several different morphological responses to waves and tides. At the Integrated Scale for Coastal Evolution an offshore morphology exists which is a response to extreme wave events with return intervals of up to 1: 250 years, while the estuarine morphology of this coast is evolving in response to residual currents in the North Sea over the inter-glacial time scale. The complex interaction both within and between these coastal scales is such that any attempt to understand or manage this coastline must initially recognise the largest scale units upon which the development of all the other units is dependent. It is hoped that this paper will provide a catalyst for the development of coastal management programmes in which effective long term solutions are seen to rely on an understanding of such large scale landforms.

References

Cambers, G. C. (1975). Sediment transport and coastal change. (East Anglian coastal research programme. No. 3). University of East Anglia.

Carr, A. (1981). Evidence for the sediment circulation along the coast of East Anglia. *Marine Geology*, 40, M9-M22.

Carr, A., King, H., Heathershaw, A., & Lees, B. (1981). *Sizewell-Dunwich Banks Field Study. Topic Report No6. Wave data observed and computed climates*. No. IOS Report No128).

Carr, A. P. (1979). *Sizewell-Dunwich Banks Field Study* IOS Report No. 89. IOS.

Carter, R. B. G. (1988). *Coastal Environments*. Academic Press.

Clayton, K., McCave, I. N., & Vincent, C. E. (1983). The establishment of a sand budget for the East Anglian coasta and its imlications for coastal stability. In Inst Civ. Engineers (Eds.), *Shoreline Protection*

Dugdale, R., Plater, A., & Albanakis, K. (1987). The fluvial and marine contribution to the sediment budget of the Wash. In P. Doody (Ed.), *The Wash and its environment*, NCC.

Eisma, D. (1981). Supply and deposition of suspended matter in the North Sea. In S.-D. Nio, R. T. E. Schüttenhelm, & C. E. Van Weering Tj (Eds.), *Holocene Marine Sedimentation in the North Sea Basin.* (pp. 415-428). Int. Assoc, Sediment. Spec. Publ.

Eisma, D., & Kalf, J. (1987). Dispersal,concentration and deposition of suspended matter in the North Sea. *Journal of the Geological Society,* 144, 161-178.

Evans, G. (1965). Intertidal flat sediments and their environment of deposition in the Wash. *Q.J. Geol. Soc. Lond.,* 121, 209-245.

Evans, G., & Collins, M. (1987). Sediment supply and deposition in the Wash. In P.Doody & B. Arnett (Ed.), *The Wash and Its Environment,* (pp. 48-63). Horncastle Lincolnshire.: Nature Conservancy Council.

Evans, G., & Collins, M. B. (1975). The transportation and deposition of suspended sediment over the intertidal flats of the Wash . In J. Hails & A. Carr (Eds.), *Nearshore sediment dynamics and sedimentation.* (pp. 273-304) Wiley Interscience.

Frostick, L. E., & McCave, I. N. (1979). Seasonal shifts of sediment within an estuary mediated by algal growth. *Est. Coast. Mar. Sci.,* 9, 569-576.

Funnell, B. M., & Pearson, I. (1989). Holocene sedimentation on the north Norfolk barrier coast in realtion to realtive sea level changes. *J Quat Sci,* 4(25-36).

Hardy, J. (1961) Coastal changes in Norfolk. PhD, University of Cambridge.

Hayes, M. O. (1975). Morphology of sand accumulation in estuaries. In L. Cronin (Eds.), *Estuarine Research.* New York: Academic Press.

King, C. A. M. (1964). The characteristics of the offshore zone and its relationship to the foreshore at Gibraltar Point, Lincolnshire. *East Midland geographer,* 3(5), 230-243.

Kirby, R. (1990). The sediment budget of the erosional inter-tidal zone of the Medway estuary, Kent. *Proc Geol. Assoc.,* 101(1), 63-77.

Komar, P. (1976). Beach Processes and Sedimentation. Prentice Hall.

Lees, B. (1981). Sediment transport measurements in the Sizewell-Dunwich Banks area, East Anglia, UK. In S.-D. Nio, R. T. E. Schüttenhelm, & C. E. Van Weering Tj (Eds.), *Holocene Marine Sedimentation in the North Sea Basin* Oxford: Blackwell.

McCave, I. N. (1987). Fine sediment sources and sinks around the East Anglian Coast (UK). *Journal of the Geological Society of London,* 144, 149-152.

NRA (1991). Shoreline Management: The Anglian Perspective Sir William Halcrow & Partners.

Pethick, J. (1980). Salt marsh initiation during the Holocene transgression: the example of the North Norfolk marshes, England. *J.Biogeog.,* 7, 1-9.

Pethick, J. (1992). Strategic planning of coastal defences. In MAFF (Ed.), *Conference of River and Coastal Engineers,* Loughborough:

Pethick, J. (1993a). The geomorphology of the Blackwater estuary English Nature.

Pethick, J. (1993b). The geomorphology of the Humber Estuary Report to NRA.

Pethick, J., Leggett, D., & Lorriman, N. (1980). Sediment budget calculations for the Holderness Coast. No. Holderness Joint Advisory Committee.

Robinson, A. H. W. (1966). Residual currents in relation to shoreline evolution of the East Anglian Coast. *Marine Geology,* 4, 57-84.

Robinson, D. N. (1970). Coastal evolution in north east Lincolnshire. *East Midland Geographer, 5*, 62-70.

Schumm, S. A., & Lichty, R. W. (1965). Time, space and causality in geomorphology. *Amer. J.Sci., 263*, 110-19.

Shih-Chiao, C., & Evans, G. (1992). Source of sediment and sediment transport on the east coast of England: Significant or coincident phenomona? *Marine Geology, 107*, 283-288.

Steers, J. A. (1927). The East Anglian Coast. *Geog J, 69*, 24-48.

Steers, J. A. (1949). The coast of East Anglia. *New Naturalist, 6.*

Steers, J. A. (1972). Coastline of England and Wales. Cambridge University Press.

Stive, M. J. F., Roelvink, D. A., & H.J., d. V. (1990). Large scale coastal evolution. In C. J. Louisse, M. J. F. Stive, & H. Wiersma (Ed.), *22nd International Conference on Coastal Engineering,* (pp. 9.1-9.13). Delft: Delft Hydraulics.

Wilmot, R. D., & Collins, M. B. (1981). Contemporary fluvial supply to the Wash. In S.-D. Nio, R. T. E. Schüttenhelm, & C. E. Van Weering Tj (Eds.), *Holocene Marine Sedimentation in the North Sea Basin* Oxford: Blackwell.

Morphology of the Wadden Sea
natural Processes and human Interference
J. Ehlers[1], H. Kunz[2]

Abstract

The Wadden Sea can be subdivided in three morphodynamic units: the tidal flats, the inlets and the barrier islands. The deeply incised tidal inlets are the most active of these units; their lateral displacement caused what has become known as 'island migration' in popular literature. The larger tidal streams of the inner German Bight (between Jade Bight and the North Frisian Islands) are equivalents of the inlets of the barrier coast, being more mobile than the latter. Coast-parallel sand transport is largely restricted to the seaward side of the barrier. Today, coast lateral sediment movement does almost exclusively take place through the tidal inlets. Overwash processes on the islands are prevented by sand dikes and dune protection. Especially the highly developed barrier islands with their holiday resorts are protected by coastal engineering structures and beach nourishment. Future development should be in closer agreement with the demands of coastal protection and nature conservation.

Introduction

The Wadden Sea (fig. 1) is an area of intensive morphological change. The geological setting of the East Frisian coast has been described by Streif (1990), Cameron *et al.* (1993, this volume) and its historical development particularly by Homeier (1962, 1969). The underlying morphogenetic processes have been summarised by Lüders (1953) and Ehlers (1988). The general influence of the tidal range on coastal morphology was pointed out by Davis (1964) and Hayes (1975). There is a close relationship between the variation of tidal range (fig. 2) and the shape of the coastline: With increasing tidal range the straight coastline of Holland is replaced by the West and East Frisian barrier islands and in the inner part of the German Bight by an open estuarine coast. North of the Elbe estuary the tidal range is decreasing again, and north of the Eiderstedt peninsula the barrier of the North Frisian Islands begins. In South Jutland again the low amplitude of tidal range has resulted in the formation of a straight coastline with coast-parallel dune ridges. The classification in fig. 2 only relates to the tidal range. Hayes (1975) has additionally considered the role of local wave climate on coastal morphology. In fig. 3 the boundary conditions

---

[1] Geologisches Landesamt, Oberstr. 88, D2000 Hamburg 13, Germany

[2] Niedersächsisches Landesamt für Ökologie, Forschungsstelle Küste (Coastal Research Station), An der Mühle 5, D2982 Norderney, Germany

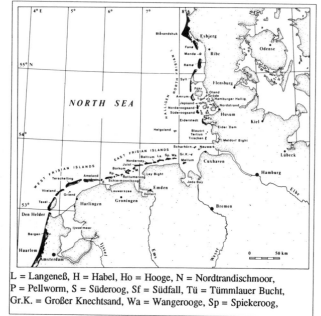

L = Langeneß, H = Habel, Ho = Hooge, N = Nordtrandischmoor,
P = Pellworm, S = Süderoog, Sf = Südfall, Tü = Tümmlauer Bucht,
Gr.K. = Großer Knechtsand, Wa = Wangerooge, Sp = Spiekeroog,

Figure 1:   Location map. The coastal dunes are shown in black
(from Ehlers, 1988).

for the East Frisian tidal inlet Otzumer Balje, as an example, have been evaluated
with respect to his classification: That part of the coast is a mixed energy tide-
dominated coast. If the landform assemblages are taken as a category for the
distinction of the different types of coastline, however, areas with tidal over 2.90 m
already belong to the macro-tidal zone (fig. 4).

Morphodynamic units

Tidal flats
The topography of the tidal flat areas is mapped and updated periodically at irregular
intervals. The tidal flats of the German Bight consist mainly of silt and fine sand,
with variable amounts of medium sand. Coarse fractions only occur in overdeepend
sections of the tidal inlets and at the bottoms of tidal streams which are eroding into
the gravel-bearing strata of the subcropping Pleistocene deposits. Mud flats are
mostly limited to tidal creeks margins and to the mainland coastal areas. The mud
flats in some cases have a slightly irregular surface; in those parts the topography is
governed by large colonies of edible mussels (*Mytilus edulis*) which act as mud
accumulators. The mussel beds can be seen as dark rises above the tidal flat surface
from great distance. They form hummocky, often net-like structures easily detectable
on aerial photographs.

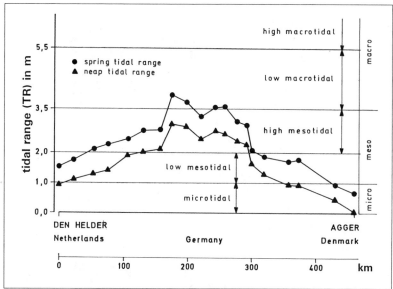

Figure 2: Tidal range (TR) variation at the Wadden Sea coast of the southern North Sea and classifications only based on TR by Hayes (1975)

The sand flats have relatively smooth surfaces which appear light coloured on aerial photographs. However, a more detailed examination reveals that the surface is not absolutely plane. The surface not only is covered by small ripples but also it is subdivided into numerous small-scale form elements. Part of the microrelief is formed by more or less radial rills created by water run-off during ebb phase. The high flats can be sculptured by the wind, and areas near tidal creeks often display densely spaced stream stripes (Ehlers, 1988). Even if drift currents occur (during strong, mainly westerly winds, and in shallow water), the actual water flow direction is to a certain degree controlled by the configuration of the tidal creeks and the tide (ebb, flood). In deep tidal creeks (channels) the wind influence is negligible; the currents are dominated by the tides.

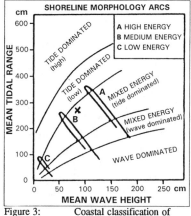

Figure 3: Coastal classification of Hayes (1975) based on Tidal Range and Wave Height; x: Otzumer Balje tidal inlet, East Frisian Wadden Sea

Fig. 6 illustrates schematically how water can be driven over the tidal flats and finally concentrate and drain through the deeper tidal creeks in another direction.

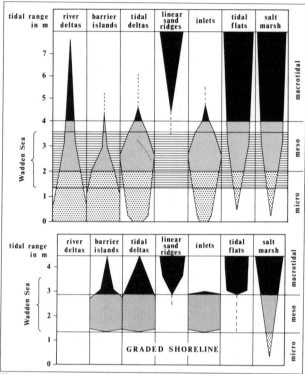

Figure 4:   Morphological variations of coastal plain shorelines
                 depending on different tidal range: (4.1) after Hayes
                 (1975), and (4.2) with regard to the North Sea coast after
                 Ehlers (1988).

Tidal currents determine the shape of the gullies and their vicinity. It is now well-established that a dynamic equilibrium exists between morphologically stable high tidal flats and waves entering from the North Sea during periods of onshore directed strong wind and storms (fig. 5.1). There is a strong linear relationship between local wave heights and water depths which also governs the development of the adjacent salt marshes (fig. 5.3, Niemeyer, 1983). Refraction studies and comparison with a map of surface sediments (Ragutzki, 1982) show that the decrease of wave energy due to dissipation on the tidal flats is also reflected by the zonation of surface sediments with a tendency to finer material (Niemeyer, 1983): Wave exposed sandy flats are followed by areas with mixed surface sediments and finally by mud flats with a high silt content; this is exemplified on fig. 5.2 for the tidal drainage basin of the Norderneyer Seegat inlet in the East Frisian Wadden Sea. Its succession of sandy, mixed and partially even muddy flats in landward direction is in accordance with the wave propagation.

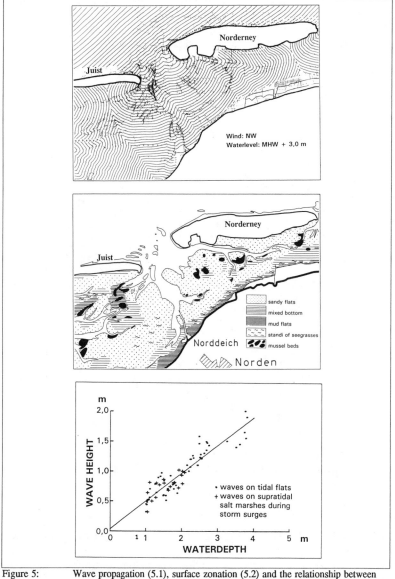

Figure 5: Wave propagation (5.1), surface zonation (5.2) and the relationship between wave height/water depth (5.3) of the Norderneyer Seegat inlet and tidal basin area (Niemeyer, 1983).

It is also apparent that salt marshes have only naturally developed in wave sheltered areas which has been demonstrated evidently with respect to tunneling effects of the tidal inlet by Niemeyer (1990) considering both effects of tidal inlet migration: saltmarsh erosion or accretion with foreland development

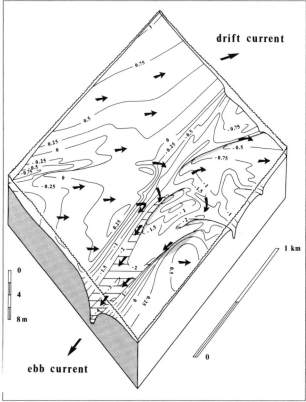

Figure 6:   Drift currents in tidal flat as function of water dept (Tidal flat south of Baltrum, Ehlers, 1988).

Tidal inlets
Tidal inlets due to the semi-diurnal tides of the southern North Sea are influenced twice daily, both by flood and ebb currents. The large water volumes of the tidal basins are forced to pass the narrow inlets at high current velocities. The resulting channels are generally eroded to depths markedly exceeding those of the adjacent sea floor.

The discharge of a tidal basin via the inlet into the sea is morphodynamically comparable to that of a major river. As a result of its high velocity the ebb current transports large volumes of sediment out through the tidal inlet. At the inlet mouth this sediment load is deposited as the current velocity decreases on the one hand and the impact of waves travelling onshore on the other hand is counter-directional, leading to a morphodynamical equilibrium resulting in sediment accumulation and formation of shoals, the so-called ebb deltas. Comparison of aerial photographs taken at different times reveals that the shoals of the ebb deltas are instable features that undergo rapid changes.

On the landward side of the barrier, the flood current also creates a tidal delta. As a result of the relatively high lying flats of the Wadden Sea area, the flood deltas are no evident morphological features. On aerial photographs, however, their brighter sandy shoals can often be easily distinguished from the darker, slightly more silty tidal flats. Major parts of the flood deltas, like the ebb deltas, are covered with megaripples. However, the megaripples of the ebb deltas are highly mobile, whereas those of the flood deltas are almost stationary, moving only a few decimetres per tide (fig. 7).

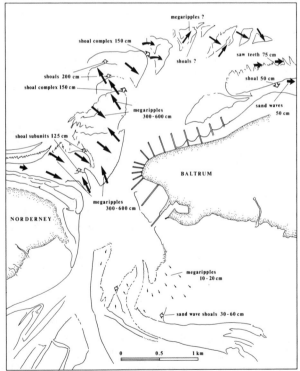

Figure 7:   Movement directions of bars and sand waves around the
                   Wichter Ee tidal inlet for a single tide; the size of the
                   arrows represents the speed of the process (from Ehlers,
                   1988).

The vast majority of coast-parallel sand transport occurs on the seaward side of the barrier islands. Sand transport from one island to the next is only imperfectly described by the often used term 'bar migration'. This expression comprises the processes of sediment transport and formation of new shoals. The 'migration rate' of shoals, which was first determined by Homeier & Kramer (1957) for the Norderneyer Seegat ebb delta (400 m/year), can only be calculated reliably for the lateral parts of the delta. Reshaping occurs very quick in the central area, so that even on aerial photographs taken at one month intervals, it is not possible in all cases to reconstruct shoal development (Luck & Witte, 1974, Ehlers, 1984).

The megaripples do not migrate along the shoals of the ebb delta, but deviate strongly from that direction by up to nearly 90° in the outermost part of the delta. This means that a considerable proportion of the sand transport occurs at right angles to the ebb delta. As the flow directions of ebb and flood currents are not directly opposite, but deviate by a few degrees, this causes -superimposed by dominant wave direction- a long-term sand migration along the ebb delta. The admixture of additional material from the sea, as well as the loss of material by deposition on the beaches and tidal flats, can significantly change the original composition of the sediments (Veenstra, 1984).

The migration of bedforms gives an indication of the sand movement directions. However this should not be mistaken for sand movement itself. Sand migrates at a faster rate as the bedforms, in the same way as small bedforms move faster than the large ones (fig. 5). A sand grain may be washed in and out of a tidal inlet several times, before it finally finds its way to the next downdrift barrier island (Hanisch 1981). The observed bedform migration rates apply to fair weather conditions. During storm periods the shoals are flooded and no measurements can be made. However, it is unterstood that the bulk of the sediment displacement and shifting of landforms occurs during such extreme weather conditions.

Tidal inlets are the gaps between the barrier islands through which waves from the North Sea enter the tidal basins. Travelling onshore they must pass the ebb deltas. Especially high energy waves are intensively attenuated by breaking on the ebb delta shoals, losing up to 90% of their energy. Furthermore this process is accompanied by a relative shift of energy to higher frequencies and the transformation of offshore single-peak to nearshore multi-peak wave spectra (fig. 8; Niemeyer, 1987).

Apart from the major, primary tidal inlets which were formed at the beginning of the Holocene transgression in areas with a low-lying Pleistocene subsurface, a number of usually smaller, 'secondary inlets', were formed later. Secondary inlets may develop where the island dune barrier is breached during storm surges, and where the resulting washover processes are so intense that a channel may form that is sufficiently deep to initiate the formation of a tidal inlet. This channel may be expanded and deepened during successive storm surges. However, a new tidal inlet only becomes established, when an appropriate tidal basin is connected to it (figs. 10-12). Since the developed parts of the islands are protected against dune erosion today, breaches may only occur at the unprotected islands end.

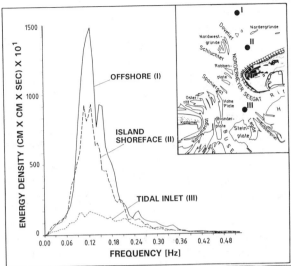

Figure 8:   Spectral wave energy dissipation due to breaking on
                   ebb delta shoals (Niemeyer, 1987).

The tidal inlets are morphodynamically the most active elements of the Wadden Sea. Their shift leads to a phenomenon which has been described in popular literature as 'island migration'. The extent of inlet migration in many cases is very limited today as a result of coastal engineering countermeasures (Luck, 1975). The historical development of the Norderneyer Seegat inlet has been reconstructed by Homeier (1969). Luck (1975) describes the situation as being in a state of dynamic equilibrum, controlled by the processes active in the tidal inlet and resulting in a passive response of the island itself (Kunz, 1991, 1993).

The Norderney Seegat is also an example for the fixation of a tidal inlet by construction of groynes which succeeded within a short period of time (fig. 9; Kunz, 1987). However, after fixing the inlets, the watersheds of the tidal basins migrated further before most of them adapted to the newly established equilibrium. Most watersheds in the East Frisian Islands are considerably displaced to the east (fig. 10).

All tidal inlets of the West, East and North Frisian Islands exhibit clearly developed ebb- and flood dominated channels. The larger tidal streams between the Jade Bight and the North Frisian Islands also consist of a number of either ebb- or flooddominated channels, separated from each other by shoals. As an example, fig. 11 shows the situaton for the Hever tidal inlet (north of Eiderstedt; fig. 1). Instead of ebb deltas, the shoals form horseshoe-shaped bars several kilometres in length. Their highest points mostly reach about -4 m below NN (NN = German ordnance datum) which means that they lie 2 - 3 m deeper than the ebb delta shoals of the East Frisian inlets.

Seismic investigations have revealed that the bars consists of cross-bedded sands (Tietze, 1983).

Old maps and charts indicate that the deep bars continuously change their positions. Early investigations in the Jade and Weser estuaries by Krüger (1911) and Poppen (1912) revealed the rapid morphological changes in deep bar regions. More recent investigations were made by Higelke (1978) along the southwestern Schleswig-

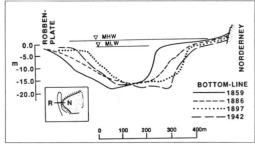

Figure 9:   Cross section between Robbenplate (R) and Norderney (N): Migration of the deep channel of the tidal inlet 'Norderneyer Seegat' since 1859 to 1942 and its fixation due to the construction of groynes about 1900 (Kramer, 1958; Kunz, 1987)

Holstein coast. In the Jade/ Weser estuaries migration rates of about 20 - 50 m per year have been measured. This migration is comparable to that of shoals of the tidal inlets, but the dimensions of the landforms involved makes the process rather look like the shifting of a whole tidal inlet. The velocity of this process, however, is 2 - 4 times faster than the average inlet migration in East Frisia, probably being a consequence of the higher tidal range and the resulting stronger tidal currents.

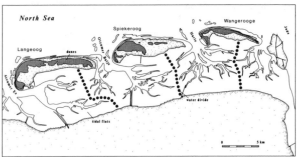

Figure 10:   Map of the tidal flat area between Langeoog and Wangerooge, showing the far eastward position of the water divides.

There have been numerous attempts to model the dynamic relationship between tidal inlets and their basins all over the world. Dieckmann et al. (1989) have considered the ratio between tidal prism of the individual basins and cross-section areas of the respective inlets for American and German Bight conditions (fig. 12). Walther (1972) also has related basin area, barrier island length and island updrift offset to tidal prism for some of the East Frisian tidal inlets (fig. 13.1). Recent investigations have shown that these relationships are still valid as could be exemplified by data for basin areas and tidal prisms from the East-Frisian Wadden Sea (fig. 13.2).

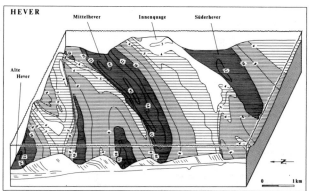

Figure 11: Block diagram of the deep bar at the mouth of the Hever tidal stream, Schleswig-Holstein. Sources: Tietze (1983) and Küstenkarte 1:25 000, Sheet 1616 K St. Peter (from Ehlers, 1988).

Islands

General subdivision

The Wadden Sea islands can be subdivided into three major groups: remnants of former mainland, barrier islands and highly mobile supratidal sand banks (shoals). Along the German North Sea coast, Sylt, Amrum and Föhr are the only islands which have Pleistocene cores.

The barrier islands

The barrier islands owe their existence to landward directed sand transport, and the accumulation of coast-parallel shoals. After these emerged, they were stabilised by dunes (Streif, 1990). Landward and coast-parallel sand transport remain active today. With respect to hydro-dynamical boundary conditions and sediment transport, the island beaches experience both erosion and accretion as a response to gradients in longshore sediment transport (Van de Graaff, et al. 1991). As an example this relationship is pointed out for the East Frisian island of Juist (fig. 14), where in updrift direction beach areas with negative, balanced and positive sediment budget exist (Kunz, 1991).

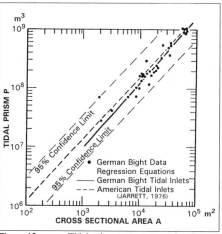

Figure 12: Tidal prisms vs. cross-sectional area for American inlets and inlets along the German Bight (Dieckmann et al. 1989).

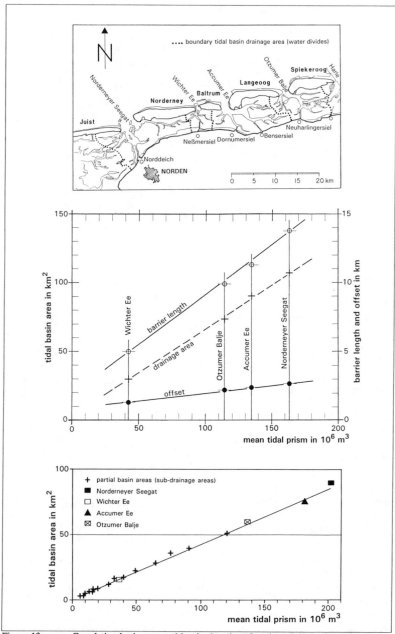

Figure 13: Correlation basin area and barrier length as function of tidal prism (data from 1975, Coastal research station Norderney).

These conditions reflect long-term or mid-term disequilibrium conditions; they do not result from occassional erosion during storm surges, from which the beaches recover within periods of several weeks (fig. 15; Homeier 1976).

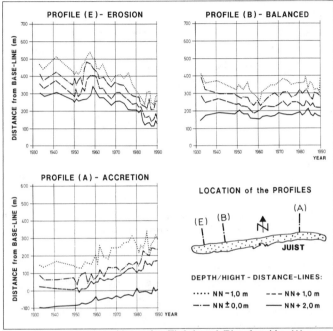

Figure 14: Beaches with negative (E), balanced (B) and positive (A) sediment budget, East Frisian island of Juist (KUNZ, 1991).

Spits have developed on both ends of these islands, as well as on the seaward end of the Eiderstedt peninsula. The latter had been a group of islands until it was connected to the mainland by dikes in 1489 (Fischer, 1956). Whereas on the North Frisian coast both ends of the islands have nearly symmetrically developed spits, along the East Frisian coastline the easterly component of coast-parallel sand transport has resulted in some modifications. Here the westerly spits are mostly underdeveloped and in some cases have migrated around the western ends of the island cores, so that they are now found on their landward side. This process has been well documented in the case of Wangerooge (e.g. Ehlers & Mensching, 1982).

Man has caused modifications of the natural system since the Middle Ages (Kunz, 1993, this volume). Whilst coast-parallel sand transport remains almost unhindered, coast-normal sand transport today is largely restricted to the tidal inlets. Therefore the natural progression which led to a slow landward migration of the islands as an adjustment to rising sea level, has been interrupted almost completely. The dunes have been fixed and washover processes are limited to a few marginal areas.

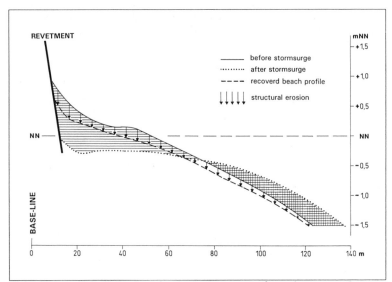

Figure 15:   Episodic beach erosion due to storm surges, afterward recovery and
                  structural erosion, East Frisian island of Norderney (Homeier, 1976).

Stabilisation of the islands has been reached as a result (Kramer, 1989), but natural
accretion has not been promoted and consequently parts of the islands need to be protected against erosion. Examples have been published by Dette & Görtner (1987) for
the North Frisian island of Sylt and by Kunz (1987) for the East Frisian island of
Norderney.

The width of the island beaches depends on the dip angle of the shoreface, the sediment budget, the tidal range and the wave climate (Wiegel, 1964). Island beaches
which experience structural erosion (Van de Graaff, et al. 1991) in general have
steeper slopes both at the beach itself and on its shoreface. As a consequence, on the
island of Sylt (fig. 16a) beaches are very narrow and have to be supplied by frequent
beach nourishments. In contrast, on Langeoog, one of the East Frisian islands (fig.
16b), the shoreface slopes gently towards the sea and the sediment budget is balanced. For that reason it has not been necessary to erect artificial structures to
protect this island. The beach there is about 200 m wide whereas at Rantum on Sylt
it is only 50 m or less. Beach width is important for the potential for both recent
dune formation and dune recovery after erosion due to storm surges.

Along the Dutch coast up to Den Helder and along the Danish coast down to
Blåvandshuk the coastal dunes form a continuous belt, interrupted only in exceptional
cases (eg. major river mouths). In contrast, on the barrier islands of the Wadden Sea
long stretches of dune have been formed. Originally the dunes on the islands formed
'dune cores', some hundred metres to -in exceptional cases- a few kilometres in diameter. Those dune cores were separated from each other by dune-free areas (Schloops).

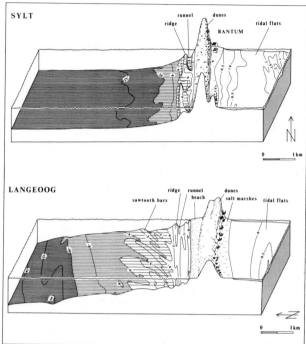

Figure 16: Block diagrams of the beach and foreshore areas of Sylt
at Rantum (a) and Langeoog (b). The dunes are
represented schematically. Sources: Küstenkarte 1:25 000,
Sheets 1114 K Ramtum and 2210 K Langeoog (from
Ehlers, 1988).

This landform assemblage came into existence when during major storm surges the
weak dune belts were breached in places, and considerable quantities of sand were
transported from the beach across the island towards the tidal flat areas. This
washover process has contributed considerably to the islands' ability to migrate in a
landward direction.

Originally Schloops were found on all West and East Frisian islands. In the 17th
century, for example, Juist consisted of three; Borkum of two dune cores. Today
almost all dune cores have been connected by sand dikes or artificially raised dune
ridges (Ehlers, 1990a).

Meanwhile almost all dunes on the islands are stabilised by vegetation. Where
necessary, parts of the foredunes are fixed by revetments to ensure protection of the
settlements. The backdunes in the interior parts of the islands, the so-called 'grey
dunes', are covered with dense vegetation. Dune destruction by pedestrian trampling
in the past has led to strong deflation and the formation of major blow-outs, which

in the worst cases has led to complete dissection of the original dune ridges into small segments. Therefore tourists are kept away from the dunes now and restricted to fenced footpaths, especially in case of the sea-facing main dune ridges (foredune) which are regarded as part of the coastal defence and also in the protected areas of the Wadden Sea National Park.

The mobile supratidal sand banks

The Wadden Sea between Jade and Eider estuaries is characterised by tidal flats, which are more extended than those bordering against the sea by major barrier islands. But in their seawardly located parts there are supratidal sand banks, which in some cases also carry small dune areas. The shape of the latter resembles that of the barrier islands. They have been interpreted as barrier islands in *statu nascendi* (Ehlers, 1988). The sand banks comprise a central sand core with recurved flood spits attached on both sides. The formation of these sand banks is relatively well known, because it has occurred in more recent times.

Trischen (north of Elbe estuary, fig. 1), for instance, is first mentioned in court documents of 1610. A description which dates back to 1689 refers to it as "a very low sand". Trischen was first mapped in 1721, still unvegetated. Permanent vegetation started to get hold on the young island since the middle of the 19th century. The salt marsh growth then proceeded so rapidly, that in 1868 reclamation works were started, and in 1872 already a salt marsh area of 16 ha had been reclaimed. Dune formation on the seaward side started in 1885, and until 1898 a 6 m high, 1.2 km long and 50 - 100 m wide dune belt had formed. In the years 1922-25 the island was diked by private initiative. But even then Trischen was not a stable island. After the severe storm flood of 1936 it became obvious that the polder would not last. In the year 1943, after the dike had been breached, the tenant had to give up and move to the mainland (Lang, 1975).

This tendency to shift has been observed at many other supratidal sand banks in the Wadden Sea. The so-called 'Außensände' (outer sands) off the Halligen in North Frisia are migrating landward at a rate of about 20 m (Norderoogsand) to almost 45 m/year (Japsand; Taubert, 1982).

The marsh islands

The islands of Pellworm and Nordstrand, the Halligen (Norderoog, Süderoog, Südfall, Hooge, Langeneß, Oland, Gröde, Habel and Nordstrandischmoor) are remnants of old marsh areas which have been cut off from the mainland in the Middle Ages. Some of the islands thus created have been poldered like e.g. Nordstrand and Pellworm, whilst others were reincorporated into the mainland (e.g. Dagebüll). These marsh islands have been repeatedly flooded and are covered now by several meters of young marine sediment, but at no time have they ceased to be islands.The complicated coastal history of the North Frisian mainland has been discussed in detail by Prange (1986).

The Halligen surface gives the impression of an absolutely flat salt marsh plain, the only relief being provided by the 'Warften' (dwelling mounds) on which the inhabitants have erected their housing and working facilities, in order to protect them from flooding during storm surges. The Deutsche Grundkarte 1:5 000, on which some Halligen are depicted with contours at decimetre intervals, does show that there is in fact a distinct microrelief. A beach ridge several decimetres high is formed on the seaward side which under natural conditions migrates eastwards at the same rate as the erosion of the Hallig margin. This beach ridge is supplemented by two, usually lower, branches along the south and north coasts. The formation of these ridges is a result of wave impact.

As already recognised by Ordemann (1912), the existing Halligen no longer represent fragments of the medieval marsh surface but considerably higher surfaces as a result of later accretion. Accumulation occurs in the 'Landunter' (= land is flooded) periods when the Halligen are flooded during storm surges. Accumulation is greatest along the salt marsh creeks which all have developed low levees, the intervening marshes forming shallow closed depressions.

The Halligen area is largely a landscape of net erosion. This can be illustrated by the fact that the land area of the present Halligen was reduced by about 80% from appr. 100 km² to 20 km² between 1650 and today. The Fallstief, a previously unimportant tidal creek, evolved into the large Norderhever tidal inlet. Between Hallig Südfall and Pellworm its depth has increased tenfold, and its width at low water sevenfold; its cross profile is now 40 times larger than 300 years ago (Nommensen, 1982).

Human interference

The island barrier adjusts itself to the respective sea level. A rising sea level results in landward migration of the barrier. Under natural conditions this is achieved by dune erosion and landward aeolian sand transport or the breaching of the dune ridges and washover processes. These processes are largely prevented at present by coastal protection countermeasures; the most exposed parts of the islands without sufficient sand supply have been fixed and fortified against erosion. This has resulted in both undernourished beaches (negative sand balance) and an increasing impact of natural forces (tide- and wave-induced currents).

The demand for coastal protection has increased considerably during the last century. The increase of tourism and the resultant improvement of the islanders' economy has led to an enormous expansion of the settlements (Ehlers, 1990b). The development of seaside resorts, starting at a larger scale in the second half of the 19th century, soon required coastal protection. Despite bad experience on Norderney, Borkum, which was originally well protected behind dune ridges, expanded towards the open beach areas. The main reasons for the coastal population to spend a tremendous amount of money on solid constructions to protect parts of the islands against erosion were: changing natural processes, growth of the population, and inadvertent positioning of buildings (Kunz, 1987). The settlement on Wangerooge, which had been completely demolished by a storm flood in 1854/55, was originally rebuilt in a protected position at the centre of the island. However, shortly afterwards hotels and boarding houses were built by the beach, forcing the construction of coastal defences.

Of the German Wadden Sea barrier islands Langeoog was the only one to remain without any sea walls and groynes (Witte, 1979).

The old village of Westerland on Sylt lay at a distance of more than 1 km from the beach. Considering the average natural coastal retreat rate in central Sylt of less than 1 m/year, this meant safety for more than 1000 years. This was sacrificed with the onset of tourism, when the area between the village and the beach was developed.

With regard to the high costs of coastal protection and the requirements of nature conservation in a national park environment, the minimum demand should be to keep the near-beach areas free of further building activities. This, however, is hardly done. The future development of the coastal area should be based on decisions that meet the requirements of an appropriate coastal zone management.

References:

Cameron, D., D. van Doorn, C. Laban & H. Streif (1993): Geology of Southern North Sea Basin; CZ'93, Coastlines of the Southern North Sea (*This volume*).

Davis, J.L. (1964): A morphogenetic approach to world shorelines. *Zeitschrift für Geomorphologie,* Sonderheft zum 70. Geburtstag Prof. H. Mortensen: 127-142.

Dette, H.H, & J. Gärtner (1987): The history of a seawall on the island of Sylt. *Coastal Sediments '87,* Proc. vol. 1, ASCI, New York, 1006-1022.

Dieckmann, R., M. Osterthun & H.W. Partenscky (1989): A comparision between German and North American tidal inlets; *21st International Coastal Engineering Conference,* Malaga, ASCE, New York; Vol. 3: 2681 - 2691.

Ehlers, J. (1984): Platenwanderung an der ostfriesischen Küste ? - Ergebnisse der Luftbildauswertung von zwei Befliegungen der Wichter Ee (zwischen Norderney und Baltrum) im Sommer 1982. *Mitteilungen aus dem Geologisch-Paläontologischen Institut der Universität Hamburg* 57: 123-129.

Ehlers, J. (1988): The Morphodynamics of the Wadden Sea. Rotterdam: Balkema. 397 pp.

Ehlers, J. (1990a): Sedimentbewegung und Küstenveränderungen im Wattenmeer der Nordsee. *Geographische Rundschau* 42: 640-646.

Ehlers, J. (1990b): Der Fremdenverkehr auf den Barriereinseln der Nordsee und seine morphologischen Auswirkungen. *Mitteilungen der Geographischen Gesellschaft in Hamburg* 80: 505-525.

Ehlers, J. & H. Mensching (1982): Erläuterungen zur Geomorphologischen Karte 1:25 000 der Bundesrepublik Deutschland, GMK 25 Blatt 10, 2213 Wangerooge, Berlin. 55pp.

Fischer (1956): Eiderstedt. Das Wasserwesen an der schleswig-holsteinischen Nordseeküste, Dritter Teil: Das Festland 3: 328 pp. Berlin: Reimer.

Graaff, J. van de, H.D. Niemeyer & J. van Overeem (1991): Beach nourishment, philosophy and coastal protection policy. In: J.v.d. Graaff, H.D. Niemeyer, J. v.Overeem (eds): Special Issue Artificial Beach Nourishments. *Coastal Engineering,* Vol. 16, No. 1.

Hanisch, J. (1981): Sand transport in the tidal inlet between Wangerooge and Spiekeroog (W-Germany). In: S.-D. Nio, R.T.E. Schüttenhelm and Tj.C.E. van Weering (eds.): Holocene Marine Sedimentation in the North Sea Basin. *Special Publication Number 5 of the International Association of Sedimentologists:* 175-185.

Hayes, M.O. (1975): Morphology of sand accumulations in estuaries. In: L.E. Cronin(ed.): *Estuarine Research*, Vol. 2: 3-22.

Higelke, B. (1978): Morphodynamik und Materialbilanz im Kostenvorfeld zwischen Hever und Elbe. Ergebnisse quantitativer Kartenanalysen für die Zeit von 1936 bis 1969. *Regensburger Geographische Schriften* 11: 167 pp.

Homeier, H. & J. Kramer (1957): Verlagerung der Platen im Riffbogen von Norderney und ihre Anlandung an den Strand. *Forschungsstelle Norderney, Jahresbericht 1956*, vol. 8: 37-60.

Homeier, H. (1962): Historisches Kartenwerk 1:50 000 der niedersächsischen Küste. *Jahresbericht der Forschungsstelle für Insel- u. Küstenschutz, Norderney;* vol. 13: 11-29.

Homeier, H. (1969): Der Gestaltwandel der ostfriesischen Küste im Laufe der Jahr hunderte - Ein Jahrtausend ostfriesischer Deichgeschichte. J. Ohling (publ.): Ostfriesland im Schutz des Deiches, Bd.II, Pewsum.

Homeier, H. (1976): Die Auswirkungen schwerer Sturmtiden auf die ostfriesischen Inselstrände und Randdünen. *Jahresbericht der Forschungsstelle für Insel- u. Küstenschutz 1975*, vol.27, 108-122

Kramer, J. (1989): Kein Deich, kein Land, kein Leben - Geschichte des Küstenschutzes. Verlag G. Rautenberg, Leer.

Krüger, W. (1911): Meer und Küste bei Wangerooge und die Kräfte die auf ihre Gestaltung einwirken. *Zeitschrift für Bauwesen* LXI: 451-464, 584-610.

Kunz, H. (1987): History of seawalls and revetments on the island of Norderney. *Proc. Coastal Sediments '87*, vol.I, ASCE, New York, 974-989.

Kunz, H. (1991): Protection of the island of Norderney by Beach Nourishment, Alongshore Structures and Groynes. *Proceedings of the 3rd International Conference on Coastal & Port Engineering in Developing Countries*, vol. I, COPEDEC, Mombasa, 29-42.

Kunz, H. (1993): Coastal Protection in the past; Coastal Zone Management in the future CZ'93, Coastlines of the Southern North Sea (*this volume*).

Lang, A.W. (1975): Untersuchungen zur morphologischen Entwicklung des Dithmarscher Watts von der Mitte des 16. Jahrhunderts bis zur Gegenwart. *Hamburger Küstenforschung* 31: 154 pp.

Luck, G. & H.H. Witte (1974): Erfassung morphologischer Vorgänge der ostfriesischen Riffbögen in Luftbildern. *Jahresbericht der Forschungsstelle für Insel- u. Küstenschutz 1973*, vol. 25,: 33-54.

Luck, G. (1975): Der Einfluß der Schutzwerke der Ostfriesischen Inseln auf die morphologischen Vorgänge im Bereich der Seegaten und ihrer Einzugsgebiete. *Mitteilungen aus dem Leichtweiß-Institut für Wasserbau der Technischen Universität Braunschweig* 47: 1-122.

Lüders, K. (1953): Die Entstehung der Ostfriesischen Inseln und der Einfluß der Dünenbildung auf den geologischen Aufbau der ostfriesischen Küste. *Probleme der Küstenforschung im südlichen Nordseegebiet;* vol. 5: 5-14.

Niemeyer, H.D. (1983): Über den Seegang an einer inselgeschützten Wattküste. *BMFT-Forschungsbericht* MF 0203, 1-264.

Niemeyer, H.D. (1987): Changing of wave climate due to breaking on a tidal inlet bar. *Proceedings of the 20th International Conference on Coastal Engineering*, Taipei, ASCE, New York, Vol. II, 1427-1443.

Niemeyer, H.D. (1990): Morphodynamics of Tidal Inlets; *Syll. Civ. Eng. Europ. Cours Progr. o. Cont. Educ.* (CEEC) - Coast. Morph., Univ. Delft, CM: 1-45.

Niemeyer, H.D. (1991): Case study Ley Bay: an alternative to traditional enclosure. *Proceedings of the 3rd Conference on Port & Coastal Engineering*, Mombasa/Kenya, Vol. I, 43-58.

Nommensen, B. (1982): Die Sedimente des südlichen Nordfriesischen Wattenmeeres (Deutsche Bucht). Ergebnisse geologisch-sedimentologischer Untersuchungen an pleistozänen und holozänen Sedimenten und an Schwebstoffen der Gezeitenströme. 268pp. Kiel: Unpublished Thesis.

Ordemann, W. (1912): Beiträge zur morphologischen Entwicklungsgeschichte der deutschen Nordseeküste mit besonderer Berücksichtigung der Dünen tragenden Inseln. *Mitteilungen der Geographischen Gesellschaft* (für Thüringen) zu Jena 30: 15-150.

Poppen, H. (1912): Die Sandbänke an der Küste der Deutschen Bucht der Nordsee. *Annalen der Hydrographie und Maritimen Meteorologie* XXXX (VI, VII, VIII): 273-302, 352-364, 406-420.

Prange, W. (1986): Die Bedeichungsgeschichte der Marschen in Schleswig-Holstein. *Probleme der Küstenforschung im südlichen Nordseegebiet* 16: 1-53.

Ragutzki, G. (1982): Verteilung der Oberfächensedimente auf den niedersächsischen Watten. *Jahresbericht Forschungsstelle für Insel- u. Küstenschutz*, vol. 32: Norderney.

Streif, H. (1990): Das ostfriesische Küstengebiet - Nordsee, Inseln, Watten und Marschen. 2nd edition. *Sammlung Geologischer Führer* 57: 376 pp.

Taubert, A. (1982): Wohin wandern die Außensände? Formänderungen der nordfriesischen Außensände und deren küstengeographische Beurteilung. *Nordfriesland* 16: 37-48.

Tietze, G. (1983): Das Jungpleistozän und marine Holozän nach seismischen Messungen nordwestlich Eiderstedts, Schleswig-Holstein. 118 pp. Kiel: Unpublished Thesis.

Veenstra, H.J. (1984): Size and shape-sorting of coastal sands in the eastern part of the German Bight (North Sea). *Geologie en Mijnbouw* 63: 47-54.

Walther, F. (1972): Zusammenhänge zwischen der Größe der ostfriesischen Seegaten mit ihren Wattgebieten sowie Gezeiten und Strömungen. *Jahresbericht der Forschungsstelle für Insel- u. Küstenschutz 1971*, vol. 23:.

Wiegel, R. (1964): Oceangraphical Engineering: 527 pp. Prentice-Hall: London.

Morphology of the Southern North Sea Coast from Cape Blanc-Nez (F) to Den Helder (NL)

B. Lahousse[1], P. Clabaut[2], H. Chamley[3], L. van der Valk[4]

Abstract

The southeastern North Sea coast of France, Belgium and the Netherlands (up to Den Helder) is characterized by coastal dunes, sandy beaches and a shallow, gentle shoreface. In the Delta area (southwestern Netherlands) the straight coastline is interrupted by a number of estuaries. The closure of several of these estuaries has resulted in significant morphological changes in this area since the past few decades. Offshore the French, Belgian and Netherlands coastlines, sets of large sand banks occur at various depths. A morphological description of the various sand bank systems is given. Their origin is not yet fully understood.

Over the last two centuries, changes in the coastal morphology and sedimentology of the southeastern North Sea are due to both natural processes and human activities.

Introduction

The southern North Sea is a shallow shelf sea with water depths generally not exceeding 40 m. The southeastern North Sea is bordered by three countries: France, Belgium and the Netherlands. The French sector extends over 60 km between Cape Blanc-Nez, southwest of Calais, and the Belgian border east of Dunkerque. In Belgium the sandy coastline is more or less straight with a lenght of approximately 65 km. The 350 km long Dutch coastline is usually divided into three parts: the Delta area between the Belgian border and Hook of Holland, the straight Holland coast between Hook of Holland and Den Helder, and the Wadden area between Den Helder and the German border (fig. 1).

The present paper is concerned with the morphology of the coastal area between Cape Blanc-Nez and Den Helder. A morphological description of the Wadden area is given by Ehlers & Kunz (this volume).

The morphology of the southeastern North Sea is the result of coastal evolution processes over several thousands of years. During the past centuries coastal evolution has for a large part been influenced by human activities. The evolution of the French

---

[1]   Harbour and Engineering Consultants, Deinsesteenweg 110 Ghent, Belgium

[2]   Consultant, 14 rue Paul Daumier, F59110 La Madeleine, France

[3]   Lille 1 University, F59655 Villeneuve d'Ascq, France

[4]   Dutch geologic Survey,P.O. Box 157, 2000 AD Haarlem, The Netherlands

and Belgian coastal areas is described by Houthuys *et al.* (this volume). The Holocene evolution of the Holland coast is, amongst others, described by Beets *et al.* (1992).

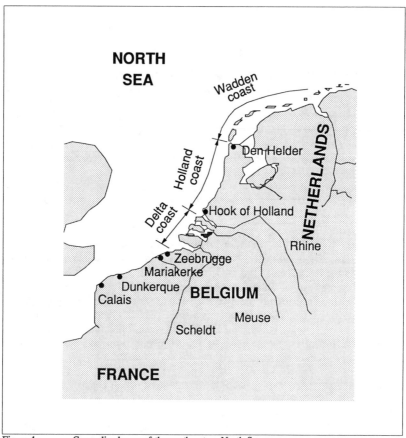

Figure 1:   Generalized map of the southeastern North Sea area.

Morphological description

In this paper aspects of the coastal morphology of the southeastern North Sea are described using a distinction in four sectors (fig. 1) with different characteristics. For the Belgian and French sectors emphasis is on a morphological description of the sand banks in the (shallow) offshore areas. For the Netherlands emphasis is placed on coastal evolution from a geological/sedimentological point of view.

Calais - Mariakerke

The coastline between Calais (France) and Mariakerke (Belgium) is characterized by the presence of large coastal dune systems, which protect the southern Flemish maritime plain against flooding by the sea.
From Calais to Dunkerque, the coastal dunes are fairly small. Beaches are up to 1,200 m wide and made up of medium-sized sand (0.200 - 0.350 mm). The dune complex east of Dunkerque is massive (width usually > 1 km), beaches are 300 to 600 m wide and exclusively composed of fine sand (0.125 - 0.200 mm). The intertidal area typically consists of an alternation of sandy bars and troughs parallel to the coastline.

The shallow offshore area is characterized by the presence of numerous sand banks which form the southern part of the Flemish bank complex (Augris et al., 1988; Beck et al., 1991). The banks are oriented WSW-ENE and run nearly parallel to the shoreline and to the direction of the tidal currents.
The four subparallel "coastal banks" are 8 to 32 km long, 1.5 to 3 km wide, and 15 m high. They are separated by 10 to 15 m deep sandy troughs. The most landward trough forms the access channel to the port of Dunkerque.
The "offshore banks" are larger (35 to 60 km long). To the west they are separated by broad troughs where coarse sediments crop out (gravel, pebble); the seperation of the eastern banks is less distinct. The offshore banks display a strong asymmetry: the southern flanks are usually much steeper than the northern flanks due to the net wave- and current-induced sand transport.
On top of the offshore banks medium-sized sandwaves (3 to 5 m high) occur. Close to the base of the banks small megaripples (2 to 3 m large) occur. Towards the crest the megaripples are larger (15 to 20 m). At the extremeties of the offshore banks more than 10 m high sandwaves are present. Sandwaves are not common on the shallower coastal banks.

The main sediment transport along this coastal sector is directed to the northeast, associated with various refraction movements and counter currents. A symmetrical but less efficient transport to the southwest is present in a more offshore position. As an example, the sediment transport pattern along the French coast between Calais and Dunkerque is shown in figure 2.

Mariakerke - Western Scheldt

The coastal morphology of this sector is strongly influenced by both the tidal mechanisms of the North Sea and the influence of the Scheldt river. The most characteristic morphological features are the sandbanks running nearly parallel to the shoreline up to 10 km offshore. The sandbanks can be devided into four groups (fig. 3).

The Flemish banks have a SW-NE orientation. They are 15 to 25 km long, 3 to 6 km wide and 25 m high. The distance between the banks range from 4 to 6 km. Further offshore, the Hinder banks have a more or less S-N direction. They are up to 35 km long and 30 m high. Finally, more to the east, the Zealand banks are oriented WSW-ENE.

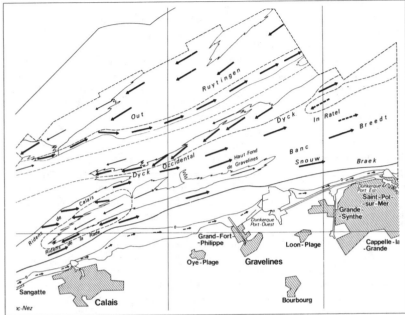

Figure 2:   Net sand transport directions along the French coast

In cross section, these banks are usually asymmetrical. For most of the Hinder banks, the eastern slope is steeper. For the Flemish banks, in the northeastern part, the northwestern slope is steeper, and in the southwestern part, the southeastern slope is the steeper one. Sandwaves and small megaripples are also common on these banks as shown on figure 4.

Sediment transport patterns are comparable to the patterns described above (fig. 2). North of Zeebrugge harbour, the residual flow is also strongly influenced by the asymmetry between ebb and flood discharges in the Scheldt estuary.

Furthermore, the area around the port of Zeebrugge is characterized by a complex sedimentological and hydrodynamic mechanism, a marine turbidity maximum area. This TMA can be explained by the existance of encounter zones of residual transports of sediments which cause a hydrodynamic trapping of the sediments.

The Delta area
The Delta area is a complex of large estuaries and tidal basins. Some of the estuaries were closed in recent times. In the classical geological meaning this area is not a delta; it can be classified as a conglomerate of tidal channels and intertidal areas. The dune coast at the western fringe of the islands is generally narrow. Seaward of the estuaries and tidal basins, ebb-tidal delta's determine the morphology of the shallow North Sea for approximately 10 km to the west of the coastline. Further offshore sand ridges (the Flemish Banks) are present (Van Alphen & Damoiseaux, 1989). The

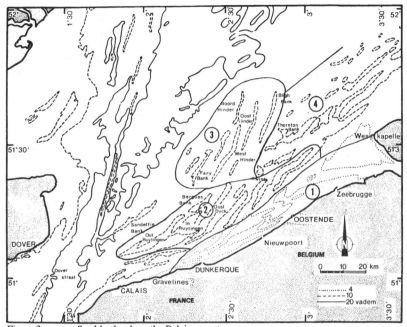

Figure 3:   Sand banks along the Belgian coast

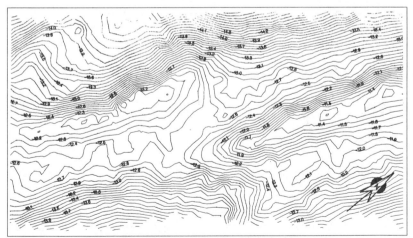

Figure 4:   Sandwave and megaripples on the West Hinder bank

morphological changes in the Delta area are very large, especially after Man has
started his actions, influencing the natural environment from the Middle Ages on.

The Holland coast

The Holland coast, between Hook of Holland and Den Helder, is characterized by an uninterrupted sandy coastal barrier with high dunes acting as sea defences. A small area in North Holland is formed by a sea-dike (the Hondsbossche Zeewering) where no coastal dunes are present. The tidal range is generally 1.7 m and much smaller than the tidal range in the Delta and Wadden areas. The 20 m depth contour in front of this part of the coast is situated much further offshore, indicating that a large sand volume is present. The 10 m depth contour is however appearing concave towards the present shore and fairly close to it (Van Alphen & Damoiseaux, 1989). This situation of the depth contours on the shoreface is related to the wave-dominated character of the shoreline.

Just offshore of Hook of Holland, a large dumping site for dredged material is present, which shows as a plateau. Another plateau, a Pleistocene high, is present in front of Petten, North-Holland (Van Alphen & Damoiseaux, 1989). The latter plateau offers considerable resistence against marine erosion.

The northern part of the Holland coast is strongly influenced by the large tidal inlet of the Marsdiep, the most southward inlet of the Wadden Sea. The morphology of the shoreface varies along the shoreline, e.g. in steepness of the coastal gradient and in grain-size. In the middle part of the Holland coast, the coastal gradient is less steep than in the vicinity of Hook of Holland and Den Helder. To the south of IJmuiden, the median grain size of the shoreface is coarser than 0.275 mm, while the median grain size to the north of IJmuiden is less than 0.200 mm.

Coastal evolution

The morphological evolution of the French coastal area has been quantified by digitizing high-resolution bathymetric data from the French Navy Hydrographic and Oceanic Service (Corbau, 1991).

In the Belgian offshore areas, the Hydrographic Service has been performing sounding operations on a regular basis since 1860. However, the older bathymetric maps are to be used with care, because of possible inaccuracies of the positioning system and of the sounding data. Since a few years, more sophisticated survey techniques are available, involving surveying by multibeam echosounder and side scan sonar, and in-situ real-time sediment transport measurements with radioactive tracers.
The Belgian nearshore areas are surveyed bi-annually through a combination of sounding and remote sensing techniques (De Wolf et al., this volume).

Monitoring of the coastline of the Netherlands is performed annually since the middle of the 19th century. Since about 1965 every year a coastal profile is measured, extending from several hundreds of metres inland (aerial photography) to about 1 km offshore (sounding techniques). Every five years the coastal profiles are extended to over 3 km offshore.
Geological mapping of the Dutch offshore areas is done through geophysical and drilling techniques.

Based on the above survey data, some aspects of coastal evolution of the southeastern North Sea are described. For details of the evolution of the Belgian coastline reference is made to De Wolf *et al.* (this volume). Coastline trends in the Netherlands are, among others, described by De Ruig & Louisse (1991).

Calais - Mariakerke

In this coastal sector, the "offshore banks" appear to be rather stable. The "coastal banks" tend to migrate landward at arate of several metres per year. In Calais and Dunkerque areas, some smaller coastal sandbanks became attached to the shore (the Nieuland bank and Schürcken bank, respectively). In both cases the attachment was due to human activities. Because in both Calais and Dunkerque harbours accumulation of marine sediement takes place, piers were built and subsequently extended. As a result, sediment accumulated along the piers and the nearest coastal bank became attached to the shore. Development of banks has been noticed as well. In the vicinity of Calais, the bank "Ridens de la Rade" came into existance less than a century ago (EDF/LNH & LCHF, 1986).

Coastline evolution is characterized by sediment accumulation between Calais and Dunkerque and by erosion (about a metre per year) between Dunkerque and the Belgian border. These contrasting developments are probably related to dike building during the past centuries. On ancient maps dikes can be recognized in the western part since the 12th century. Since that time the coastline has shifted seaward by about 2 km.

Mariakerke - Western Scheldt

The banks off the Belgian coast seem to be rather stable sand bodies. During storms, some of the megaripples and sandwaves are eroded, but they are build up again soon afterwards. The long-term morphological evolution is governed by the system of tidal gullies (Van Cauwenberghe, 1966; Maenhaut van Lemberge, 1993). Flood gullies are shallow and relatively close to the shore; ebb gullies are further offshore and deeper.

The coastal evolution of the Zeebrugge area is strongly influenced by harbour construction activities and dredging operations. As a result of the extension of the harbour moles by some 3 km, the residual transport patterns changed significantly. The TMA, which used to be located west of Zeebrugge harbour, is now eastward of the harbour entrance. The area near Zeebrugge is investigated in detail frequently (fig. 5).

The Delta area

From the onshore geological record it is known that the Delta area was dissected by tidal inlets until some 5000 years ago. After this the area was very probably separated from the sea by a coastal barrier, apart from an inlet associated with the River Scheldt. Behind the barrier wide-spread peat growth took place until Medieval times. By then the coastal area including the peat region was dewatered to a high degree by invading tidal channels draining the peat. This led to wide-spread shrinkage of the peat, which caused large-scale flooding. In historic time, large surfaces of the intertidal area were reclaimed causing confinement and deepening of the tidal

Figure 5:   Sandwave and megaripples in the Zeebrugge area.

channels. This process was still active when the Deltaplan for coastal protection against storm surges was carried out as a reaction to the 1953 flooding which took 1835 lives and caused much damage.

The coastal development during the past century or so may be summarized as follows. Man had constrained the intertidal area by reclaiming large areas. This led to confinement of the tidal channels, which in turn caused a considerable deepening of the channel thalwegs. Consequently the ebb-tidal deltas have grown. At the coastal dune headlands erosion prevailed mainly due to storm-surge wave attack. Eroded dune- and beach sands were transported longshore away from the headlands. Deposition of these sands occurred in the lee-side areas away from the headlands, locally giving rise to extensive coastal progradation by formation of beach plains (Schouwen; Goeree). In the northern part of the area, west of Rotterdam, human influence has almost obliterated natural coastal development.

Recent changes associated with the execution of the Deltaplan involving complete or partial closure of the tidal inlets (exept for the inlet of the Western Scheldt due to shipping) are considerable. Due to the closure of the inlets, the in- and outgoing tides have been blocked. The sediment exchange between the ebb-tidal delta and the estuary has been cut off or largely dimished. Under the present hydraulic conditions, the landward sediment transport by shoaling waves is no longer compensated by the seaward transport by the ebb-tide. The changed conditions cause the removal of sediment from the ebb-delta front and from the tidal bars in the mouth of the estuary. At the same time, the deep channels became oversized due to the diminishing of the tidal volume. Consequently, large volumes of sand (a.o. from the adjacent bars) are being deposited in these overfit channels. In the ebb-tidal delta, the orientation of the shoals is changing from more or less oblique to the shore to a more shore-parallel orientation. Rates of geomorphological change are declining, after rapid (15 years) initial changes. However, general patterns of coastal change as observed during the last century are expected to continue. These patterns indicate erosion of the coastal dune headlands and deposition of this eroded sediment in areas northeast of the eroded coastal sections due to the overall net northward sediment transport along the Delta coast (Louters *et al.*, 1991).

The Holland coast

The coastal evolution of the western Netherlands during the last 6000 years is strongly related to the rate of sea level rise and the formation of beach ridges (Beets *et al.*, 1992). The last estuary to be closed was a former outlet of the river Rhine, in the 12th century. Around that time the coastal profile steepened, the sand was blown ashore in the form of a dune sand sheet with transverse dunes. Later, in the 14th to 15th century, when the coastal dunes were stabilized by vegetation, these transverse dunes transformed into rows of parabolic dunes, which still can be recognized in the present-day landscape. The cause of the steepening of the coastal profile is still debated, but is seems probable that there may have been a period around the 9th to 11th century with more and stronger winds than today. The last 150 years the coastal dunes have been maintained very effectively as a sea defence by Rijkswaterstaat and the Waterboards, which allowed very few dynamic changes. Along parts of the Holland coast, notably in the eroding southern and northern parts, groines were placed. These groins were effective in limiting wave influence, but could not stop coastal erosion.

The upper shoreface is a multiple-barred system, while the lower part of the shoreface is dominated by storm sedimentation, down to a depth of about 16 m. At greater depths tidal currents play a significant role along with storm waves, keeping fine-grained sediment in suspension (fig. 6).

In front of the Holland coast a broad field of sand ridges is present, situated below the shoreface in a waterdepth of 14-23 m. These "shoreface connected ridges" differ from the Zeeland ridges further to the south. They are generally lower: only 3-7 m high. The ridges closest to the coast are shoreface-attached. Although their origin is still uncertain, they were probably formed in an open marine environment, quite similar to the present one. Sedimentological observations have shown that at present the ridges are still being reworked by waves and tidal currents (J.W.H. van de Meene, pers. com.). Further offshore at waterdepths greater than 20 m, a large field of sand waves is present with wave heights of 2-10 m and spacings between 60 to 600 m.

Sand losses of the foreshore are small in front of the Holland coast when compared to the Zeeland and Wadden areas. To the north of Hook of Holland and around the IJmuiden piers of the arificial Amsterdam harbour entrance, locally considerable coastal progradation (of several m/year) has taken place. Nevertheless, from a recent coastal survey it is clear that yearly fluctuations in the coastal area are strongly related to the hydrodynamic processes. In spite of local differences and deviations, the coast of Holland can be considered to be in a state of dynamic equilibrium to the processes active in the area.

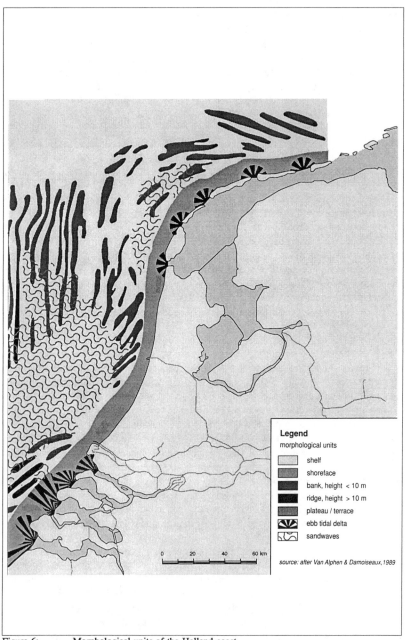

Figure 6:   Morphological units of the Holland coast

References

Augris C., Vicaire O. & Clabaut P. (1988) Carte des sédiments superficiels au large du Nord - Pas-de-Calais. *Feuille de Calais-Dunkerque.* Editions Ifremer.

Beck C., Clabaut P., Dewez S., Vicaire O., Chamley H., Augris C., Hoslin R. & Caillot A. (1991) Sand bodies and sand transport paths at the English Channel - North Sea border: morphology, hydrodynamics and radioactive tracing. *Oceanologica Acta*, vol. sp. 11.

Beets, D.J., Van der Valk L., Stive M.J.F. (1992) Holocene evolution of the Coast of Holland. *Marine Geology* 103, 423-443.

Corbeau C. (1991) Bilan sédimentaire pluri-décennal du littoral dunkerquois. DEA, Univ. Lille I.

De Ruig, J.H.M. & Louisse, C.J. (1991) Sand budget trends and changes along the Holland coast. *J. Coastal Res.* 7, pp. 1013-1027.

De Wolf, P., Fransaer, D. Van Sieleghem J. & Houthuys, R. (1993) Morphological trends of the Belgian coast shown by 10 years of remote sensing based surveying. *This volume.*

EDF/LNH & LCHF (1986) Catalogue sédimentologique des côtes françaises. Côtes de la Mer du Nord et de la Manche. Eyrolles, Ed.

Ehlers, J. & Kunz, H. (1993) Morphology of the Wadden Sea. *This volume.*

Houthuys, R., De Moor, G. & Sommé, J. (1993) The shaping of the French-Belgian North Sea coast throughout Recent geology and history. *This volume.*

Maenhaut van Lemberge V. (1993) Southern North Sea Project-Borehole BH89/1; Technical report Geologie, *Bull. Belg. Veren. Geol.*

Louters, T., Mulder J.P.M., Postma, R. & Hallie, F.P. (1991) Changes in coastal morphological processes due to the closure of tidal inlets in the SW Netherlands. *J. Coastal Res.* 7, 635-652.

Van Alphen J.S.L.J. & Damoiseaux, M.A. (1989) A geomorphological map of the Dutch shoreface and the adjacent part of the continental shelf. *Geol. & Mijnbouw* 68, pp. 433-443.

Van Cauwenberghe C. (1966) Hydrografische analyse van de Scheldemonding ten Oosten van de meridiaan 3.05' tot Vlissingen. *Het Ingenieursblad*, no. 17.

Coastal morphological Modelling for the southern North Sea

H.J. de Vriend[1], R.C. Steijn[2]

Abstract

The paper gives a inventory of mathematical model concepts for coastal morphological applications, from sophisticated short-term process models based on first physical principles, via process-based medium-term morphodynamic models, to semi-empirical long-term dynamic models and descriptions of the equilibrium state of coastal morphological systems.

Next, it briefly outlines a number of applications to coastal problems in the southern North Sea, which cover the whole range of model types.

Finally, the perspectives of the integration of morphodynamic models into CZM-tools are discussed.

Introduction

Mathematical models for the prediction of coastal morphology have a role to play in Coastal Zone Management, in spite of their possible limitations when it comes to the description of real-life situations. Whenever we want a quantitative prediction of the coastal behaviour, we need a model. Models can also be useful for a diagnosis of observed coastal behaviour, e.g. in order to predict how the system will respond to future changes. Depending on the kind of information we want out of this model, it can be crude and simple, or it has to be sophisticated and complex.

In general, the choice of a model should be based on two pieces of information, viz.:

- what information do we want from the model, in what detail and with what accuracy?
- which type of model is able to provide this information against the lowest price (money-wise or according to other standards)?

Other arguments, such as the capabilities of the hardware platform or the software environment, should be essentially immaterial.

The assessment of what model type is needed is far from trivial, and it is certainly not related directly to the space and time scales of the problem. Besides, the gamut of model concepts is covering a wide area of applications, often with not too large overlaps, in spite of the progress which has been made in the modelling of coastal dynamics in recent years. Hence we seldomly have a choice between a variety of concepts of different detail and accuracy for the same problem.

---

[1] Senior researcher, Delft Hydraulics, P.O. Box 152, 8300 AD Emmeloord, The Netherlands

[2] Researcher, Twente Technical University, P.O Box 217, 7500 AE Enschede, The Netherlands

In the following we will give a review of the various model concepts which are in use for the coasts of the southern North Sea, especially in The Netherlands.

Model concepts

A large class of coastal morphodynamic models is based on more or less well-established descriptions of the constituent physical processes (waves, currents, sediment transport), which are coupled via a bottom evolution module based on the sediment balance.

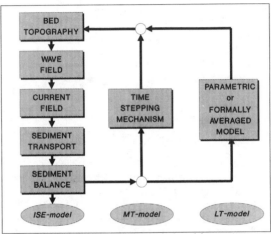

Figure 1:   Compound morphological model concepts

Figure 1 represents three basic model concepts of this type, viz.:

(1)   "initial sedimentation/erosion" (ISE) models, which go through the sequence of constituent models only once, and have the initial rate of bathymetric change as a final result; in fact, it boils down to a hydrodynamic and sediment transport computation based on the assumed invariance of the bed topography,

(2)   "medium-term morphodynamic" models, in which the new bottom topography is fed back into the hydrodynamic and sediment transport computations; this yields a looped system which describes the dynamic time-evolution of the bed; the time scale of this "morphodynamic" simulation is not far from the hydrodynamic time scale (duration of a storm, tidal period), and

(3)   "long-term morphological" models, in which the constituent equations are not describing the individual physical processes at the hydrodynamic time scale, but integrated processes at a higher level of aggregation; these equations can be derived from process models via formal mathematical operations (time-splitting, time-averaging), physical reasoning and/or empiricism (closure hypotheses).

In the next sections, each of these concepts will be discussed in further detail. In line with the historical development of coastal morphodynamic modelling, we will start at the process end and build up gradually towards larger scales.

Coastal area ISE-models

ISE-models are seldomly applied to situations with a predominant variation in one horizontal direction, such as the evolution of the cross-shore profile of a coast which is uniform alongshore, or the evolution of a coastline if the cross-shore profile is always in equilibrium. In those cases, medium-term dynamic models are just as easily accessible and give much more information.

If the horizontal directions are equivalent, medium-term dynamic models are much more complicated and less accessible, if it were only because they are rather demanding with respect to user's skill and computing power. As long as this situation lasts, coastal area ISE-models will have their place on the market.

ISE-models are usually compounded from existing operational models for waves, currents and sediment transport, followed by a computation of bathymetric change rates from the sediment balance. Thus we can make use of existing knowledge on how to model the constituent processes, although we have to be careful there: a model which is good enough to describe a process in isolation, is not necessarily also good enough to describe the interaction with other processes (Dingemans et al., 1987).

Applications of coastal area ISE-models have been described extensively in the literature. Applications to the southern North Sea have been described, among others, by De Vriend and Ribberink (1988) and Steijn et al. (1989), for the Delta area in south-west of The Netherlands, by Ribberink and De Vroeg (1992), for the Eijerlandse Gat inlet between Texel and Vlieland, and by Steijn and Louters (1992), for the Friesche Zeegat, another inlet of the West-Frisian Wadden Sea. Besides, the concept has been applied to a number of inlets in the East-Frisian and North-Frisian Wadden Sea (e.g. Van Overeem et al., 1992).

In general ISE-models give a lot of insight into the short-term processes, but the translation of their results into medium- or long- term predictions requires a good deal of interpretation.

Which physical phenomena should be included in an ISE-model depends to a large extent on the objective of the application. In general, the system becomes more complicated and sensitive as it approaches its equilibrium state, or as it concerns longer time scales. 3-D tidal and density-driven residual currents, for instance, play a much larger role in the long-term sediment balance of a coast than in the short-term dynamics of the upper shoreface.

In preparation of the forthcoming Coastal Defence Study of The Netherlands, the implementation of a 3-D coastal ISE-model is under investigation. 3-D tidal models, which include the effects of wind and density gradients, and corresponding models for the transport of fine sediment, are already available, e.g. for the Rhine outflow area. For these models to be applicable to the shoreface, however, sea wave effects have to be included, both in the current model (De Vriend and Kitou, 1990) and in the transport model (Van Rijn and Meijer, 1991).

Medium-term dynamic models

The first medium-term dynamic model concept for coastal morphology was probably the coastline model (Pelnard-Considère, 1956), which basically assumes a fixed shape of the cross-shore profile and lumps all processes on that profile into a single point, which represents the position of the coastline. Hence the level of aggregation of these models is rather high, and they work best for predictions in the longer term (as compared with the predominant morphological time scale).

Medium-term coastal profile models do work at this predominant time scale. Based on process decriptions at the hydrodynamic time scale, they are supposed to include phenomena such as dune erosion, beach formation and bar dynamics (e.g. Broker Hedegaard et al., 1992, and Roelvink and Broker Hedegaard, 1993).

An example of the use of coastal profile models is the prediction of the cross-shore profile response to sand mining and shoreface nourishments (Roelvink, 1988; Van Alphen et al., 1990), in the framework of the 1990 Coastal Defence Study of The Netherlands.

An interesting application of the models in the near future concerns a number of full-scale experimental shoreface-nourishments, in Denmark, Germany, The Netherlands and Belgium, which will be monitored, modelled and analysed in the framework of the EC-programme Marine Science and Technology (MAST).

Medium-term coastal area models for nearshore applications are still in an early stage of development (e.g. De Vriend et al., 1993a). In contrast to coastal profile models, they are still based on depth-integrated formulations, whence they are unable to describe the dynamics of the coastal profile.

This means, that the applicability of coastal area models in their present form is limited to problems where "cross-shore" transport mechanisms are of minor importance, such a scour due to flow contraction, the dynamics of offshore sediment deposits and shipping channels, or, to some extent, the dynamics of tidal inlets and their ebb-tidal deltas. Another possibility is to include a parametric profile evolution model (Watanabe et al., 1986; Horikawa, 1988), in cases where the profile evolution proceeds much faster than the morphological processes of interest.

Multi-dimensional tidal morphodynamic models are somewhat further advanced (Wang et al., 1991; Wang, 1992). These models have already been extended to quasi-3D, i.e. with a parametric representation of the vertical structure of the flow and the sediment concentration field. Thus they include the effects of secondary flows due to curvature, coriolis and wind, and those of lag effects in the sediment concentration. In situations with a strongly non-uniform bathymetry, these effects can be very important. As long as this concept does not include wave effects, however, it can only be applied in sheltered areas, such as tidal basins.

Formally averaged long-term models

In order to avoid unnecessary and expensive computations of short-term processes which in the long run have no residual effect, we can attempt to formally average the mathematical system over the smaller time scales. As this system contains a number of non-linear terms, in the hydrodynamic modules, as well as in the sediment transport module, this averaging operation inevitably yields residual effects of the short-term motion (cf. the radiation stress terms which are found when averaging the flow equations over the waves). These terms have to be expressed in terms of the averaged variables in order to close the mathematical system. ISE-models can be of good help to derive those closure relationships (De Vriend *et al.*, 1993b).

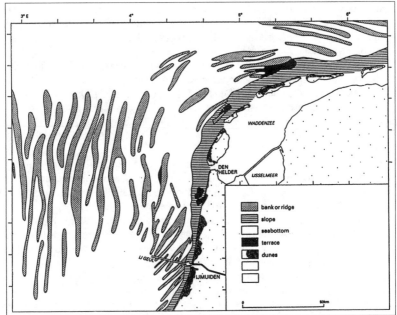

Figure 2:   Linear sand banks and shoreface-connected ridges near the Dutch coast (after Van Alphen and Damoiseaux, 1989)

Another approach concerns rhythmic features of the sea bed, many of which are found in the southern North Sea, such as large offshore sand bank systems (Pattiaratchi and Collins, 1987; also see fig. 2), shoreface-connected ridges (Van de Meene, 1991), sandwave fields (Stride, 1982), nearshore bar systems (De Vroeg *et al.*, 1988), rhythmic features of the beach (Short, 1992), etcetera. Attempts to describe such features as small-amplitude free instabilities of a basic state of the sea bed have been rather successful in describing the wave length, crest orientation and propagation speed of the orientation of the fastest growing mode (Hino, 1974; Huthnance, 1982; De Vriend, 1990; Hulscher *et al.*, 1992). In order to predict the amplitude, external or inherent limiting factors have to be introduced. Huthnance (1982) describes how wave action can limit the growth of linear sand banks, Vittori and Blondeaux (1992) show how the non-linearity of a finite-amplitude bottom wave system can limit its amplitude and can select certain preferential patterns.

The applicability of the resulting free-instability models is restricted exclusively to the feature they describe: every new rhythmic phenomenon has to be analysed in its own right, in physical and mathematical terms. Besides, the instablities have to be free, i.e. they must not be influenced by boundaries or other external factors which force them into a prescribed mode.

On the other hand, there is quite some practical demand for the prediction of rhythmic sea bed behaviour, e.g. in relation to pipelines and shipping channels. A better understanding of the formation and the behaviour of nearshore bar systems is certainly needed to predict the effectiveness of shoreface nourishments on barred coasts. The analysis of rhythmic features on the sea bed and the shoreface is therefore not only scientifically interesting, but also of practical use.

Behaviour modelling

In their present state of development, process-based models as described in the foregoing have a limited capability for long-term predictions, because

(a)   the physical processes which are included in these models have been selected for their relevance to the short- and medium-term coastal behaviour; this warrants neither their relevance, nor their sufficiency to describe long-term evolutions;

(b)   medium-term dynamic models are designed to describe the medium-term dynamics of the coast, which often dominates the long-term trend; they require too much computer time to cover a period of time which exceeds the time scale of these highly dynamic processes by an order of magnitude, and the numerical procedures are insufficiently controlled to yield reliable estimates of the weak long-term trend;

(c)   the modelling approach is still quite deterministic: we try to predict the bottom level as a deterministic function in space and time; thus we produce only one realization of a highly stochastic non-linear process, and we would have to make more runs with different input series in order to have an indication of the probability distribution of this process; in most practical applications, medium-term dynamic models would be prohibitively time-consuming for this approach;

(d)   we still don't know how to handle the uncertainty which is inherent to the forcing (e.g. by wave action) of the complex non-linear coastal system; like in meteorology, this may even limit predictability.

Since we cannot do without long-term prediction methods, however crude, we therefore have to find ways around process-based long-term prediction models, at least for the time being. To that end, we have to make use of all knowledge available, theoretical as well as empirical, on phenomena as well as processes (Terwindt and Battjes, 1990).

One way to achieve this is behaviour modelling. The idea is simple: for a certain class of situations, the coastal behaviour as observed from field surveys and possibly a few medium-term model runs is mapped onto a simple mathematical system which exhibits a similar behaviour, without being derived rigorously from first physical principles. The parameters in this system are adjusted such that the model hindcasts the observed behaviour, and they are fixed when using the model to forecast future evolutions.

This essentially phenomenological approach has the advantage of describing only the phenomena of interest, usually against very reasonable costs. Important disadvantages are the weak physiscal basis and, related to this, the limitation of each model to a narrow class of situations (in principle even site-specific).

Coastal profile behaviour models

A simple example is a coastal profile behaviour model, which lies very close to the existing cross-shore transport concept in multi-coastline models (Bakker, 1968; Perlin and Dean, 1983). It was derived from numerical experiments with a medium-term coastal profile model, which was run with realistic inputs to simulate the evolution of a shoreface nourishment (Stive *et al.*, 1991). The low-pass filtered profile behaviour, together with the empirical observation from many datasets that the shape of the upper part of the profile is more or less invariant, lead to an asymmetric diffusion-type model formulation.

When expressed in terms of the cross-shore position, X, as a function of the vertical coordinate, z, this model reads (fig. 3)

$$\frac{\partial X}{\partial t} - \frac{\partial}{\partial z}\left[D(z)\ \frac{\partial X}{\partial z} \right] = S(z) \tag{1}$$

in which t denotes time. The diffusion coefficient $D(z)$ is the parameter which has to be adjusted in order to reproduce the observed behaviour, and $S(z)$ is a source function which serves to introduce interferences (e.g. nourishments or extractions), or the effect of longshore transport gradients. The diffusion coefficient $D(z)$ will usually decay from a high value in the upper part of the profile (the active zone), to a much lower value on the lower shoreface, which is is line with the observed distribution of morphological activity over the profile.

$D(z)$ can be estimated off-line by trial-and-error, but there are also very efficient parameter identification methods for this type of equation (Capobianco, 1991). The main problem there is to find a sufficiently large and consistent dataset from the site to be studied. So far, this had to be generated artificially with a medium-term dynamic model. Thus the behaviour model becomes in fact a map of the medium-term model.

The diffusion concept turns out to work reasonably well over a wide range of time scales (Stive *et al.*, 1992). An example of its potential use is the evaluation of a hypothetical coastal maintenance policy by repeated beach and profile nourishments, in order to compensate for the effect of sea level rise (fig. 4; Stive *et al.*, 1991).

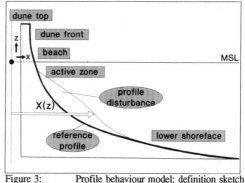

Figure 3:   Profile behaviour model: definition sketch

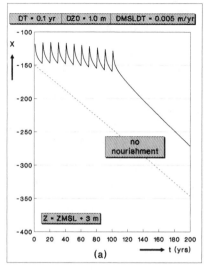 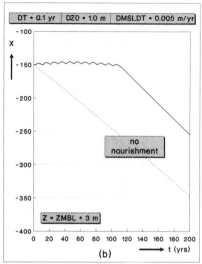

Figure 4:   Long-term evolution of a repeatedly nourished profile (a) beach nourishments, (b) shoreface nourishments (from Stive et al., 1991; courtesy Coastal Engineering)

The model shows that repeated nourishment tends to steepen the profile and builds up a sand deficit at the lower shoreface. If the nourishments are stopped, this leads to an accelerated erosion of the upper shoreface, which is stronger in the case of beach nourishments (shoreface nourishments "lose" more sand to the lower shoreface, whence the sand deficit remains smaller).

The model also showed that the amount of sediment needed for shoreface nourishments is higher than for beach nourishments, but not to the extent that shoreface nourishments should be rejected a priori for economical reasons.

Tidal inlet behaviour models

A large part of the coasts of the southern North Sea consists of series of barrier islands with tidal inlets in between, which connect a shallow back-barrier basin with the sea. The West- and East-Frisian Wadden Sea is the largest of these barrier island coasts.

The Wadden Sea is ecologically extremely valuable, but also quite vulnerable. Its ecosystem rests to a large extent on the large intertidal area, which is sensitive to changes in the mean sea level and the tidal regime (e.g. Eysink, 1990, and Eysink *et al.*, 1992). The prediction of the morphological response to such changes, either natural or man-made, is therefore very important.

On the other hand, the system is extremely complicated, with a pronounced topography and a key role for small residual effects of the tidal motion and a complex interaction with wave effects on the outer deltas and the adjacent island coast. Therefore, the long-term prediction capacity of process-based models is limited and behaviour-oriented modelling seems worth trying, especially because empirical knowledge is well-established here (e.g. Eysink *et al.*, 1992).

An interesting case to start with is the Friesche Zeegat inlet, between Schiermonnikoog and Ameland. In the early seventies a large part (approximately one third) of the basin area was reclaimed and the system's response to that major interference has been monitored in detail (Biegel, 1991). Thus we have obtained a well-documented and almost complete time-history of the establishment of a new equilibrium state of this heavily disturbed system.

The area has been studied extensively from a variety of disciplines, also with detailed process-based numerical models (Steijn, 1992; Wang, 1992). The process-knowledge which was obtained from these models was used to set up a simple and more or less site-specific behaviour model of the basin, which was able to reproduce the morphological evolution in terms of aggregate large-scale quantities (e.g. total channel cross-section and total flats' area per subsection of the basin) with a small number of adjustable parameters (Van Dongeren and De Vriend, 1992).

A similar approach is being taken to model the large-scale behaviour of the outer delta and the adjacent coastlines. Here we use a modified version of a coastline model based on the two-line concept (fig. 5).

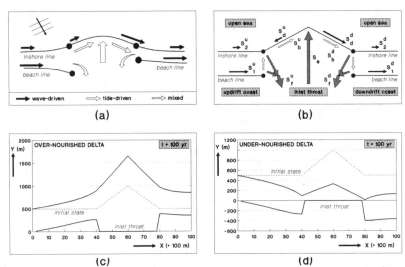

Figure 5:   Two-line model for the outer delta of a tidal inlet (after De Vriend and Bakker, 1992) (a) model concept, (b) equilibrium state, (c) response to overnourishment from the inlet, (d) response to undernourishment from the inlet

Although this concept is still in an early stage of development, it looks as though it is capable of producing useful results and can easily be integrated with two-line models for the long-term prediction of the barrier island coasts and the Holland coast. A next step on the way to long-term morphological modelling of the Wadden Sea is the combination of tidal models (linear network, non-linear network, 2-D) with similar morphological concepts as described above. This would enable us to model complex multi-inlet systems where the tidal motion is non-trivial, without having to go through sediment transport computations and morphological time loops at the time scale of the tide (Di Silvio, 1991).

Long-term sediment balance models

If we step to historical time scales, i.e many decades to centuries, we can further reduce the spatial resolution of our model. For the coastal profile, for instance, it is no longer necessary to use a numerically discretized continuum model, such as the diffusion model: the profile can now be described as a small number of interlinked rigid sections, the position of which follows from sediment balance considerations (Cowell and Roy, 1988; Stive *et al.*, 1990, and De Vriend *et al.*, 1993b).

Figure 6 shows an example of such a "panel-type" system. The upper panel, which represents the dune front, the subaerial beach and the active zone, can only move parallel to itself and has a fixed position with respect to the mean sea level. This reflects the observed invariance of the shape of the upper shoreface when following the mean sea level. The horizontal position follows from the sediment balance, similar to the "Bruun rule" (Bruun, 1962), but taking into account how much sediment is supplied by the lower parts of the shoreface.

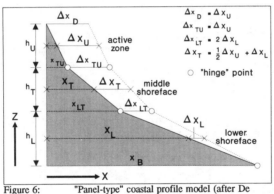

Figure 6:   "Panel-type" coastal profile model (after De Vriend and Roelvink, 1989)

The lower panel represents the lower shoreface and hinges at the sea floor, which is assumed to be morphologically inactive. The central panel is hung up between the two others and represents the middle shoreface as a transition zone.

The cross-shore sediment transport in this system can be eiter "autonomous", i.e. independent of the morphological state of the profile, or a function of the slope of the relevant panels. The latter provides a feedback mechanism from the morphology to the transport, which makes the system dynamic and introduces an equilibrium mechanism.

Since we are dealing with very slow large-scale evolutions, the longshore coastal evolution is not very likely to be an order of magnitude slower than the cross-shore one. Therefore, the panels have to be given a longshore dimension, and a number of cross-shore sets have to be put in parallel and interlinked. Longshore transports can also be "autonomous" (e.g. tidal), or related to the orientation of the panels with repect to the wave climate.

Thus we obtain a horizontally two-dimensional panel-model, which is somewhat similar to a three-line model (Perlin and Dean, 1983). Stive *et al.* (1990) applied an earlier version of this model to predict the long-term evolution of the Holland coast. Stive (private communication; also see De Vriend *et al.*, 1993b) has tested the above concept on the same case, with reasonable success.

Perspectives of integration in CZM-tools

One of the tools of coastal zone management (CZM) is a computational framework, which encompasses a variety of modules which quantify different aspects (e.g. physical, biological, socio-economical) of the system to be managed. Very often, a morphological module is needed there. That module has to fit in with the other parts of the system, i.e. it should produce adequate information against reasonable costs, without unnecessary detail or redundancy. This simple requirement presents either side of the "market" with a problem:

-       the "demand side", because they have to formulate what they want to know from the morphological module, including the ranges of uncertainty and unreliability of the information,

-       the "supply side", because they have to know what information a certain concept can produce in a given situation, against what costs and with what reliability.

However obvious these questions may seem, they can seldomly be answered accurately enough to make the two sides meet without iteration.

Clearly, it should not be the costs of a module which determine the choice. If we need information on a complex system, but cannot afford to include the appropriate model into the computational framework, we have to bring the results of that model into a more amenable form, e.g. by tabulation, or by mapping them onto a simpler mathematical system. The latter complies very well with the behaviour modelling approach which was described in the foregoing.

Even if physical modelling is ahead of the other disciplines in the framework when it comes to quantitive prediction, sooner or later the more complex models will be called upon. At that moment morphological modellers should have a cleare picture of

(a)   the "deterministic" reliability of the information produced by their models, i.e. conceptual correctness, effects of simplifications, numerical accuracy, sensitivity to inputs,

(b)   the uncertainty of the model output due to the stochastic nature of part of the input,

(c)   the predictability limits, if any, of the physical system and its model representation.

At the last two points, there is a lot to be learned from other disciplines, such as Systems Theory and Applied Mathematics.

Conclusion

The material presented herein shows that coastal morphological modelling is developing along two interlinked lines, one based on the description of the underlying physical processes, and the other based upon the description of observed coastal behaviour. These two lines have produced a variety of model concepts, ranging from a simple box-type sediment balance approach to complicated 3-D dynamic models, and from refined small-scale dynamics to crude aggregated large-scale behaviour.

Most of these concepts were tested and applied in the southern North Sea, thanks to substantial investments into the innovation of coastal defence by the bordering countries.

Clearly, there is still a long way to go in coastal modelling, but the intermediate products provide a good potential to exploit the existing knowledge.

Acknowledgement

This paper was written as part of the author's involvement in the Netherlands Centre for Coastal Research, a cooperation between Delft University of Technology, the University of Utrecht, Rijkswaterstaat and Delft Hydraulics.

The supply of information on the Texel Coastal Defence Project by dr. J.S. Ribberink and on the long-term prediction of the Holland coast by dr. M.J.F. Stive is gratefully acknowledged.

References

Bakker, W.T., 1968. The dynamics of a coast with a groyne system. *Proc. 11th ICCE*, London, England, p. 492-517.

Biegel, E.J., 1991. Equilibrium relationships in the ebb tidal delta, inlet and back barrier area of the Frisian inlet system. Rijkswaterstaat, *Report GWAO-91.016*, 79 pp

Broker Hedegaard, I., Roelvink, J.A., Southgate, H.N., Péchon, Ph., Nicolson, J. and Hamm, L., 1992. Intercomparison of coastal profile models. *Proc. 23rd ICCE*, Venice (in press).

Bruun, P., 1962. Sea-level rise as a cause of shore erosion. *J. Waterways and Harbors Div.*, ACSE, 88(WW1), p. 117-130.

Capobianco, M., 1991. Parameter identification for a class of partial differential equations in coastal profile modelling. *CIRM Conf. "Mathematical Problems in Environmental Protection and Ecology"*, Trento, Italy.

Cowell, P.J. and Roy, P.S., 1988. Shoreface transgression model: programming guide. *Unpublished report*, Coastal Studies Unit, University of Sydney, 23 pp.

De Vriend, H.J., 1990. Morphological processes in shallow tidal seas. *In: R.T. Cheng (ed.): "Residual Currents and Long-term Transport", Coastal and Estuarine Studies*, Vol. 38, Springer-Verlag New York, p. 276-301.

De Vriend, H.J. and Bakker, W.T., 1993. Sedimentary processes and morphological behaviour models fro mixed-energy tidal inlets. *Submitted for publication.*

De Vriend, H.J. and Kitou, N., 1990. Incorporation of wave effects in a 3D hydrostatic mean current model. *Proc. 22nd ICCE*, Delft, The Netherlands, p. 1005-1018.

De Vriend, H.J. and Ribberink, J.S., 1988. A quasi-3D mathematical model of coastal morphology. *Proc. 21st ICCE*, Malaga, Spain, p. 1689-1703.

De Vriend, H.J. and Roelvink, J.A., 1989. Innovation of coastal defence: along with the coastal system. Rijkswaterstaat/Delft Hydraulics, "Coastal Defence after 1990", *Techn. Rept. no. 19*, 39 pp.

De Vriend, H.J., Zyserman, J., Péchon, Ph., Roelvink, J.A., Southgate, H.N. and Nicholson, J., 1993a. Medium-term coastal area modelling. *To be published in Coastal Engineering*.

De Vriend, H.J., Latteux, B., Capobianco, M., Chesher, T., Stive, M.J.F. and De Swart, H.E., 1993b. Long-term modelling of coastal morphology. *To be published in Coastal Engineering*.

De Vroeg, J.H., Smit, E.S.P. and Bakker, W.T., 1988. Coastal Genesis. *Proc. 21st ICCE*, Malaga, Spain, p. 2825-2839.

Dingemans, M.W., Radder, A.C. and De Vriend, H.J., 1987. Computation of the driving forces of wave-induced currents. *Coastal Engineering*, Vol. 11, no. 5&6, p. 539-563.

Di Silvio, G., 1991. Averaging operations in sediment transport modelling: short-step versus long-step morphological simulations. *Preprints Int. Symp. Transp. Susp.* Sed. Mod., Florence, Italy, p. 723-739.

Eysink, W.D., 1990. Morphologic response of tidal basins to changes. *Proc. 22nd ICCE*, Delft, The Netherlands, p. 1948-1961.

Eysink, W.D., Biegel, E.J. and Hoozemans, F.J.M., 1992. Impact of sea level rise on the morphology of the Wadden Sea in the scope of its ecological function: investigations on empirical relations. *Delft Hydraulics*, ISOS*2 report H 1300.

Hino, M., 1974. Theory of formation of rip current and cuspidal coast. *Proc. 14th ICCE*, Copenhagen, p. 901-919.

Horikawa, K., 1988. Nearshore dynamics and coastal processes. *University of Tokyo Press*, 522 pp.

Hulscher, S.J.M.H., De Swart, H.E. and De Vriend, H.J., 1992. A dynamical model of offshore tidal sand banks. *Proc. ICCE '92*, Venice, Italy (in press).

Huthnance, J.M., 1982. On one mechanism forming linear sand banks. *Est., Coastal and Shelf Sci.*, Vol. 4, p.79-99.

Pattiaratchi, C. and Collins, M., 1987. Mechanisms for linear sand bank formation and maintenance in relation to dynamical oceanographical observations. *Prog. Oceanol.*, 19, p. 117-176.

Perlin, M. and Dean, R.G., 1983. A numerical model to simulate sediment transport in the vicinity of coastal structures. CERC, Vicksburg, *Misc. Rept. no. 83-10*, 119 pp.

Ribberink, J.S. and De Vroeg, J.H., 1992. Coastal protection Eijerland (Texel). Hydraulic and morphologic effect study. *Delft Hydraulics*, Report H 1241.

Roelvink, J.A., 1988. Shoreface nourishment, an alternative coastal defence technique? *Delft Hydraulics*, Reprt H 825.30 (in Dutch).

Roelvink, J.A. and Broker Hedegaard, I., 1993. Cross-shore profile models. *To be published in Coastal Engineering*.

Short, A., 1992. Beach systems of the central Netherlands coast: Processes, morphology and structural impacts in a storm driven multi-bar system. *Marine Geology*, 107, p. 103-137.

Steijn, R.C. and Louters, T., 1992. Hydro- and morphodynamics of a mesotidal inlet in the Dutch Wadden Sea. *Proc. "Oceanology Int. '92" Conf.*, Brighton, U.K.

Steijn, R.C., Louters, T., Van der Spek, A.J.F. and De Vriend, H.J., 1989. Numerical model hindcast of the ebb-tidal delta evolution in front of the Deltaworks. In: *Falconer, R.A. et al.: "Hydraulic and Environmental Modelling of Coastal, Estuarine and River Waters"*, Gower Technical, Aldershot, p. 255-264.

Stive, M.J.F., Roelvink, J.A. and De Vriend, H.J., 1990. Large scale coastal evolution concept. *Proc. 22nd ICCE*, Delft, The Netherlands, p. 1962-1974.

Stive, M.J.F., Nicholls, R. and De Vriend, H.J., 1991. Sea-level rise and shore nourishment: a discussion. *Coastal Engineering*, Vol. 16, p. 147-163.

Stive, M.J.F., De Vriend, H.J. and Nicholls, R.J. and Capobianco, M., 1992. Shore nourishment and the active zone: a time-scale dependent view. *Proc. ICCE'92*, Venice, Italy (in press).

Stride, A.H. (ed.), 1982. Offshore tidal sands: processes and deposits. *Chapman and Hall*, London.

Van Alphen, J.S.L.J. and Damoiseaux, M.A., 1989. A geomorphological map of the Dutch shoreface and adjacent part of the continental shelf. *Geologie en Mijnbouw*, 68(4), p. 433-444.

Terwindt, J.H.J. and Battjes, J.A., 1990. Research on large-scale coastal behaviour. *Proc. 22nd ICCE*, Delft, p. 1975-1983.

Van Alphen, J.S.L.J., Hallie, F.P., Ribberink, J.S., Roelvink, J.A. and Louisse, C.J., 1990. Offshore sand extraction and nearshore profile nourishment. *Proc. 22nd ICCE*, Delft, The Netherlands, p. 1998-2009.

Van Dongeren, A.R. and De Vriend, H.J., 1993. A model of morphological behaviour in tidal basins. *To be published in Coastal Engineering*.

Van de Meene, J.W.H., 1992. Shoreface connected ridges along the Dutch coast. *Proc. "Coastal Sediments '91"*, Seattle, Washington, p. 512-526.

Van Overeem, J., Steijn, R.C. and Van Banning, G.K.F.M., 1992. Simulation of the morphodynamics of tidal inlets in the Wadden Sea. *Proc. Int. Coastal Congress*, Kiel, Germany.

Van Rijn, L.C. and Meijer, K., 1991. Three-dimensional modelling of sand and mud transport in currents and waves. *Preprints Int. Symp. Sed.*, Florence, Italy, p. 683-708.

Vittori, G. and Blondeaux, P., 1992. Sand ripples under sea waves. Part 3: brick pattern ripple formation. *J. Fluid Mech.*, 239, p. 23-45.

Wang, Z.B., 1992. Morphodynamic modelling for a tidal inlet in the Wadden Sea. *Delft Hydraulics, Coastal Genesis Report H 840.50-II*, 37 pp.

Wang, Z.B., De Vriend, H.J. and Louters, T., 1991. A morphodynamic model for a tidal inlet. *Proc. CMOE'91*, Barcelona, Spain, p. 235-245.

Watanabe, A., Maruyama, K., Shimizu, T. and Sakakiyama, T., 1986. Numerical prediction model of three-dimensional beach deformation around a structure. *Coastal Eng. Japan*, Vol. 29, p. 179-194.

Long Term Water Level Observations and Variations

J. Jensen[1], J. L.A. Hofstede[2], H. Kunz[3], J. de Ronde[4], P.F. Heinen[4], W. Siefert[5]

Abstract

Numerous tide gauges have been recording data along the coastlines of the southern North Sea for more than a century, at Amsterdam even since 1700. Every tide gauge locality has its own unique sea level history; the data are more or less influenced by several factors. Some examples of how these factors affect the trend analysis of water levels (high, low, mean) are presented.

Statistical aspects with relation to predictions of the future water level trend are addressed. The evaluation of the Mean Sea Level (MSL) has become important with respect to the expected global climate change. Its relative trends are not constant, neither with regard to locality nor with regard to time. The MSL rise in the southern North Sea varies between 10 and 20 cm/100 years with an overall mean of about 15 cm/100 years. A rise of the relative Mean High Water (MHW), which has accelerated over the past decades, is found at all of the stations. The MLW sank slightly during the beginning of the second half of this century but subsequently has stabilised or shows a light positive trend. This results in a MTR increase for the tide gauges along the Dutch and German coast since 1950.

Introduction

The eustatic variations of the global sea level are strongly influenced by worldwide changes in climate. There is no longer any doubt that the "greenhouse effect" and other human activities have a definite impact on our climate. In the North Sea, bounded by Norway (N), Denmark (DK), Germany (D), The Netherlands (NL) and Great Britain (GB), and especially in the German Bight, an enhanced rise of the sea level will also lead to far-reaching alterations.

[1] University GH Siegen, P.O. Box 101240, D5900 Siegen, Germany

[2] Landesambt für Wasserhaushalt und Küsten, Saarbrükenstrasse 38, D2300 Kiel 1, Germany

[3] Coastal Research Station, An der Mühle 5, D2982 Norderney, Germany

[4] Ministry of Transport, Public Works and Water management, Tidal Waters Division, P.O. Box 20907, 2500 EX Den Haag, The Netherlands

[5] Strom und Hafenbau, Dalmanstrasse 1/3, D2000 Hamburg 11, Germany

In this context, not only changes in MSL are important, especially alterations in annual values of the mean tidal high water (MHW), mean tidal low water (MLW) and mean tidal range (MTR) will play a decisive role (all references to trends in the MSL, MHW etc. are considered to be *relative* changes between the land and sea level). Changes in tidal dynamics of the flat coastal regions affect erosion, degree of storm surge risk, ground water level and shipping. The evaluation of time series of annual data (MHW and MLW) and semi-diurnal tidal high water (THW) and tidal low water (TLW) shows an increase of the THW along the German and Dutch coastline, whereas the TLW decreases or stays constant (e.g. Jensen, 1984). This leads to the assumption, that during the last three decades a change in the tidal dynamics in the North Sea may have taken place. If this is a long term trend needs to be proven because of the influence of long term periodic changes (Jensen *et al.*, 1990).

Sea level rise between AD 1000 and 1850

Sea level curves covering the Holocene epoch are mostly based upon geological data and reach back to the last ice age about 18,000 years ago. Curves, which are based upon tidal gauge data, only cover the period from about AD 1850. Hardly any curve exists, which links the Holocene to the modern curves. For this reason, a probable MHW-curve for the southern North Sea covering the period from about AD 1000 onwards was established (fig. 1).

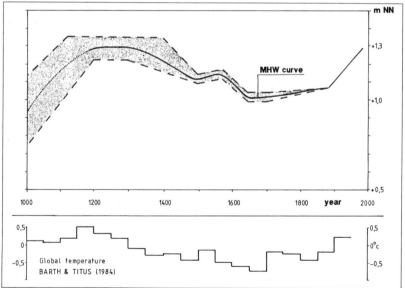

Figure 1:   Probable MHW-curve for the southern North Sea since about 1000 (adapted from Hofstede, 1991)

Based upon dated organic samples from the East-Frisian barrier islands ‚Streif (1989) suggested that the local MHW-level might have been between NN +1.22 and 1.35 m for a certain time interval between 1125 and 1395. This would mean that the modern MHW-level in this area was already reached about 500 to 800 years ago. According to Barth and Titus (1984), the 50-year global mean temperature for the time span 1200/1250 (Medieval Climate Optimum) was a little higher than the 1900/1950-mean (fig. 1). The MHW-maximum between 1125 and 1395 according to Streif (1989) correlates well with these climatological data.

The Medieval Climate Optimum was followed by a severe deterioration. According to the temperature curve of Barth & Titus (1984) the period between 1600 and 1700 was the coldest since 900 at the latest. The temperature difference between 1200/1250 and 1650/1700 was about 1.5 °C. Also, the first great expansion of glaciers in the Alps during the Little Ice Age occurred between 1600 and 1640 (Maisch, 1989). So it seems realistic to suggest that the MHW-minimum of the Little Ice Age was reached around 1650 (Gornitz et al., 1982; Wigley & Raper, 1987; Oerlemans, 1989).

From 1700 on, sea level has been recorded at the tide gauge of Amsterdam (Netherlands). So from this time on more or less accurate data on relative sea level are available. Continuous German tide gauge records reach back to 1826 (Travemünde, Baltic Sea) and 1843 (or fixed to tidal datum 1855) (Cuxhaven, North Sea). The Baltic Sea is connected to the North Sea by three small entrances only. The consequence is a sea with a tidal range that can be neglected. The MSL-curves for these three gauges are presented in fig. 2.

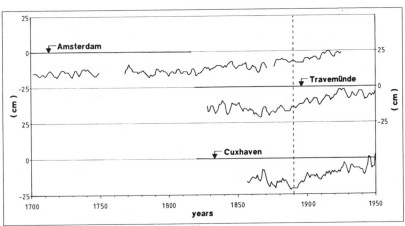

Figure 2:   MSL-rise at three long-term tide gauges

According to these curves no significant fluctuations occurred between 1700 and about 1890. However, a stable MSL does not necessarily mean that the MHW remained constant. Independent changes in the tidal range might induce a MHW change as well. However, first results of Dutch investigations on tidal range fluctuations along the Dutch North Sea coast since 1700 seem to suggest that no significant change took place during this period.

According to Flohn (1985) a small temperature rise occurred between 1700 and 1850. During this period glacier variations appear to have been within a limited range, while after 1850 a world-wide retreat began (Oerlemans, 1989). So once again, the climatological data correspond well with the reconstructed MHW-curve of figure 3.

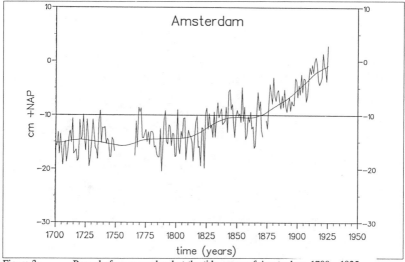

Figure 3:   Record of mean sea level at the tide gauge of Amsterdam, 1700 - 1925

Based upon 21 long-term tide gauges Gornitz & Solow (1991) found weak statistical evidence for a common turning point around 1895 in the European MSL records which also can be seen in figure 4. It corresponds with the world-wide glacier retreat and temperature rise since about 1870. According to Oerlemans (1989), these processes might have contributed about 9.5 cm to the MSL rise since 1890.

It appears that, at least since AD 1100, sea level changes along the southern North Sea coast were controlled mainly by thermal eustatic processes.

Tide gauge records along the Dutch coast

The Netherlands have a dense network of tide gauges along their coastline, most of them have a record length of more than 100 years. The oldest tide gauge record of the world is that of Amsterdam which started in 1700 and lasted until 1930 (van Veen, 1954) at which time unfortunately (for the Amsterdam record) the former "Zuider Zee" was closed and Amsterdam was disconnected from the sea.

The record of Amsterdam is given in figure 3, including a filtered signal calculated with SSA (Singular Spectrum Analysis) (Heinen, 1992 and Vautard et al., 1989). The record of Amsterdam shows quite clearly a change in relative sea level rise. Before about 1850, the measurements indicate no sea level rise, after about 1800 there is a distinct rise of the sea level. From 1800 to 1930 the rise has been 17 cm.

Figure 4:   Location of mentioned tide gauges in the Netherlands

To explain this change in the rate of sea level rise one can look at the link with the sudden decline of most glaciers around the same time. This change is probably due to relatively lower levels of volcanic dust in the atmosphere (Oerlemans, 1988). One could also say that these changes are evidence of the end of the Little Ice Age. One thing at least is clear, it is not due to the greenhouse effect.

All tide gauge records along the the Dutch coast show a clear trend of relative sea level rise, which is shown in figure 5. A map of the Netherlands with the location of the tide gauges is given in figure 4.

The term *relative* sea level rise has been used here because the changes measured are due to two phenomena, namely sea level rise and subsidence. The Dutch benchmarks have their foundation in the Pleistocene layer several tens of meters deep. The Pleistocene under the Netherlands is subsiding at a rate of 2 - 8 cm per 100 years.

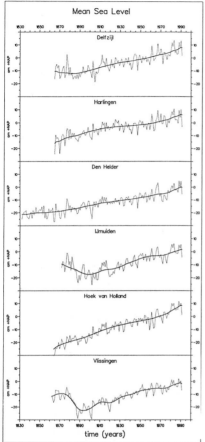

Mean Sea Level

From 1900 onwards the records in figure 5 show more or less the same trend; before 1900 some of the records have a more fluctuating behaviour due to unknown causes. The rate of relative sea level rise is given in Table 1 for the periods 1900-1990 and 1940-1990.

	1900 - 1990	1940 - 1990
Vlissingen	23	19
Hoek van Holland	26	28
IJmuiden	21	20
Den Helder	17	20
Harlingen	16	19
Delfzijl	21	23
Mean value	21	22

Table 1:   Mean relative sea level rise in cm per 100 years.

The fluctuations between the rates of relative sea level rise of the different tide gauges cannot be neglected. These differences can be partly explained by different rates of subsidence of the Pleistocene and partly by local changes near the gauges (dredging and harbour works). The local changes also can greatly influence the tidal range and the MSL.

Figure 5:   Records of 6 Dutch tide gauges up to 1991; the smooth lines have been derived with SSA.

In figure 5 the filtered signal calculated with SSA for every tide gauge has been plotted as well. This method has also been used for a combined time series of the six tide gauges (fig. 6). During the last 15 years a small increase in relative sea level rise can be noted at several gauges as well as in the combined signal.

In order to investigate this, first it has been tried to reduce the noise level of the signals. A part of the signals can be explained by the changes in air pressure and by the nodal tide. With a multiple linear regression technique using the nodal signal and air pressure data of six stations (Edinburgh, Bergen, De Bilt, Stykkhisholmur, Charlotte Town and Lisbon) it was possible to reduce the noise level. E.g. for the combined signal of the six Dutch tide gauges the noise level was reduced by 40%.

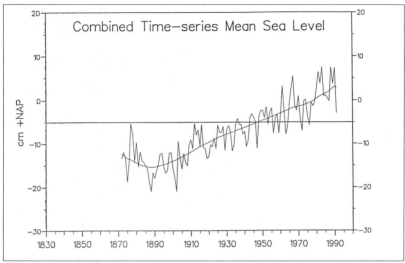

Figure 6:   Combined record of six Dutch tide gauges up to 1991; the smooth lines have
been derived with SSA

Of the total variance 75% could be explained by the trend, 11% by air pressure and
0.06% by the nodal tide, leaving 14% for the noise. The time series for this case are
shown in fig. 7.

When SSA is used after reduction of the noise level by using air pressure data, one finds that the resulting filtered signal show remarkable differences compared with the earlier ones (fig. 8 and 9). The small increase in relative sea level rise at the end of the records has disappeared and the SSA-lines shown are smoother than before. Along the Dutch coast not only relative sea level is increasing but also the tidal range is showing an increase. Increasing tidal ranges are also found along the Belgium and German coasts, but not along the British coast.

In Table 2 the rates of rise of MHW, MLW and MTR for the six tide gauges are given for the period 1940-1990.

Period 1940-1990			
Station	Mean High Water	Mean Low Water	Mean Tidal Water
Vlissingen	29	15	14
Hoek van Holland	(44)	22	(22)
IJmuiden	32	16	16
Den Helder	22	12	10
Harlingen	31	14	17
Delfzijl	(49)	(-10)	(60)
Mean value without ()	29	16	14

Table 2:   Mean rise in cm per 100 years of
mean high water, mean low water
and mean tidal range during the
period 1940-1990
( )=there is no uniform trend

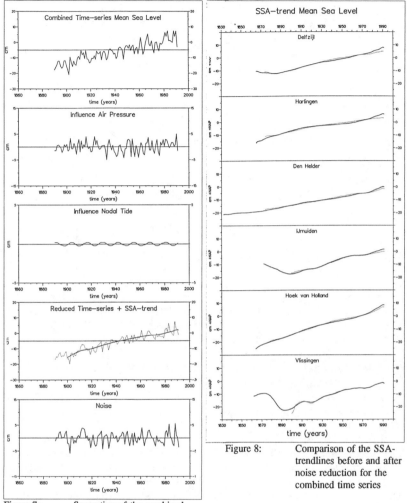

Figure 7: Separation of the combined record of Dutch mean sea level in the parts induced by air pressure, nodal tide, SSA-trend and noise

Figure 8: Comparison of the SSA-trendlines before and after noise reduction for the combined time series

All gauges show an increase of the MTR. The greatest increase can be seen in Delfzijl and can be explained largely by dredging in the Ems estuary. In spite of sea level rise, MLW is decreasing. The second largest increase is in Hoek van Holland. The MTR of this station shows large fluctuations and there is no uniform trend. The other stations show a more or less uniform trend in the MTR.

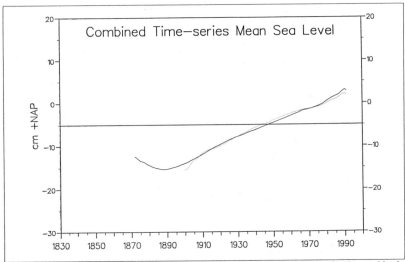

Figure 9:   Comparison of the SSA-trends before and after noise reduction for the combined
                time series

The causes of these changes must be partly due to dredging, harbour works and delta works (the closing of several major tidal inlets in the southern part of the Netherlands) and partly due to some up to now unknown external sources which seem to change the tidal amphidromic system in the North Sea.

As an example, fig. 10 shows the changes in relative MSL, MHW and MTR of the tide gauge Vlissingen.

Tide gauge records along the German coast

Figure 11 gives a general view of the German Bight, the southeastern part of the North Sea, and the sites of the main gauges along and in front of the German coast from the Dutch border in the west to the Danish border in the north. Most of these gauges were established during the second half of the last century. The oldest gauge on the German coast is the one in Cuxhaven that started recording in 1788, with continuous records since 1843 and with a reliable data connection to the German datum (NN = Normal Null, fixed to the Amsterdam gauge and the Dutch datum [NAP]) since 1855 (e.g. Gaye, 1951, Rhode, 1977). The rate of subsidence/uplift for the tide gauge locations is not well known. For the western part (East-Frisia) subsidence is likely in a range of about 5 cm/100 years.

A first idea of the regional changes in the southeastern North Sea is given by the comparison of mean tidal ranges and propagation velocities for a period around 1925 and 1980 by Lassen and Siefert (1991). The first survey of the MSL conditions of the German North Sea coast was made by Lassen (1989).

For the evaluation of long term trends on relative sea levels, gauge locations are necessary which can be expected to be free of any influence by human activities such as dredging, training works etc.; furthermore continuous series of records from as long ago as possible must be available. 12 such stations have been selected in the German Bight; these are located on the islands (Borkum, Norderney, Helgoland, List on Sylt, Wittdün on Amrum), at the mouths of large estuaries (Emden, Wilhelmshaven, Bremerhaven and Cuxhaven) or in small harbours (Büsum, Husum, Dagebüll) (fig. 11).

Additional data from seasonally operating and intermediate gauges in the North Sea were included and connected to the German datum, which led to the result that the southeastern North Sea was almost completely covered with more than 230 gauges.

Special correlations between data of seasonal and of main gauges provide reliable information on annual means, even for discontinuous records (Siefert and Lassen, 1985).

From these gauges, the MHW and MLW are used as the arithmetic averages of all tides of each individual hydrological year (the German hydrological year starts on Nov. 1st). The sequences of these values describe the time history of these annual water levels.

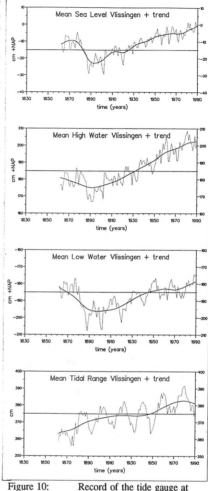

Figure 10:   Record of the tide gauge at Vlissingen for MSL, MHW, MLW and tide range. Smooth lines with SSA

The German Bight as part of the North Sea is governed by semi-diurnal tides with nearly sinusoidal curves in front of the coast. MTR varies between 2.0 m and 3.5 m for the locations indicated in figure 12.
MTR is decreasing to the northwest, where amphidromic points for the M2 and S2 tides are situated (fig. 12). Spring and neap tides do not show remarkable dif-ferences from mean tides. Their tidal ranges are usually within 10 to 20 % of the mean.

Figure 11:   General view of the German Bight with the main gauging stations

Mean time differences between the occurrence of THW and TLW at different stations and the reference time (transition of the moon through the Greenwich meridian) become smaller in the last decades. The evaluations of THW and TLW occurrence times lead to the assumption that in the last decades the propagation velocity of the tide has increased in the southern North Sea (Jensen *et al.*, 1990).

In these very simple tidal conditions it is convenient to use a substitute parameter for MSL, that usually lies some centimetres below MSL:

$$T\tfrac{1}{2}w = \tfrac{1}{2} * (THW + TLW)$$

Nevertheless MSL-changes and T½w-changes are more or less identical, so that it is quite easy to evaluate MSL changes by simply analyzing annual THW and TLW developments.

A long term trend s_T can be described by a linear function:

$$H = H(t) = H_0 + s_T * t$$

With respect to an evaluation after the method of least squares, s_T represents the mean slope of the function related to a period of 100 years (the effect of the nodal tide with a period of $T = 18.6$ years must be reduced).

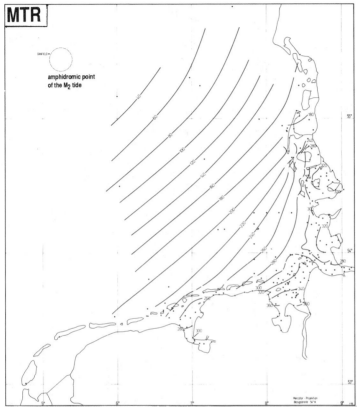

Figure 12:   MTR 1975-1985 in the southeastern North Sea

As a summary of the time series of annual tidal high and low water presented in fig. 13, table 3 shows, related to the base of 1991, the changes in cm/100 years for MHW, MLW, MTR and ≈MSL from the time series N = 100 years and the extrapolation of N = 37 years (2 cycles of the nodal tide) (stations from table 3 as well as Helgoland and Wittdün).

For the 10 gauges, the time series of the annual values for the MHW show an increasing trend superimposed by annual fluctuations. These are mostly due to large scale meteorological effects and occur simultaneously (e.g. 1947) in the time functions of all gauges. In the time series for the MLW the same annual fluctuations as in the MHW can be detected (e.g. 1947). The long term trend, however, is not as clear as that for the MHW (Töppe, 1992).

Taking the equally weighted values of all 12 stations, mean values for the complete area of the German Bight can be computed (table 3; Jensen, 1984; Führböter and Jensen, 1985).

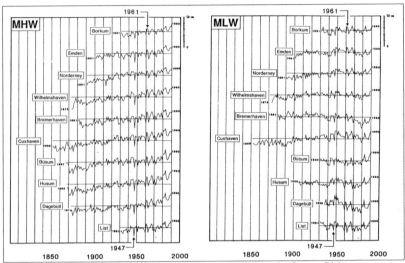

Figure 13:   Time series of annual values for the MHW along the German Bight

	Secular Trends in cm/100 years calculated from N=37 and N=100 years							
	Mean High Water		Mean Low Water		Mean Tidal Range		Mean Sea Level	
Locations	1955 1991 (37a)	1892 1991 (100a)	1955 1991 (37a)	1892 1991 (100a)	1955 1991 (37a)	1892 1991 (100a)	1955 1991 (37a)	1892 1991 (100a)
Borkum	31	-	15	-	15	-	23	-
Emden	43	-	-22	-	65	-	10	-
Norderney	27	-	5	-	22	-	16	-
Wilhelmshaven	27	27	-1	1	28	26	13	14
Bremerhaven	24	24	-39	-18	65	42	-7	3
Cuxhaven	34	25	-9	13	43	12	13	19
Helgoland	24	-	2	-	21	-	13	-
Büsum	27	19	35	-	11	-	41	-
Husum	56	31	6	-	49	-	31	-
Wittdün	44	-	1	-	43	-	23	-
Dagebüll	60	29	-17	-	77	-	21	-
List	32	-	4	-	28	-	19	-
Mean value	36	26	0	0	39	27	18	12
Stand.deviation	+/-12	+/-4			+/-21	+/-12	+/-11	+/-7

Table 3:   Secular Trends s_T of MLW, MHW, MTR and ≈MSL for different stations

Whereas the standard deviations for the MHW are much lower than the absolute values (which clearly show an increasing tendency within the last 37 years), the mean deviations for the MLW are higher than the absolute values themselves.

These numbers indicate a trend of:

+12 to +18 cm/100 years

for MT½w (and roughly for MSL as well) on the basis of the last 100 or the last 37 years respectively.

The different developments of MHW and MLW seem to be the result of a nonlinear reflection process in the shallow water areas of the tidal flats surrounding the German North Sea coast.

In order to gain a better insight into the temporal changes of the trend, the trend s_T is computed only for a period of N = 37 (100) years starting at the beginning of the data, this results in a time dependent function $s_{T \, (N \, = \, 37 \, (100) \, years)}$ as $s_{T(t)}$.

Figure 14 shows this time-function (trend of a linear regression model over 37 years (2 cycles of the nodal tide)) for Cuxhaven. The first period of 37 years at Cuxhaven

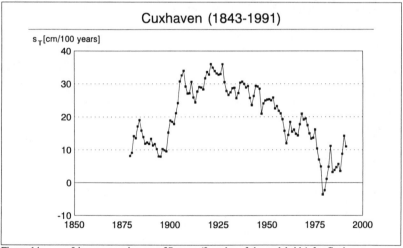

Figure 14:   Linear regression over 37 years (2 cycles of the nodal tide) for Cuxhaven

starts in 1843 (from 1855 reliable data). The secular trends (in cm/100 years, as running means over 37 years) in Cuxhaven vary between 0 and + 35 cm/100 years, with a mean value of about + 20 cm/100 years. Trend investigations should be based on time windows with a length of cycles of the nodal tide (T = 18.6 years) for the calculation of running means.

The increase of the trend in MHW can clearly be seen, the trends of MLW are much smaller, at some tide gauges MLW is even decreasing. Due to this development the trends of MTR increase faster and those of MSL rise much slower than the trends of MHW. Conspicuous alterations have taken place in the last decades.

Evaluations made by Jensen, Mügge and Schönfeld (1990 and 1992) reveal that there is no uniform trend in the tidal regime of the German Bight. The secular rise strongly depends on the location of the tidal gauge and on the period of time considered. One obtains different results depending upon whether MSL (predominantly used parameter) or THW and TLW are computed. Based on the historical development of MSL in the German Bight, no extraordinary rise can be detected when time periods of more than 30 years are taken into account. Only computations of trends over shorter periods reveal an enhanced increase of the MSL in the German Bight. More interesting developments have occurred with respect to the extreme values of the tidal regime. The rise of MHW compared with constant or decreasing MLW leads to an enhanced increase of the MTR. The rise of MTR is of great importance for tidal dynamics. Enhancement of tidal energy will lead to increased current velocities and sediment transports. This has great impact on the morphological structure of the coastal regions and is of high importance for coastal engineering purposes.

In order to gain a better insight into the development of the extreme high (e.g. storm surges) and low water levels the following quantiles: $THW_{99\%}$, $THW_{95\%}$ and $THW_{50\%}$ ($\approx$MHW), $TLW_{50\%}$ ($\approx$MLW), $TLW_{5\%}$ and $TLW_{1\%}$ were investigated (Table 4).

	THW-trend [cm/100a]			TLW-trend [cm/100a]		
	THW 99%	THW 95%	THW 50%	TLW 50%	TLW 5%	TLW 1%
Borkum	43	36	14	15	3	18
Emden	81	63	39	-21	-43	-37
Norderney	48	45	24	2	-11	4
Wilhelmshaven	51	47	30	-5	-11	-5
Lt. Alte Weser	51	58	20	5	-11	-8
Cuxhaven	52	69	33	-14	-29	-19
Helgoland	45	52	20	-2	-13	-4
Büsum	64	74	46	28	24	34
Husum	78	91	51	-3	-15	-31
Wittdün	86	80	43	-8	-13	-7
Dagebüll	103	88	54	-24	-54	-76
List	77	57	27	-4	-6	-5

Tabel 4:   Trends in the time series (TLW- and THW-Quantiles) from 1954 to 1991 of selected tide gauges along the German Bight

The trend of extreme high water levels (e.g. $THW_{99\%}$) is much steeper than the trend of the mean high water levels (e.g. $THW_{50\%}$). The same can be seen for the most low water levels and extreme low water levels ($TLW_{1\%}$).

Depending on the period, its length and the treatment of the available data slightly different results for MSL variations can be found, given for three representative gauges (fig. 13) in Table 5.

Location	Period	Treatment	Secular Trend cm/100 years
Cuxhaven	1925-1974	hourly values	20+/-6
	1855-1987	yearly means	14+/-6
	1945-1987	yearly means	-1+/-3
	1906-1986	mean tide curves	12+/-3
	1955-1991	yearly means, (MT½w)	13+/-6
	1892-1991	yearly means, (MT½w)	19+/-8
Borkum	1933-1988	mean tide curves	15+/-3
	1955-1991	yearly means, (MT½w)	23+/-5
Helgoland	1953-1986	hourly values	1+/-4
	1916-1986	mean tide curve	4+/-1
	1955-1991	yearly means, (MT½w)	13+/-6

Table 5: Different estimations (data basis, length of the time series and method) for MSL changes for three representative gauges

The almost tide-free Baltic Sea may be regarded as a damped gauge of the North Sea (Jensen and Töppe, 1986). The centennial change of the MSL in the Baltic Sea (Travemünde gauge) amounts to 16 cm/100 years during the period 1826 to 1990 (N = 165 years). The trend increases when shorter periods are investigated (last decades). In comparison to the behaviour of the North Sea, the mean water levels at the Travemünde gauge show a behaviour similar to the MSL of the North Sea.

Time series and trend-corrected time series of annual extreme water levels (storm-tide water levels) allow us to make assessments concerning storm surge intensity based upon probability theory of the frequency of occurrences of storm tides (Jensen, 1985).

Selection of data sets and its impact on results

Many investigations apply statistical tools and data selection criteria that deal with tide gauge data. This will be addressed by examples.

Trend investigations can be based on different time windows for the calculation of running means. An example showing the impact of the window length is given in figure 15. Displayed are MHW as well as MLW registered by the tide gauge Bremerhaven. The large differences are obvious, although the length of the windows only differs from 20 to 30 years. Similar results have been published by Lohrberg (1989; see also Jensen, 1984; Führböter and Jensen, 1985). A part of the curves could be explained by dredging, training works etc. in the Weser estuary and the influence of the nodal tide. The trend curves can be described by a superpositioning of periodic parts (Jensen *et al.* 1991).

Extreme events may be neglected to get more significant information on trends of mean high or low water levels. With regard to MHW, a result is shown in figure 16.

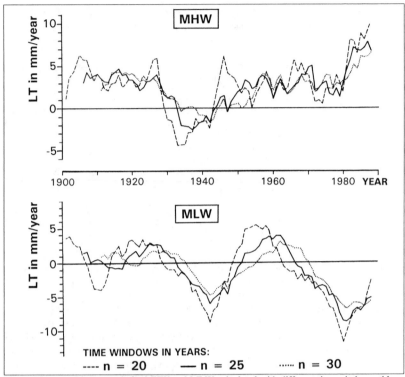

Figure 15:   Linear trends for MHW and MLW calculated with different time windows; tide
                gauge Bremerhaven (Kunz & Niemeyer, 1993)

The upper graph compares linear trends of the tide gauge Wilhelmshaven (time
window of 20 years) gained by investigations which were based on all tides (solid
line) and on tides below the level of storm surges (dashed line; see also tables 4 and
5).

The probability method (DNA, 1979; Rohde 1979) used defined specific frequencies
of occurrence differentiated by three grades of severeness (f < 10; 0.5; 0.05 with f
= occurrences per year). The differences between the two graphs can be seen in the
figure lower: they reach from about +3 to -4 cm; a kind of periodic pattern can be
recognised. Investigations concerning MLW lead to comparable results.

It is also essential to consider the long term trend of the MHW (secular rise). Data
have to be corrected with respect to a non-horizontal reference level, neglecting the
secular rise will lead to falsified results; an artificial trend of increasing storm surge
frequency is created which does not exist in reality. An example is plotted in figure
17 for the tide gauge Norderney.

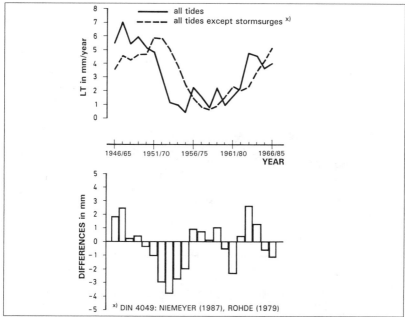

Figure 16:   Linear trends of MHW-level rise for different data sets; tide gauge
              Wilhelmshaven (Niemeyer, 1987)

To get valid information on trends of storm surge climate, it is recommended to base
investigations on time series with a length of at least two return periods. Reported
frequencies differ quite a bit: 7 years (Siefert, 1989) to 60 years (Lüders, 1936).

There is no definite proof for a long term tendency of increasing storm surge frequency; extraordinary storm surges occurred in former times as well (Niemeyer,
1987, Halcrow/NRA ,1991), but some authors claim that such tendency exists (e.g.
Lamb, 1982).

Conclusions

A rise in the MHW ($\approx$25 cm/100 years), which has accelerated over the past decades,
is found at all of the stations (all references to trends in the MSL, MHW etc. are
considered to be *relative* changes between the land and sea level). The MLW sank
slightly during the beginning of the second half of this century but subsequently has
stabilised or shows a light positive trend. This results in a MTR increase ($\approx$20 cm/100
years) for the tide gauges along the Dutch and German coasts since 1950. In spite of
this, the MTR along the British coast seems to decrease. The MT½w, as an
approximation for MSL, has therefore not risen as fast as the MHW.

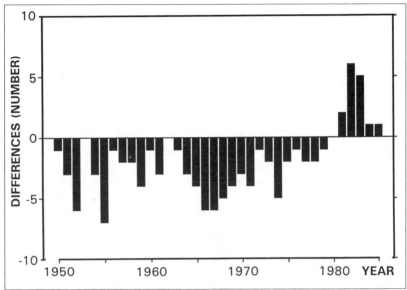

Figure 17:   Over- and underestimation of yearly storm surge frequencies (number) by
                neglecting the linear trend of the MHW; tide gauge Norderney (Niemeyer, 1987)

The rise of MHW compared with constant or decreasing MLW leads to an enhanced increase of the MTR. The rise of MTR is of great importance for tidal dynamics. Enhancement of tidal energy will lead to increased current velocities and sediment transports. This has great impact on the morphological structure of the coastal regions and is of great importance for coastal engineering purposes. The causes for this development must be partly due to dredging, harbour works and delta works (e.g. the closing of several major tidal inlets in the southern part of the Netherlands) and partly due to some up to now unknown external sources which seem to change the tidal amphidromic system in the North Sea.

MSL in a coastal region forms a relatively complicated sphere. The trends of the MHW, MLW, MTR and MSL are not constant, neither with regard to locality nor with regard to time. Depending on data set, evaluation, period and length of period, MSL rise in the southern North Sea varies between 10 and 20 cm/100 years with an overall mean of about 15 cm/100 years. The MSL rise shows a higher tendency in last years, which should not be misunderstood as a significant change in the long term rend. For the interpretation of these results a subsiding in a range from 2 to 8 cm per century must be taken into account.

References

Barth, M.C., Titus, J.G., (eds), 1984: Greenhouse effect and sea level rise, Van Nostrand Reinhold, New York

DNA 1979: DIN 4049 - Begriffe der quantitativen Hydrologie; Deutscher Normen ausschuß (DNA)

Flohn, H., 1985: Das Problem der Klimaänderungen in Vergangenheit und Zukunft, Erträge der Forschung 220

Führböter, A., Jensen, J., 1985: Longterm Changes of Tidal Regime in the German Bight (North Sea). *Proc. 4th Symp. on Coastal and Ocean Management,* Baltimore

Gaye, J., 1951: Wasserstandsänderungen in der Ostsee und der Nordsee in den letzten 100 Jahren, *Die Wasserwirtschaft, Sonderheft:* Vorträge der gewässerkundlichen Tagung 1951 in Hamburg

Görnitz, V., Lebedeff, S. and Hansen, J., 1982: Global sea level trend in the past century, *Science* 215, 1611-1614

Görnitz, V, Solow, A., 1991: Observations of long-term tide-gauge records for indications of accelerated sea-level rise

Halcrow/NRA, 1991: Shoreline Management: The Anglian Perspective, *in:* The future of shoreline management - conference papers: report of the National River Authority and Sir William Halcrow Partners, GB

Heinen, P.H., 1992: Singuliere Spectrum Analyse toegepast op tijdreeksen (in Dutch), Rapport DGW-92.031

Hofstede, J.L.A., 1991: Sea Level Rise in the Inner German Bight Since AD 600 and its Implications upon Tidal Flats Geomorphology, *in* Brückner and Radtke (*Ed.*): Von der Nordsee bis zum Indischen Ozean. Franz Steiner Publ., Stuttgart

Jensen, J., 1984: Änderungen der mittleren Tidewasserstände an der Nordseeküste. *Mitt. des Leichtweiß-Inst.,* Vol. 83, Braunschweig

Jensen, J., 1985: Über instationäre Entwicklungen der Wasserstände an der Nordseeküste, *Mitt. Leichtweiß-Institut der TU Braunschweig,* Vol. 88

Jensen, J., Töppe, A., 1986: Zusammenstellung und Auswertung von Original-aufzeichnungen des Pegels Travemünde, *DGM,* Vol. 4

Jensen, J., Mügge, H.-E., Schönfeld,W., 1990: Development of Water Level Changes in the German Bight, an Analysis based on Single Value Time Series. *Proc. 22nd ICCE,* Delft, ASCE, New York

Jensen, J., Mügge, H.-E., Schönfeld,W., 1992: Wasserstandsentwicklung in der Deutschen Bucht, *Die Küste,* Vol. 54

Kunz, H., Niemeyer, H.D., 1993: Der Einfluß des Datenkollektivs sowie gewählter Bezugshorizonte auf Ergebniss statistischer Untersuchungen zum Wasserspiegelanstieg an der niedersächsischen Küste, *Jahresbericht Forschungsstelle Küste,* Norderney

Lamb, H. H., 1982: Climate, History and the Modern World, S. Methuen & Co. Ltd., London and New York

Lassen, H., 1989: Örtliche und zeitliche Variationen des Meeresspiegels in der südlichen Nordsee. *Die Küste,* Vol. 50

Lassen, H.,Sieffert, W., 1991: Mittlere Tidewasserstände in der südöstlichen Nordsee - säkularer Trend und Verhältnisse um 1980. *Die Küste*, Vol. 52

Lohrberg, W., 1989: Änderungen der mittleren Tidewasserstände an der Nordseeküste, *DGM*, Jahrg. 33, Vol. 5/6

Maisch, M., 1989: Der Gletscherschwund in den Bündner-Alpen seit dem Hochstand von 1850, *Geographische Rundschau* 41(9), 474-485

Niemeyer, H.D., 1987: Zur Klassifikation und Häufigkeit von Sturmtiden; *Jahresbericht 1986, Forschungsstelle Küste*, Vol 38, Norderney

Oerlemans, J., 1988: Simulation of historic glacier variations with a simple climate-glacier Model, *Journal of Glaciology*, Vol. 34, No. 118, 333-41

Oerlemans, J., 1989: A projection of future sea level, *Climatic Change* 15, 151-174

Rohde, H., 1977: Sturmfluthöhen und säkularer Wasserstandsanstieg an der deutschen Nordseeküste, *Die Küste*, H.30

Rohde, H. 1979: Die neue DIN 4049 - Teil 1 - Hydrologie, Begriffe, quantitativ, *Wasser und Boden*, 31, Vol. 12

Schönfeld, W., Jensen, J., 1991: Anwendung der Hauptkomponentenanalyse auf Wasserstandszeitreihen von Deutschen Nordseepegeln, *Die Küste*, Vol. 52

Siefert, W.,Lassen, H., 1985: Gesamtdarstellung der Wasserstandsverhältnisse im Küstenvorfeld der Deutschen Bucht nach neuen Pegelauswertungen. *Die Küste*, Vol. 42

Siefert, W., 1989: Mean Sea Level changes and storm surge probability, *Proc. XXIII IAHR/AIHR-Congress*, Ottawa

Streif, H., 1989: Barrier islands, tidal flats and coastal marshes resulting from a relative rise of sea level in East-Frisia on the German North Sea coast, *Proc. KNGMG Symp. "Coastal lowlands, Geol. and Geotechn."*, 1987

Töppe, A., 1992: Zur Analyse des Meeresspiegelanstiegs aus langjährigen Wasserstandsaufzeichnungen an der Deutschen Nordseeküste, Promotion, *Mitt. Leichtweiß-Institut der TU Braunschweig.*

Vautard, R. and Ghil, M., 1989: Singular Spectrum Analysis in nonlinear dynamics, with application to paleoclimatic time series, *Physica* D35 (1989)

Veen, J. van, 1954: Tide gauges, subsidence gauges and flood-stones in the Netherlands, *Geologie Mijn.* NS 16

Wigley, T.M.L. and Raper, S.C.B., 1987: Thermal expansion of sea water associated with global warming, *Nature* 330

Long Term Wave Statistics

J.A. Ewing[1], R.A. Flather[2], P.J. Hawkes[1] and J.S. Hopkins[3]

Abstract

Long term wave statistics are needed for many applications in coastal engineering and oceanography in the southern North Sea. Three sources of wave data are available: visual observations, wave measurements and hindcast data from numerical wave models. The availability of these sources and their reliability are discussed. A review is made of applications of long term wave statistics to coastal problems in the southern North Sea.

Introduction

Long term wave statistics are important in many research and engineering studies. We may define long term wave statistics as the distribution of wave characteristics over periods of years rather than hours. In the short term of about 1 hour, waves are characterised by a spectrum giving the energy of the waves as a function of frequency. The area under the spectrum is m_o which provides the definition of significant wave height, H_s, given by $H_s = 4m_o^{1/2}$. A corresponding wave period can be associated with this wave height and this is often chosen to be the mean zero-crossing period, T_z, or the spectral peak period, T_p. Finally, a wave direction, θ_p, associated with the direction at the spectral peak, is often used to characterise the direction of the wave system. A sea state can thus be defined in the short term by its height, period and direction.

In contrast, long term wave statistics are associated with the probability distribution of wave height, period and direction taken over many years.

Three types of information on long term wave statistics are available in the southern North Sea. These are derived from visual observations, wave measurements and hindcast data synthesised from numerical wave models. The largest data base consists of visual observations from ships and from offshore platforms in the southern North Sea.

Visual observations have been made for many years and various compilations of these statistics have been produced by different countries in the form of atlases.

---

[1] Hydraulic Research Ltd Wallingford, Oxfordshire OX10 8BA, U.K.

[2] Proudman Oceanographic Laboratory, Bidston, Meseyside L43 7RA, U.K.

[3] Marine Advisory Service, Meteorological Office, Bracknell RG12 2SZ, U.K.

Although individual observations are not as precise as measurements, long term statistics derived from the large data sets of visual observations are believed to provide a useful representation of the climate of a sea area. Measured wave data from instruments in the sea are the most direct and accurate means of describing waves but, due to the cost of deploying and monitoring wave recorders, very few data sets are available for periods exceeding more than a year. Finally, an increasing and useful source of information is now becoming available from numerical wave models operated routinely by national meteorological agencies. This source of wave data, whose accuracy is limited by the physical processes represented in the wave models, offers the possibility of providing detailed wave information on a regular spatial grid, and of extending short period records by correlating available measurements with model hindcasts.

Long term wave statistics can provide information on extreme values of wave height which are used in the design of offshore structures. Wave statistics from a comparatively short data set covering a few years are then extrapolated to provide estimates for return periods of 50 years, for example. Wave climate statistics are also used in fatigue calculations and in the estimation of the probability of weather windows for work on the towing and installation of offshore platforms.

Many coastal engineering applications rely on extreme wave conditions for use in design, particularly of sea walls, harbours and breakwaters. Extremes derived from long term wave data are obviously more reliable than those derived from short data sets. One of the authors (Dr Hawkes) is a member of the International Association for Hydraulics Research Working Group on Extreme Wave Statistics. It is due to report in 1993 upon such things as pre-processing and fitting of data, and confidence in the predicted extremes as a function of goodness of fit and length of data set.

Long term data in coastal waters is necessary to establish mean and extreme conditions, year-to-year variability, and evidence of longer term trends. Net littoral drift on sand and shingle beaches is very sensitive to small changes in wave direction (as well as to changes in wave heights and periods). The natural inter-annual variability of wave climate may mean that a beach which is almost in balance will drift slightly northward in one year and slightly southward the next. Long term wave data is necessary to estimate potential mean rates of drift reliably, for management of mobile beaches and to estimate the natural inter-annual variability of drift and any long term trends. It can be used, in conjunction with long term water level data, to determine the correlation between high waves and high water levels, to derive joint probability extremes for use in design.

The basic information on long term wave statistics is contained in the 'scatter diagram' or bivariate distribution giving the probability of occurrence of wave height and period. This presentation is now standard for wave data. The wave heights are bounded above by a line related to the "steepness" ($2\pi H_s/g T_z^2$) of the waves with a limiting steepness of about 1:15 corresponding to actively generated storm seas. The occurrences of low wave heights at long wave periods represent swell waves.

A difficulty exists in coastal regions where the waves change their character rapidly due to the local water depth, fetch and possibly tidal currents. Visual observations and also wave measurements are only able to provide information at specific points so that recourse must be made to numerical hindcast data or possibly in the future, remote-sensed data to give information over large areas.

Sources of visual wave observations in the southern North Sea

Under the auspices of a World Meteorological Organization scheme, many national meteorological services collect, exchange and archive visual observations of winds (speed and direction) and waves (height, period and direction for each identifiable wave train) made by vessels under way. Although made subjectively, the reports are carried out under internationally agreed procedures and are subject to rigorous quality control checks, so there is a fair degree of consistency over space and time. Clearly, the density of observations depends on the pattern of shipping activity; observations are densest off Iberia and the eastern seaboards of North America and Asia, and sparsest in the southern Oceans.

From these data archives (e.g. Shearman, 1983), observations can be readily retrieved and summarised for any specified area and period, to provide a reasonable guide to the local wave climate. The US Navy Marine Climate Atlases (e.g. 1981) provide a good overview of large scale variations in wave conditions. Korevaar (1989) has compiled data from the North Sea to illustrate smaller scale variations and extremes.

In another major collation of global wave data, Hogben and Lumb (1967) compared visual observations with data from Ocean Weather Ships equipped with wave recorders. They concluded that visual estimates of wave height are biased only a little in comparison with measured heights and display a correlation coefficient of over 0.8, and so can be considered reasonably reliable. Visual estimates of wave period, however, correlated rather poorly (0.5) with measured values, and so should be used with caution. In an extension of this work, known as 'Global Wave Statistics', Hogben et al. (1986) developed parametric relationships between wind speeds, wave heights and periods, in order to maximise the information provided by the larger numbers of wind speeds available from most oceans. Detailed information from the European Data Base is given for six areas in the southern North Sea, as shown in figure 1. However, by its very nature the method produces results which are representative of rather large areas, and so this work probably represents the limit to which parametric methods can be taken for the definition of wave climate from visual observations. It seems likely that numerical modelling based on the energy balance equation will provide better site-specific guidance in future.

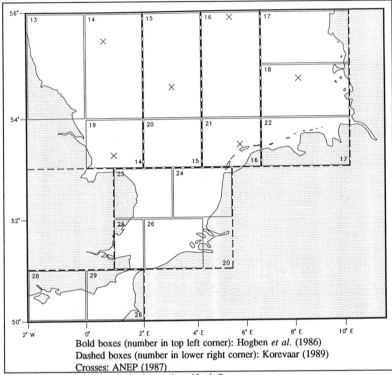

Bold boxes (number in top left corner): Hogben *et al.* (1986)
Dashed boxes (number in lower right corner): Korevaar (1989)
Crosses: ANEP (1987)

Figure 1:   Data sources in the southern North Sea

Wave measurements in the southern North Sea

The British Oceanographic Data Centre (BODC) at the Proudman Oceanographic
Laboratory, Bidston, maintains a global inventory of measured wave data on behalf
of the International Oceanographic Commission. (The BODC was formerly the Data
Banking Service of MIAS: the Marine Information and Advisory Service of the Institute of Oceanographic Sciences). In the southern North Sea the latest catalogue (dated
March 1982) includes a list of data holdings mainly from oil platforms and light-
vessels: it is hoped to update the catalogue with recent entries. The measurements
were taken with a variety of instruments including wave staffs, the Datawell 'Wave-
rider' and the Shipborne Wave Recorder. A review of instrumental wave sites around
the east coast of England has been given by Brampton and Hawkes (1990).

In recent years the UK Department of Energy have commissioned directional wave
measurements with the Datawell 'WAVEC' by the Institute of Oceanographic
Sciences. These measurements at Flamborough Head (December 1984 - May 1986)
and Cromer (December 1985 - January 1987) are reported by Clayson and Ewing
(1990 and 1988).

Measured wave data have been collected by national agencies in countries around the southern North Sea. The Office of Coastal Ports in Ostend, Belgium maintains a data base with wave data from various sites along the coast near Zeebrugge and Antwerp. Extensive monitoring of waves at sites in the area has been co-ordinated by the KNMI, Rijkswaterstaat and Directorate-General of Shipping and Maritime Affairs in the Netherlands: the sites include the stations Euro, K13 and ELD. Wave measurement programmes are also active in Danish and German waters.

Measurements with directional wave buoys are becoming routine. The information in the scatter diagram can now include information on wave direction in the form of a directional 'rose' at each combination of wave height and period. For a 30 minute wave record, the standard deviation of measured values of H_s and T_z (due to sampling variability) are both about 5% of H_s and T_z. The mean wave direction has a standard deviation of about 5°. The accuracy of measured wave statistics is thus adequate for most engineering applications. Of greater significance is the length of the data base of measurements since this governs the confidence limits of extremal estimates and how representative the data is for climatic purposes. At least one complete year of measurements is required but ideally about 5 years or more of recordings are needed for reliable estimates of extreme values. Wave spectral information from instruments are used in validating numerical (hindcast) models and in studies where the energy content of the waves and its variation with frequency is important, such as in the design of wave energy convertors.

There have been indications of an increase in both the average and extreme values of wave height in the north-east Atlantic from measurements (over about 20 years) carried out by the Institute of Oceanographic Sciences at ocean weather stations 'India' and 'Juliett' (Carter and Draper, 1988). The increase in wave height has not been accompanied by a corresponding increase in wind speed at the weather ships. A possible explanation for the increase has been given by Hogben (1989) as due to the influence of swell levels caused by changes in the patterns of the prevailing winds. The situation in the southern North Sea is unknown since most wave measurements have only been made over a few years. However the implications of an increase in wave height are serious for coastal defences and their management (Hawkes *et al.*, 1991).

Remote-sensed wave data from earth satellites can also give useful information over the oceans. A vertically-mounted radar altimeter in the satellite can sense the roughness of the sea surface and hence provide an estimate of the significant wave height over a 'footprint' of about 7km diameter. A programme of work (Carter *et al.*, 1988) is underway to use information from ERS-1 (launched in July 1991) to derive wave climate information in European waters. Figure 2 shows the paths covered by ERS-1 in the southern North Sea. However, coastal wave information cannot be obtained within one 'footprint' of the land. This source of information will become of increasing importance in wave climate statistics.

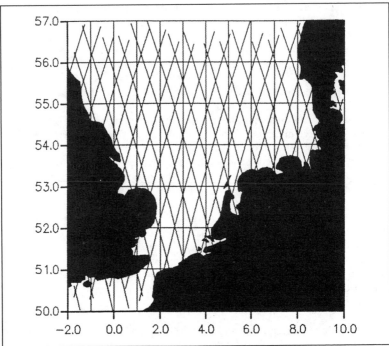

Figure 2:   Path covered by radar altimeter in ERS-1 (35 day repeat)

Wave modelling in shallow water

The numerical wave models presently used in the North Sea are based on the integration of the energy balance equation:

$$\frac{DE}{Dt} = S = S_{in} + S_{nl} + S_{ds} + S_{bot} \tag{1}$$

where E ($\underline{x}$, t; f, θ) is the directional wave spectrum, $\underline{x}$ is the position in space, t is the time and (f, θ) are the frequency and direction of the waves. S is called the Source Function which includes physical processes to describe the growth, interaction and decay of waves. S_{in} describes the wind input to the waves, S_{nl} represents nonlinear wave-wave interactions, S_{ds} represents the dissipation due to wave breaking and, finally, S_{bot} describes the dissipation in shallow water due to bottom processes. (For a general discussion of the Source terms see WAMDI, 1988).

Second-generation wave models include a parametric representation of the S_{nl} Source Term, so as to keep computing requirements within practical limits. However, third-generation models, not generally used in engineering applications, include a rigorous evaluation of S_{nl}.

The Source term for bottom friction is usually modelled on the results from the JONSWAP study in the German Bight (Hasselmann *et al.*, 1973) in the form:

$$S_{bot} = \frac{-\Gamma}{g^2} \cdot \frac{(2\pi f)^2}{\sinh^2 kh} \cdot E \tag{2}$$

where k is the wave number, h is the water depth, g is the acceleration due to gravity and $\Gamma = 0.038 m^2 s^{-3}$. For applications in other areas of the North Sea it may be necessary to tune this Source term to take into account the nature of the bottom and its roughness (Shemdin *et al.*, 1978).

In local areas with banks and shoals, wave breaking is important. The dissipation due to wave breaking in shallow water can be modelled on the use of a breaking wave height coefficient (for example, Weggel, 1972) together with an estimate of S_{ds} derived for a tidal bore, as proposed by Battjes and Janssen (1978).

At the present time S_{ds} is not included in wave models operated by the meteorological agencies but this term will be necessary in calculations where a fine mesh is needed to resolve banks and shoals off the east coast of England, for example.

Simplified methods are often used to predict waves at coastal locations (Shore Protection Manual, 1984). The methods are usually site-specific but are economical and enable a long series of wind data to be used in the calculations of the nearshore wave climate. The HINDWAVE model (Hawkes, 1987), for example, takes hourly sequential wind data from a single coastal anemograph and efficiently converts it to hourly sequential wave data for deep water coastal locations. The model is based on the JONSWAP (Hasselmann *et al.*, 1973) parametric wave prediction formula, as modified by Seymour (1977) for use in restricted wave generation areas.

Around most of eastern England, there are about 20 years of wind data available for use in simplified wave hindcasting models. In conjunction with site-specific wave transformation models, and perhaps long term sequential water level data, the simulated deep water wave data can be brought inshore to sites of coastal engineering interest. A disadvantage of simplified wave hindcasting models is that they neglect wave energy arriving from outside the immediate area of interest (i.e. swell). However an advantage is that they can be calibrated and validated against local wave measurements. The 20 years of sequential data can be used to examine wave climate, extreme waves, littoral drift rates, structural responses, and any natural variability or trend in the above.

Interactions between waves and tide-surge motion
There has been considerable interest in recent years in the interactions between waves and tide-surge motion, mainly stimulated by the need for better forecasts for coastal flood warnings. Wolf, Hubbert and Flather (1988) examined the various interaction mechanisms, several of which were found to be significant. In particular, additional wave refraction arises from the changing water depths and currents associated with

tides and storm surges. Depth-dependent Source terms will also change. The tide-surge motion is also modified by the waves through changes in the sea surface roughness and hence surface wind stress (which generates storm surges), and the bottom stress (which removes energy from the tide-surge motion) is modified by near-bed wave orbital velocities.

In the presence of currents, the fundamental conserved quantity is wave action N, defined as N=E/f, and not wave energy. Equation (1) is replaced by the "action balance equation" (Bretherton and Garrett, 1968). A number of spectral wave models have been established based on the action balance equation.

Hubbert and Wolf (1991) reformulated the third-generation (WAM) model and investigated the influence of directional resolution on the simulation of wave refraction by currents and changes in water depth. Tolman (1991a, b) has also established a third-generation wave model and applied it to some storm events in the southern North Sea, using a 24km x 24km spatial grid and spectral resolution of 26 frequencies x 24 directions. He found small but significant changes in wave parameters due to interactions with tide and surge currents; in particular H_s changed by 5%-10%. These studies have led to the development of fully coupled wave-tide-surge models to investigate and predict these processes. At the Proudman Oceanographic Laboratory (POL), such work is well advanced, based on the WAM wave model and the POL tide-surge model, which runs routinely at the Meteorological Office to provide operational flood warnings in the UK. An outline of the tide-surge model is given in Flather, Proctor and Wolf (1991). Hindcast simulations of the storm of 26 February 1990, which caused the failure of the sea wall and flooding at Towyn in North Wales, have recently been carried out using the wave-tide-surge model on a grid 1/9° in latitude by 1/6° in longitude (about 12km) and wave resolution of 26 frequencies x 24 directions covering the Irish Sea, Celtic Sea and English Channel. Preliminary results (Wu and Flather, 1992) show significant changes in wave parameters caused by the interactions.

Sources of hindcast wave data in the southern North Sea
A wave climatology has been produced for ten locations in the North Sea over the period 1968-1983 (ANEP, 1987). The wave model used for the calculations (HYPAS) is a second-generation model developed by Gunther and Rosenthal (1984). Six of the ten locations for the hindcasts are south of latitude 56°N. The conditional distributions of wave height with wave period, wind speed and wind direction are presented in ANEP (1987) for the four seasons and also annually.

National meteorological agencies began to operate spectral wave prediction schemes in the mid-1970's, to meet the needs of the offshore industry for sea state forecasts. The UK Met Office introduced a 50km resolution model covering the North Sea and surrounding waters in July 1977, using analysed and forecast wind fields from the operational atmospheric models to provide energy input, and representing wave energies in 6 frequency bands and 12 directional bands (Golding, 1983). Increased computing power and improved understanding of wave physics have allowed many of these original schemes to be developed to give much improved spatial and spectral resolutions. The current (1992) UK Met Office model operating over the North Sea has remained essentially unchanged since July 1986. It is a second-generation depth-dependent model operating on a grid 0.25 degree latitude x 0.4 degree longitude with resolution of 13 frequencies x 16 directions. The hindcast wind and wave fields have been archived from all versions of the model in use since 1977.

Some agencies have now developed third-generation wave models, but these are computationally much more expensive to run and so are currently operated at fairly coarse spatial resolution; significant advantages over the higher resolution second-generation models have yet to be clearly demonstrated.

The accuracy of hindcast values of wave height depends on the area of application, reliability of the wind fields and the modelling procedures. For most numerical models in use in the southern North Sea the root mean square error in significant wave height ranges from 0.5m to 1.0m.

The most comprehensive hindcasting exercise undertaken over the North Sea is NESS (North European Storm Study), funded by a group of oil companies and government agencies (Francis, 1987). The project required the preparation of high quality wind fields to drive the wave and surge models, and extensive validation of the model output against available measurements from offshore locations. The period covered was 25 years (1964-1989); continuous hindcasts of winds and waves were carried out at 3 hour intervals for each winter (October-March) period, three summer periods (1977,78,79) were also continuously analysed, and 40 additional summer storms were also studied. In parallel with the wave hindcasting work, water levels and depth-integrated currents were also modelled for 250 storm periods over the 25 years.

The NESS wind fields were analysed at 50km resolution, and those for the most important storms were subjectively modified by an experienced meteorological analyst, taking into account the storm development and time/space continuity of areas of strongest winds. The 50km winds were then interpolated onto a 30km grid, and the hindcast models were run at this resolution. The wave model used was the HYPAS second-generation model of GKSS (Gunther and Rosenthal, 1984), and the surge model was a slightly modified version of System 21 of the Danish Hydraulic Institute. As a supplement to the main NESS project, an area of the southern North Sea south of approximately 55°N was analysed on a 10km spatial resolution over 132 storm periods within the 25 years.

As well as providing space and time continuity, the NESS archive allows simultaneous specification of wind, wave and current values for the estimation of resultant forces on an offshore structure (Shaw, 1992). The archive also has the advantage of being derived from models which remained unchanged over the full 25 year hindcast period. Models which are run operationally by meteorological services are subject to improvement from time to time, and so hindcasts from these cannot be considered strictly homogeneous over such long periods.

The NESS archive is held by the UK Met Office but is confidential to the funding participants of NESS for a period of 10 years from the completion of the project in 1991. However, requests for access to the data will be considered by the NESS User Group, who can be contacted at MATSU, Culham Laboratory, Abingdon, UK.

Applications of long term statistics in coastal regions

Nearly all coastal engineering situations require some knowledge of the local wave climate, and it is always better to have measurements or predictions spanning a long period of time if possible. However, there are some applications, best illustrated with reference to examples, where long term wave data is essential.

Variability of derived extreme wave heights in Lyme Bay

Although not in the southern North Sea, this example indicates the potential error which may arise when predicting extremes from too short a data set. 15 years of wave data (1974-1989) were hindcast off Lyme Bay (southwest England), using consistent input wind data from Portland and consistent wave prediction methods. The results were grouped into 2, 5, 10 and 15 year (July to June) batches, each of which was extrapolated to extreme significant wave heights in a consistent manner. The results are listed in Table 1. Predictions of the 100 year return period event have a range of 2.9 metres (5.41-8.29m), and even the 1 year predictions have a range of 1.6 metres (4.17-5.77m). This example suggests that even 5 years of data may not be enough to obtain a reliable prediction of extreme conditions, due to the natural variability of conditions from one year to another.

Return period (yrs)	Period of data (in years)													
	74-76	76-78	78-80	80-82	82-84	84-86	86-88	74-79	79-84	84-89	74-84	79-89	74-89
0.1	3.5	4.3	4.5	3.7	3.8	3.5	3.6	4.0	4.1	3.5	4.0	3.8	3.9
1	4.5	5.7	5.8	4.7	4.7	4.2	4.5	5.2	5.2	4.2	5.1	4.8	4.9
5	5.1	6.6	6.7	5.3	5.2	4.6	5.0	6.0	5.9	4.7	5.8	5.5	5.5
10	5.4	7.0	7.0	5.6	5.4	4.8	5.3	6.3	6.2	4.9	6.1	5.7	5.8
50	6.0	7.9	7.9	6.2	5.9	5.2	5.8	7.1	6.9	5.4	6.8	6.3	6.5
100	6.3	8.3	8.2	6.5	6.1	5.4	6.0	7.4	7.2	5.5	7.1	6.6	6.7

Table 1:   Extreme wave heights derived from Lyme Bay hindcast data
                Years are taken from 1/7-31/6

Variability of wave heights off Great Yarmouth

17 years of wave data (1973-1990) were hindcast in deep water off Great Yarmouth (Norfolk), using consistent input wind data from Gorleston and consistent wave prediction methods. The results were grouped into 1 year (July to June) batches. The mean significant wave height was determined for each batch, together with the values exceeded 1% and 10% of the time. The results are plotted as a function of time in figure 3. Some trends are apparent in the data. There is no long term trend in the mean values, although some of the annual values are up to 25% greater or 30% lower than the average over the 17 year period. The 1% and 10% values appear to show

long term trends, with the former decreasing by about one third of one percent per year and the latter increasing at about the same rate. Even with 17 years of data it is hard to tell if these are genuine changes, or just the result of a natural variability of the annual averages, which have a scatter of about ±20% about the overall average values.

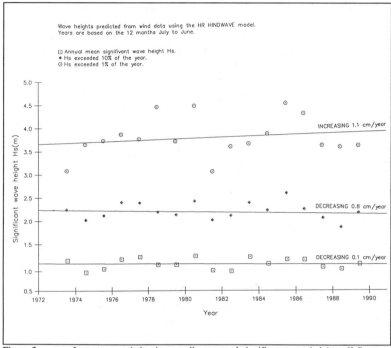

Figure 3:   Long term variation in annually averaged significant wave heights off Great Yarmouth

Joint probability of high waves and high water levels at Dowsing

Two of the most important variables in the design of coastal and offshore structures are extreme wave heights and extreme water levels. Even if it were possible to predict both types of extreme with accuracy, it would still be necessary to make an assumption about whether or not the two extremes would occur simultaneously. One might expect a correlation between high wave heights and high surges since in the southern North Sea both would tend to occur during times of high winds from the north. However, there is no obvious reason to expect the same level of correlation with astronomical tidal levels (which are the major component of the total water level). The degree of correlation between high waves and high water levels can be estimated intuitively from such things as the ratio of extreme surge to Highest Astronomical Tide, and the general exposure of the area. However, correlation is best determined from an analysis of long term sequential wave and water level data, following which extreme values of combined waves and tides can be predicted.

As an example of the importance of long term wave data and the derived joint probabilities, consider predictions based on 9 years of data at Dowsing, off Lincolnshire. For convenience, assume that the key design variable is the "total level" comprising water level plus significant wave height. This parameter represents an offshore platform clearance level, being approximately equal to the crest elevation of the highest individual wave above the still water level. The "best" 50 year return period total level predicted is 16.9m CD, using the correlation between high waves and high water levels derived directly from the data. This value compares with 16.0m CD assuming independence between the two variables or 18.9m CD assuming dependence. These figures indicate the importance of a correct assessment of joint probability and the need for long term wave and water level data.

Conclusions

Long term wave statistics in the southern North Sea are available from three sources: visual observations, wave measurements and hindcast data from numerical wave models.

Visual observations are plentiful but have limited accuracy in coastal regions where wave conditions change rapidly. Wave measurements are the most direct and reliable source of wave information but are not available at sufficient locations. It is hoped that wave data will be available from satellites in the near future to provide detailed wave climate information. At present, carefully validated hindcast data is the most useful source of wave information in the southern North Sea.

In engineering applications the user should make use of all available data since complete reliance on one source can be misleading. Long term wave statistics are important, particularly in design applications, where extreme values are required. Extrapolation from too short a data set can lead to serious errors. Long term data are essential for assessment of trends, which may have significant practical implications for coastal management planning. Again, extrapolation from too short a data set can lead to erroneous conclusions. Several years of wave data, either measured or simulated, are necessary to obtain the conditional extreme wave statistics which are particularly important in coastal applications. These include direction-dependent extreme wave heights and the joint probability of high waves and high water levels.

Acknowledgement
Ms. R. Tokmakian of the James Rennell Centre for Ocean Circulation kindly supplied Figure 2.

References

ANEP (1987). "Seasonal climatology of the North Sea". *Allied Naval Eng.* Publication 14, NATO.

Battjes, J A and Janssen, P A E M (1978). "Energy loss and set-up due to breaking of random waves". *Proc. 16th Conf. Coastal Eng.*, ASCE, Hamburg, 569-587.

Brampton, A H and Hawkes, P J (1990). "Wave data around the coast of England and Wales. A review of instrumentally recorded information". *Hydraulics Research Ltd.*, Wallingford, Report SR 113.

Bretherton, F P and Garrett, C J R. (1968). "Wavetrains in inhomogeneous moving media". *Proc. R. Soc.*, London, A 302, 529-554.

Carter, D J T, Challenor, P G and Srokosz, M A (1988). "Satellite remote sensing and wave studies into the 1990s". *Int. J. Remote Sensing*, 9, 1835-1846.

Carter, D J T, and Draper, L (1988). "Has the north-east Atlantic became rougher?" *Nature*, 322, 7 April 1988.

Clayson, C H and Ewing, J A (1988). "Directional wave data recorded in the southern North Sea". *Inst. Ocean. Sciences Deacon Lab.*, Wormley, England. Report No. 258, pp70.

Clayson, C H and Ewing, J A (1990). "Directional wave data recorded off Flamborough Head". Ibid, Report No. 273, pp45.

Flather, R A, Proctor, R and Wolf, J (1991). "Oceanographic forecast models". pp 15 - 30 in *Computer Modelling in the Environmental Sciences* (Eds D G Farmer and M J Rycroft). IMA Conference Series 28, Clarendon Press Oxford.

Frances, P E (1987). "The North European Storm Study (NESS)". *Advances in Underwater Technology, Ocean Sci. and Offshore Eng.* 12, *Modelling the Offshore Environment*, 61-70.

Golding, B (1983). "A wave prediction system for real-time sea state forecasting". *Quart. J. R. Met. Soc.*, 109, 393-416.

Gunther, H and Rosenthal, W (1984). "A shallow water surface wave model based on the Texel-Marsen-Arsloe (TMA) wave spectrum". *Proc. 20th Congress IAHR*, Moscow.

Hasselmann, K et XV al. (1973). "Measurements of wind-wave growth and swell decay during the Joint North Sea Wave Project (JONSWAP)". *Dt. Hydrogr. Z.*, A8(12), pp95.

Hawkes, P J (1987). "A wave hindcasting model". Advances in Underwater Technology, Ocean Science and Offshore Eng. 12, *Modelling the Offshore Environment*.

Hawkes, P J, Jelliman, C E and Brampton, A H (1991). "Wave climate change and its impact on UK coastal management". *HR Wallingford*, Report SR 260.

Hogben, N (1989). "Increases in wave heights measured in the North-Eastern Atlantic: a preliminary reassessment of some recent data". *J. Soc. Underwater Tech.*, 15, 2-4.

Hogben, N and Lumb, F E (1967). "Ocean wave statistics". Her Majesty's Stationery Office, London.

Hogben, N, Dacunha, N M C, and Olliver, G F (1986). "Global wave statistics". Unwin Brothers for British Maritime Tech. Ltd.

Hubbert, K P and Wolf, J (1991). "Numerical investigation of depth and current refraction of waves". *J. Geophys. Res. - Oceans,* 96(C2), 2737-2748.

Korevaar, C G (1989). "Climatological data for the North Sea based on observations by voluntary observing ships over the period 1961-1980". *KNMI,* de bilt, Netherlands, Scientific Report 89-02.

Seymour, R J (1977). "Estimating wave generation on restricted fetches". *J. Wat. Port. Coastal and Ocean Divn.,* ASCE, 103, 251-264.

Shaw, C J (1992). "The North European Storm Study (NESS)". *The Impact of Tech. Developments on Safety Cases.* Conference in London, 19 March 1992.

Shearman, R J (1983). "The Met. Office main marine data bank". *Marine Observer,* 53, 208-217.

Shemdin, O *et al.* (1978). "Nonlinear and linear bottom interaction effects in shallow water". In *Turbulent fluxes through the sea surface, wave dynamics and prediction,* Eds. A Favre and K Hasselmann, Plenum Press, 347-370.

"Shore protection manual" (1984). U.S. Army Coastal Eng. Research Center. U.S. Govt. Printing Office, Vol 1.

Tolman, H L (1991a). "A third generation model for wind waves on slowly varying, unsteady, and inhomogeneous depths and currents". *J. Phys. Oceanogr.* 21,766-781.

Tolman, H L (1991b). "Effects of tides and storm surges on North Sea wind waves". *J. Phys. Oceanogr. 21,* 776-781.

U.S. Naval Oceanography Command Detatchment (1981). "Marine climatic atlas of the world". Vol. IX: Means and standard deviations, Asheville, N.C.

WAMDI (1988). "The WAM model - a third generation ocean wave prediction model". *J. Phys. Oceanogr.,* 18, 1775-1810.

Weggel, J R (1972). "Maximum breaker height". *J. Wat. Harb. Coastal Eng. Divn.,* ASCE, 98, WW4.

Wolf, J, Hubbert, K P and Flather, R A (1988). "A feasibility study for the development of a joint surge and wave model". *Proudman Oceanographic Laboratory,* Report No.1: 109pp.

Wu, X and Flather, R A (1992). "Hindcasting waves using a coupled wave-tide-surge model". pp.159-170 in *3rd International Workshop on Wave Hindcasting and Forecasting,* Montreal, Quebec, 19-22 May 1992. Preprints. Ontario: Environment Canada.

Water Quality Management in the southern North Sea

J.P.G. van de Kamer[1], K.J. Wulffraat[1], A. Cramer[1], M.J.P.H. Waltmans[1]

Abstract

The inhomogeneous spatial distribution of the water and (suspended) sediments quality in the southern North Sea is determined mainly by the specific transport and mixing of water masses. Tidal averaged, anti-clockwise circulation patterns consist of Atlantic water, entering from the south through the Dover Strait and from the north along the British east coast. Fresh water is contributed by continental and British rivers. The northward directed outflow is concentrated along the Danish and Norwegian coasts.

Variations in riverine inputs, dumping of dredged materials and atmospheric deposition as well as physical, chemical and biological processes superimpose temporal inhomogenities or trends.

As far as nutrients are concerned, a comparison with background values clarifies that in the continental coastal zone present winter concentrations are threefold for phosphate and nitrate. As a consequence, the risk of nuisance algae blooms is relatively high. From an ecotoxicological point of view contaminant concentrations are too high as well. Adverse effects on the reproduction of, among others, seals and sea stars have been reported for the Wadden Sea and the Scheldt estuary.

For this reason riparian states have resolved to implement protective measures to reduce substantially (in the order of 50 %) the inputs of nutrients and contaminants over the period 1985-1995. The inputs from the river Rhine have since decreased markedly. Concentrations in the coastal zone, however, have yet not dropped significantly. On a national level The Netherlands has set a concrete objective in terms of the abundance of several biological species in order to attain a sustainable use of the water system. Water quality standards are forthcoming.

Further policy measures are being prepared using a systems analysis approach. In order to eventually comply with the standards, alternative strategies are compared using model calculations.

Apart from water quality improvement, additional measures will be or have been taken to meet the sustainable use objective. Sand and gravel extractions, also on behalf of beach nourishments, as well as fisheries, offshore industry, shipping and recreation affect the ecosystem. Mitigation of the impact for the Dutch part of North Sea is realised by various restrictions, for the area as a whole or for some defined protected area.

---

[1] Ministry of Transport, Public Works and Water management, Tidal Waters Division, P.O. Box 20907, 2500 EX The Hague, The Netherlands

Human activities on the southern North Sea

Sustainable use

The shallow North Sea is wedged in by the coasts of the United Kingdom, Norway, Sweden, Denmark, Germany, the Netherlands and Belgium. The North Sea has an area of about 575.000 square kilometres. It contains roughly 55.000 cubic kilometres of salt water: less than one hundredth of one per cent of all the salt water on the planet.

About 200 million people live in the heavily industrialised North Sea river basins and they are second only to the North Americans in their use of natural resources and consumer goods. However, Western Europe's environmental impact focuses on a shallow, highly productive shelf sea instead of deep ocean water.

The North Sea offers a thousand uses to the people living nearby: fishing, nature reserves, shipping, offshore industry, sand and gravel extraction and recreation. Even a certain amount of waste is discharged into it. If mankind is too greedy, these uses will come into conflict. It seems that this is the hub of the problem in the North Sea. Due to overfishing, oil spills, overloading with waste and physical disturbances, the North Sea is no longer able to accommodate a fully-fledged system. Additionally, a destruction and separation of habitats has been caused by sea defence measures and embankments. A signal of distress is the disappearance of animals at the top of the food chain, such as dolphins, sharks and certain fish-eating birds, from the coastal waters. To solve the North Sea problem, we have to base our future policy on sustainable use.

Chemical compounds

In this chapter we will focus on the enhanced concentrations of chemical compounds and the sustainable use of the North Sea in this respect. Two types of compounds are considered: nutrients (eg. nitrogen and phosphorus) and contaminants (eg. PCBs and the heavy metal cadmium). Eutrophication (overloading with nutrients) and contamination, however, only partly explain why the North Sea ecosystem is unable to hold its own. There are other important factors, eg. fishing and the disturbance it causes. Every year mankind catches 25 per cent of the total North Sea fish stock. Bottomtrawls and dredges towed over the seabed have an effect on the sediment and lead to physical disturbance. This type of intensive exploitation inevitably leads to changes in populations: short-living organisms, such as worms or herring, that reproduce rapidly and in large numbers are favoured. Long-living predators from near the top of the food chain, such as rays, sharks and porpoises, disappear.

In the case of fishery sustainable use may be defined as collecting the "interest" of the fish stock and respecting the rest of the ecosystem. In the case of discharging waste it means that toxic levels should be avoided, now and in the future.

Inputs from the ocean

Elevated levels of nutrients, PCBs, heavy metals and pesticides are at the root of the North Sea problem. Well known are the polluted rivers flowing into the North Sea, thereby sealing its fate. Northern Atlantic inflow and the English Channel are, however, by far the largest inputs of chemical compounds. Although concentrations are rather low, the observed waterflow accounts for enormous loads. A balanced view is necessary, however, since nutrient concentrations in both Northern Atlantic inflow and the English Channel are in the range of background concentrations (Bentley *et al*, in prep; Laane, 1992). The same holds for heavy metals in the north, but in the English Channel metal concentrations are enhanced (Statham *et al*, submitted; Balls, 1985; Kersten *et al*, 1990). The anthropogenic factor is estimated to be 50 to 75 %. A second need for balancing Atlantic inputs is the fact that only a fraction of the Northern Atlantic inflow reaches the Southern Bight of the North Sea, below 56 degree north latitude, where concentrations of contaminants are enhanced and environmental problems arise.

Atmospheric deposition
Atmospheric deposition is of greater importance for the eutrophication and contamination of coastal waters and the Southern Bight (GESAMP, 1989). Atmospheric deposition might even be a trigger for algae blooms. After heavy rains nitrogen concentrations in the upper layer of the water column may be doubled (van Boxtel *et al*, 1991). Deposition is difficult to monitor, especially at sea. Samples are easily contaminated by seaspray. Therefore extrapolation of measurements at coastal stations is widely used. Application of computer modelling is promising. Based on atmospheric emissions in Europe, under average meteorological conditions and assuming a deposition velocity, atmospheric deposition can be calculated (Warmenhoven *et al*, 1989). Both measurements and modelling do not take into account re-emission from sea water. Heavy metals end up in the sediments of deposition areas like the Wadden sea and the Wash. The larger part of PCBs will evaporate (Klamer, 1989).

Rivers
Rivers carry excess water from land to sea. They are a major source of nutrients and contaminants. The quantity of contaminants being discharged into the marine environment has been the subject of research in North Sea riparian states. River loads are a touchstone for their environmental policy. Annual reviews of river loads into the North Sea are presented by the Paris Commission. Since monitoring stations are located at fresh water-salt water boundaries or at tidal limits, discharges in estuaries are summed up by the Paris Commission for a grand total per river basin. In this way riverine input will be overestimated, as estuaries act as a filter for contaminants (Klamer, 1989; NERC, 1991). In estuaries biological degradation takes place; suspended particulate matter and adsorbed contaminants are deposited, while volatile compounds evaporate from the water column. In this way over 50 per cent of the heavy metals and some 90 per cent of the PCB load is filtered out. Actually estuaries are the hot spots of environmental pollution.

Monitoring strategy, detection limits and number of contaminants to be monitored differ a great deal in the various North Sea riparian states. Only for a few heavy metals a North Sea total can be given (Hupkes, 1990). It is shown that river loads of heavy metals decreased significantly over the last decade. Net cadmium loads from the Nieuwe Waterweg, representative for the river Rhine, decreased from over 20 tons annually (1980) to 3 tons in 1990, which still is six times the background level

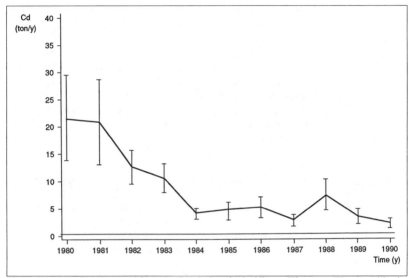

Figure 1:   Cadmium load from the river Rhine via the Nieuwe Waterweg. The natural load
                   amounts to about 0.5 ton/year.

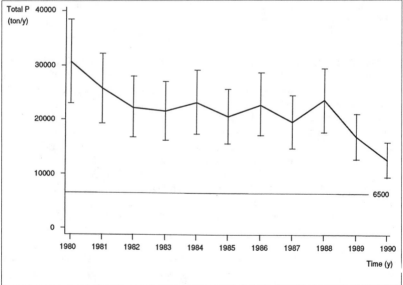

Figure 2:   Phosphorus load from the river Rhine via the Nieuwe Waterweg. The natural load
                   amounts to about 6500 ton/year.

(fig. 1). The net phosphorus load from the river Rhine into the North Sea shows almost a fifty per cent reduction over the last decade. Reduction greatly improved from 1988 onwards, by successfully banning phosphorus from washing powders (fig. 2). Present river loads are at least double the natural input, mainly because of discharges by the artificial fertilizer industry in the Rotterdam harbour area.

Dredged materials
Main ports are situated along the North Sea coast. As a result of density currents, most of these harbours and their entrance channels suffer from severe silt build up. Moreover, docks and harbours are artificially constructed, resulting in restricted water movements which, in turn, promotes settling of sediments. For this reason ports and entrance channels have to be dredged from time to time (Nieuwendijk et al, 1991). Every year several million tonnes of sand and silt are disposed of in the North Sea. As sediments in harbour areas are contaminated by local discharges or from upstream rivers, dumping of dredged material counts for an extra input of contaminants into the North sea (Donze, 1990). Since discharges of waste water have been tackled and the most contaminated dredged material are stored at containment areas, the input via dumping of dredged material has decreased over the last decade (of course storing at containment areas is not to be seen as a sustainable activity).

Incineration
Incineration and dumping of industrial waste has been banned successfully. Only the United Kingdom still carries out dumping of sewage sludge, fly ashes, colliery wastes and some liquid industrial waste, of which sewage sludge imposes a severe burden on a small area (Nihoul, 1991; Compaan et al, 1992; Crogan, 1984). Both shipping and the offshore industry account for substantial discharges of oil and polycyclic aromatic hydrocarbons. However they do not contribute significantly to the inputs of nutrients, heavy metals and PCB, although some authors suggest PCB discharges by offshore industry (Schulz-Bull, 1991).

Review of the inputs
A review of the inputs of cadmium, PCB, phosphorus and nitrogen into the North Sea (tonnes per year) is given in table 1. Figures in the table apply for the mid 80's.

Transport of chemical compounds through the North Sea

Hydrodynamics
Apart from processes like adsorption and desorption of compounds on suspended matter, exchange of this material between bottom and water by sedimentation and resuspension, chemical transformations and the uptake of compounds by organisms, the transport of chemical compounds through the North Sea is largely determined by the movement and mixing of its water bodies (van der Giessen, 1989).
The movement and mixing of water bodies is elegantly described and depicted in the North Sea Atlas (ICONA, 1992). The water in the North Sea circulates according to a fixed pattern: Atlantic water enters via the English Channel in the south and along the Scottish coast in the north. Outflowing water leaves the North Sea along the

	cadmium	PCB	phosphorus	nitrogen
English Channel	54-102	0.04-0.08	$10\text{-}16.10^4$	$7\text{-}12.10^5$
Northern Atlantic	300-500	1.8-3.2	$40\text{-}150.10^4$	$30\text{-}80.10^{5\ A)}$
Atmosphere [B)	8-30	8-200	«	$16\text{-}30.10^3$
dredged materials	$23^{\ C)}$	$0.7^{\ D)}$	$14.10^{3\ D)}$	NI
incineration	0.14	$«^{\ E)}$	«	«
industrial wastes	«	«	NI	35.10^3
sewage sludge	2.5	0.02-0.4	3.10^3	12.10^3
rivers + discharges	$18\text{-}27^{\ F)}$	$0.05^{\ F)}$	$25\text{-}47.10^3$	$37\text{-}68.10^{4\ F)}$

« no significant contribution, N.I. no information
A) NO_3-N only.
B) estimated via computer modelling.
C) 18 ton cadmium North Sea total, excluding Belgium which is estimated to be 5 ton
     (average of 1986 till 1989)
D) Both PCBs and phosphorus are analyzed only by the Netherlands (20.10^6 m^3
     sediment annually). Doubling this amount represents a rough estimate for total
     North Sea (40.10^6 m^3 sediment annually)
E) total emission of organo-chlorides estimated to be 5 tonnes.
F) estuarine retention taken into account

Table 1:   Review of inputs

Norwegian coast. The inflow is caused by the predominantly westerly winds, which
push Atlantic water towards the North sea, and by the tidal wave. Ebb tides do not
completely neutralise the water movement of the flood tides. A counter-clockwise
residual current remains in the North Sea basin. When the wind blows from a
northerly or easterly direction for a long time, the circulation pattern may be
temporarily reversed. Averaged over the whole year, the residual current along the
Netherlands coast amounts to a few centimetres per second. The North Sea water
travels a distance of about 1500 kilometres per year. It is refreshed every one or two
years (fig. 3).
The different water masses meet and partly mix. Close to the coasts of the United
Kingdom and the continent the seawater has a high concentration of river water,
which is relatively light and spreads slowly from the rivers over the salt seawater. As
a reaction to this movement more saline seawater is transported coastward along the
bottom. Due to its slow seaward spread and the strong residual current along the
coast, the river water with its load of contaminants lingers along the coast for a long
time. Contaminants consequently become concentrated in a zone a few dozens
kilometres wide along the coast (fig. 4).
In the deeper parts of the North Sea stratification occurs in summer.

Transport and fate of chemical compounds
The hydrodynamics mainly determines mass transport of nutrients and contaminants.
Due to this type of transport, chemical compounds are dispersed in the North Sea, eg.
along the Netherlands coast the nutrient concentration of original river water
decreases gradually by dilution with ocean water from the English Channel. However,
during this transport various processes affect the concentration of compounds.

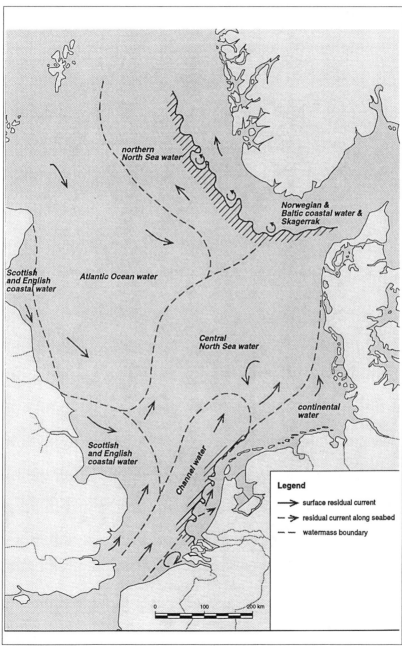

Figure 3:   Residual currents and water masses in winter (ICONA, 1992).

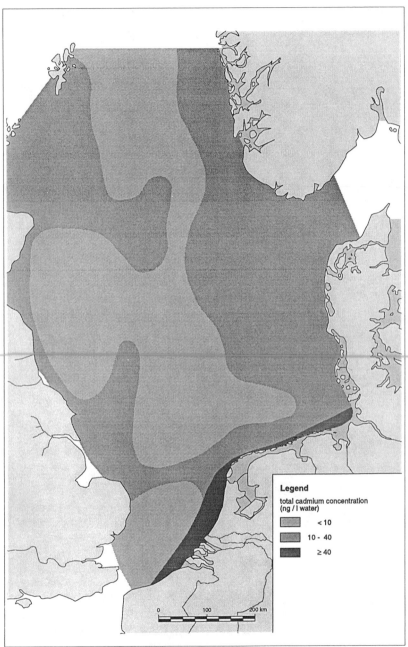

Figure 4:   Total cadmium concentration, based on measurements in 1986 (ICONA, 1992).

Heavy metals like cadmium and organic micropollutants like PCBs attach to suspended matter. This matter is deposited at places with low current velocities and little wave action, such as the margins of the Wadden sea and in the Silver Pit, the Wash, the Oystergrounds, the Elbe-Rinne and the Norwegian Trench (fig. 5). Thus contaminants accumulate in these areas and threaten bottom life. After reduction of the direct anthropogenic inputs (see the second section of this paper) these areas will turn into sources of contamination because of resuspension and desorption.

Decay, evaporization and transformation of compounds are main chemical processes causing change of concentrations in the compartments water and sediment. The time and spatial scales of these processes differ a lot. Nitrogen compounds may become gaseous after denitrification, while PCBs evaporate.

Metals and organic micropollutants are accumulated by organisms directly from the water or via the foodweb and form a major risk to marine life (See next section). By the uptake of carbon and nutrients phytoplankton biomass is formed. Elevated concentrations of nutrients may give rise to excessive algae blooms or to toxic algae species.

The impact of chemical compounds

Effects of enhanced concentrations
Concern about rising concentrations of various chemical compounds in the North Sea and their adverse effects on organisms has led to political action. The ministers of riparian states agreed to reduce the emission of 37 compounds by 50% or more in the period 1985-1995. For the nutrients phosphorus and nitrogen the reduction is fixed at 50%, for cadmium at 70% and for PCBs at 100%.

As far as the nutrients are concerned the concern is based on signals of distress like the bloom of the toxic algae species *Chrysochromulina* along the Norwegian coast (ICES, 1991), the increased level of chlorophyl (a measure for algae biomass) along the Dutch coast as well as an increased duration of the *Pheaocystis* bloom (North Sea Task Force, 1992). These phenomena cause mortality of fish, depletion of oxygen and foam on beaches. By reducing the riverine and atmospheric emissions, the chance, that these effects may happen diminishes, whereas the concentrations will decrease and nutrients are potential limiting factors for algae production. The current concentrations of phosphorus and nitrogen in the continental coastal zone exceed the coastal background levels by a factor of about 3 and 2.5 respectively (North Sea Task Force, 1992).

As far as the micropollutants are concerned the concern is based on signals of distress as well. Cadmium or PCB exposed sea stars produce defective offspring in long term exposure experiments conducted under semi-field experiments. The internal concentrations in the experimental animals were largely comparable to those in the Western Scheldt (den Besten, 1991). The PCBs, or some of their breakdown products or impurities, and probably the DDT family are found to cause reproductive failure, disrupt the metabolism of steroid hormones and affect the immune system of seals (Reijnders, 1988). Organometallic contaminants were reported to affect the whelk *Buccinum undatum* (ten Hallers-Tjabbes *et al*, 1992). This mollusc has been exterminated in the continental coastal zone, is still fairly abundant in open sea and shows imposex in the transition area (fig. 6). The current (1986) concentrations of cadmium

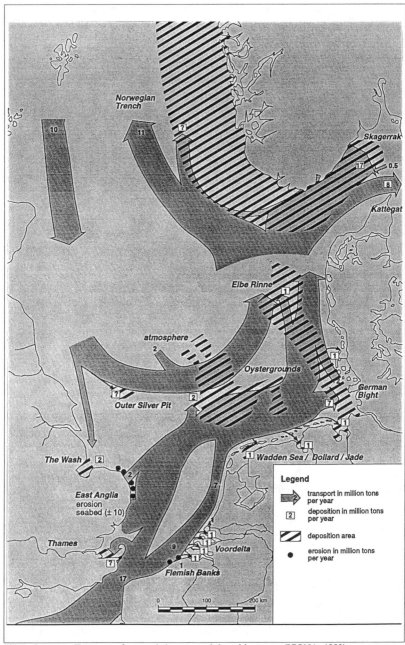

Figure 5:   Transport of suspended matter and deposition areas (ICONA, 1992).

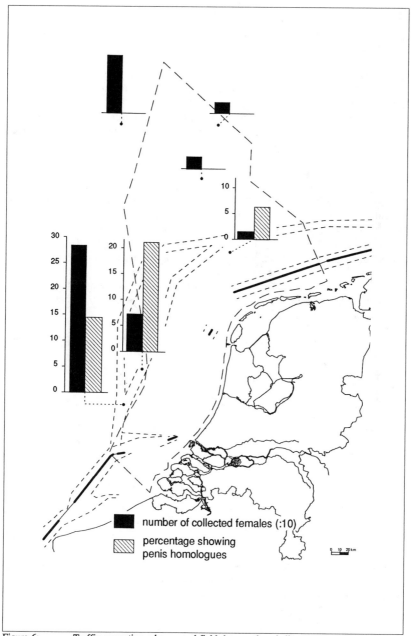

in the sediments in the Dutch continental sector exceed the background level by a factor of about 3 (North Sea Task Force, 1992). For PCBs the background levels are zero, witch means that in all reported cases (North Sea Task Force, 1992) these levels are exceeded.

Temporal trends
Chemical compounds enter the marine environment via different pathways, of which rivers, disposal of dredged materials and atmospheric deposition are the most important. Mainly because of political action and economical reasons, the inputs of many compounds to the North Sea via rivers and sludge have shown a decrease (see the second section of this paper).

In spite of the remarkable reduction (except for nitrogen) of those inputs via the Dutch rivers and via the disposal of sludge along the Dutch coast, the concentrations of phosphorus, cadmium and PCBs do not show a similar trend. As an example the distribution of phosphorus in the continental coastal zone is considered (North Sea Task Force, 1992). Apart from the inputs, the distribution depends mainly on the transport and mixing of water masses and on biological processes. To exclude biological processes it is common practice to use winter phosphorus concentrations for comparing levels over time and over large areas. Dissolved concentrations are maximal during this period. To exclude variations in transport and mixing due to variations in river discharges, a normalization procedure on salinity is generally used. Results for the Dutch coast are shown in figure 7.

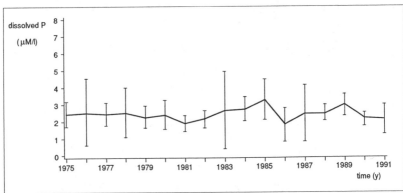

Figure 7:   Temporal trend of dissolved phosphorus.

Regression analysis of winter phosphate concentrations on salinity has been used to calculate the concentrations and their confidence limits at 30-salinity. No clear trend is to be seen in spite of the remarkable reduction of the river load (fig. 1).
The explanation of the poor relation between the riverine load and the concentration in the coastal zone is fourfold:
  *   the datasets cause wide confidence limits;
  *   phosphorus is emitted in the Rotterdam area;

* phosphorus is desorbed from the sediments of river, estuary and sea (after the adsorption in the sixties and seventies);
* most of the coastal water originates from the English Channel and is polluted by eg. the Scheldt and the Seine.

As yet the conclusion is that the 50% reduction measures should be completed for all inputs.

Environmental quality standards

In order to guarantee a sustainable use of the North Sea, environmental quality standards for contaminants are being developed in the Netherlands. Based on a concrete objective in terms of the abundance of several biological species and based on an evaluation of ecotoxicological data, two risk levels are determined: the Maximum Tolerable Risk (MTR) level and the Negligible Risk (NR) level (Jonkers et al, 1992). For cadmium the MTR has been fixed at 2.5 µg/l; the NR at .025 µg/l. According to van Eck et al (1985) the background concentration for cadmium is about .02 µg/l. Present concentrations in the Dutch coastal zone are about .04 µg/l. All figures refer to the dissolved fraction.

The risk levels will eventually guide the policy on reduction of inputs. To calculate the effect of alternative reduction measures, water quality models are used. When current reduction measures result in concentrations which exceed risk levels, additional measures will be necessary.

The effect of managerial actions

Reduction of inputs

To analyze the effect of different measures with respect to the uses and ecology of the North Sea, a management support system has been developed within the project Management Analysis North Sea (MANS). The MANS system considers items as calamitous spills, contaminants, eutrophication, physical planning, ecosystem effects and economy. Various models describe these items and their interrelations. A subset of models deals with contaminants, eutrophication and economy in an integrated way. This subset calculates the economical and ecological effects of reduction of indirect (eg. rivers) and direct (eg. dredged materials) inputs. In order to estimate economical effects of reduction of indirect inputs, the main sources of land based pollution had to be inventoried. Traffic, industrial, agricultural and domestic activities all contribute to the riverine and atmospheric input. In order to estimate the ecological effects various physical, chemical and biological processes have been incorporated in the models. In figure 8 the modelling approach to quantify the impact of contaminants on the water quality is schematically presented (Roos et al, 1990). The approach for nutrients is analogous.

The various steps in the analysis of the ecological effects of contaminants, shown in figure 8, are briefly outlined below. The contaminant model is fed by data representing the actual inputs of compounds into the North Sea or the inputs after managerial actions. These inputs may be directly into the sea or indirect via river discharge. Retention processes in estuaries and harbours are simulated. Dissolved and suspended mass transport is simulated, based on the results of hydrodynamic models for averaged meteorological and astronomical conditions. Based on the knowledge of the behaviour of the compounds in marine environments, the water and sediment concentrations are calculated. Sedimentation fluxes of both organic and inorganic matter and adsorption of these compounds play a major role in this respect. Finally the calculated concentrations are used for comparison with standards, no effect levels and other ecotoxicological data. Alternatively the uptake of contaminants by organisms can be calculated in order to perform a direct risk assessment.

Before application for managerial purposes, the model should be calibrated using actual input data and data on salinity, suspended matter and contaminants. These field data result from monitoring activities.

Simulations with the eutrophication model pointed out that a reduction of the input of the nutrients phosphorus and nitrogen by 75% will cut in half the algae excess in the Dutch coastal waters. This kind of results has indeed been a support for the formulation of the Dutch water policy.

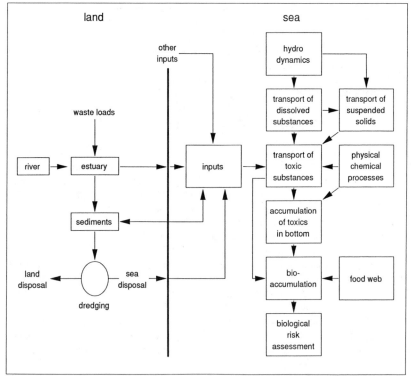

Figure 8:   The model approach to quantify the impacts on contaminants of the water quality.

Experimentally closed areas

Dutch water policy recognizes that clean water is not enough. In order to meet sustainable use measures other than reduction of inputs of chemical compounds are needed as well. Further, elaborating the example of the adverse effects of fishing, measures regarding this type of exploitation of the North Sea may imply spatial zoning. For particular areas the Council of Fishery Ministers of the EC already determines annually the Total Allowable Catches for the commercially important fish species. Further restrictions apply to the beam and otter trawls within the 12 miles zone and the so called plaice box and cod box. The aim of these measures is the protection of the stocks, spawning grounds and nurseries of commercial fish species. In order to protect other organisms as well, additional measures are being considered: areas closed for all types of fisheries. To support this kind of managerial action, an area of 12.000 km^2 north of the Wadden sea will probably be closed for research purposes. The research will focus on the ecological restoration of the area and adjacent areas.

The Dutch government designated a part of the Netherlands Continental Shelf to be an environmental zone. Its ecological value requires a high level of protection. Spatially, the implementation of the most far reaching measures will therefore be concentrated in this zone. As yet the only exclusion applies to the extraction of sand and gravel (on the landward side of the line of 20 meter below MSL) because of the rich variety of the benthic fauna. In addition extraction should be prohibited to ensure the safety of the sea defences. It may be expected that more exclusions and restrictions will follow, eg. with respect to fishing after evaluation of the research in the experimentally closed area.

Clearly not only water quality but also the physical and biological condition of the North Sea needs improvement to achieve a sustainable use. Related managerial actions are aimed at optimal (not maximal) harvesting, with all uses getting their share. The success of these actions will be reflected by a stable and diverse flora and fauna of the North Sea: a basin of 55.000 cubic kilometres of salt water on whose borders 200 million people live.

References

P. Balls (1985). Trace metals in the Northern North Sea. *Marine Pollution Bulletin* vol 16 no5 pp 203-207.

D. Bentley, R. Lafite, N.H. Morley, R. James, P.J. Statham, J.C. Guary (in prep). Flux de nutriments entre la Manche et la Mer du Nord, situation actuelle et evolution depuis 10 ans.

P.J. den Besten (1991). Effects of cadmium and PCBs on reproduction of the sea star *Asterias rubens. Dissertation.* University of Utrecht.

A.M.J.V. van Boxtel, M.O. von Königslöw, F.M. Tossings (1991). Atmospheric deposition of nutrients into the North Sea: assessment of possible effects on algae growth. *Geosens.*

H. Compaan, R.W.P.M. Laane (1992). Polycyclic aromatic hydrocarbons in the North Sea, an inventory. *Institute for Environmental Sciences (IMW)*, Delft. TNO-report 92/392.

W.C. Crogan (1984). Input of contaminants to the North Sea from the United Kingdom. *Institute of offshore engineering*, Heriot-Watt University. Edinburgh.

M. Donze ed. (1990). Aquatic pollution and dredging in the European Community. *DELWEL publishers.* The Hague.

G.Th.M. van Eck, H. van 't Sant, E. Turkstra (1985). Voorstel referentiewaarden fysisch chemische waterkwaliteitsparameters Nederlandse zoute wateren. *VROM-DGMH, RWSRIZA/DDMI* nota 1984.

C.C. ten Hallers-Tjabbes, J.P. Boon, H. Lindeboom (1992). Ecoprofile of the whelk, *Buccinum undatum* l

Joint Group of Experts on the Scientific Aspects of Marine Pollution, GESAMP (1989). The atmospheric input of trace species to the world ocean. World Meteorological Organization. *GESAMP* report No 38.

A. van der Giessen (1989). Water quality modelling for the North Sea. *Rijkswaterstaat Tidal Water Division.*

H. Hupkes (1990). Pollution of the North Sea imposed by west European rivers. *International Centre for Water Studies.* Amsterdam.

ICES (1991). The chrysochromulina polylepsis Bloom in the Skaggerak and the Kattegat in May-June 1988: Environmental Conditions, Possible Causes, and Effect. *ICES cooperative research report* 175.

ICONA (1992). North Sea Atlas. *Stadsuitgeverij Amsterdam.* ISBN 90-5366-047-Xgeb.

D.A. Jonkers, J.W. Everts (1992). Seaworthy; deviation of risk levels for microcontaminants in North Sea and Wadden Sea. *Report Ministries of Housing, Physical planning and the Environment and Transport, Public Works and Water Management,* The Hague (in Dutch with English summary).

M. Kersten, W. Kienz, S. Koelling, M. Schröder, K. Förstner (1990). Schwermetal belastung in Schwebstoffen und Sedimenten der Nordsee. *Vom Wasser* 75 pp 245-272.

J.C. Klamer (1989). PCBs in de Noordzee, Bronnen en Verspreiding. *Rijkswaterstaat, Dienst Getijdewateren.* Report GWAO-89.2001.

R.W.P.M. Laane (ed) (1992). Background concentrations of natural compounds in rivers, sea water, atmosphere and mussels. *Rijkswaterstaat, Tidal Water Division.* Report 92.033.

NERC (1991). An estuarine contaminant simulator, user manual by Plymsolve. *NERC.* Plymouth.

C. Nieuwendijk, A.M.J.V. van Boxtel (1991). Baggerspecieproblematiek in België, Bondsrepubliek en Nederland. *Water,* vol 61 pp 238-245.

C.C. Nihoul (1991). Dumping at sea. Ocean & Shoreline Management 16 pp 313-326.

North Sea Task Force (1992). *Quality status report of the North Sea,* subregion 4. Preprint.

P.J.H. Reijnders (1988). Environmental impact of PCBs in the marine environment. In: *P.J. Newman & A.R. Agg. Environmental Protection of the North Sea,* Heinemann, Oxford: 85-98.

A. Roos, J.F.M. van Vliet, J.A. van Pagee, T.A. Nauta, M.B. de Vries (1990). An integral approach to support managerial actions on micropollutants in the Southern North Sea. In: North Sea Pollution, Technical Strategies for Improvement. *Proceedings of Aquatech,* Amsterdam: pp 101-123.

D.E. Schulz-Bull, G. Petrick, J.C. Duinker (1991). Polychlorinated biphenyls in North Sea water. Marine Chemistry,36 pp 365-384.

P.J. Statham, Y. Aujer, J.D. Burton, P. Choisy, J.C. Fischer, R.H. James, N.H. Morley, B. Oudane, F. Puskaric, M. Wartel (*submitted to Oceanological Acta*). Fluxmanche - Fluxes of Cd, Co, Cu, Fe, Mn, Ni, Pb and Zn through the Straits of Dover into the Southern North Sea.

J.P. Warmenhoven, J.A. Duiser, L.Th de Leu, C. Veldt (1989). The contribution of the input from the atmosphere to the contamination of the North Sea and the Dutch Wadden Sea. *TNO Division of Technology for Society*, Delft. TNO-rapport R 89/349A.

Water Quality of the Wadden Sea

F. de Jong[1]

Abstract

A brief overview will be given of the different sources of input to the Wadden Sea (rivers, land run-off, atmosphere, dumping). The specific problems encountered in comparing the different sources will be addressed (i.e. computation of riverine loads, estimation of atmospheric inputs). For some substances temporal trends in riverine inputs will be presented. Trends in inputs will be compared to trends in concentrations of nutrients and metals for the period 1980-1990. An evaluation of 10 year environmental monitoring on the above presented information will be given.

The water quality policies of the Wadden Sea states will be summarized and evaluated. A prognosis for the future policy will be presented taken the decisions -and in particular the decision that ecological quality objectives for the Wadden Sea be developed- to the 6th Trilateral Wadden Sea Conference (Esbjerg, 1991) as a starting point.

Introduction

The Wadden Sea is a shallow sea extending along the North Sea coasts of the Netherlands, Germany and Denmark (fig. 1 and 2). It is a highly dynamic wetland ecosystem with tidal channels, sandbars, mudflats and saltmarshes. An island barrier of 23 islands with sand dunes and 14 high sands separate the area from the southeastern North Sea. The Wadden Sea covers an area of about 9000 km^2, of which 10% belongs to Denmark, 60% to Germany and 30% to the Netherlands.

The Wadden Sea is Europe's largest marine wetland. An overview of the national and international protection regimes is given by Enemark (1993). Reineking (1993) describes the natural values of the Wadden Sea ecosystem.

Human activities have an adverse impact on the Wadden Sea ecosystem. Of the various threats pollution is considered one of the most serious. The relatively high level of contamination of the Wadden Sea is caused by three main factors:

1.   A number of rivers, whose catchment areas are highly industrialized and agrominized, debouches into the Wadden Sea (fig. 2). The catchment area adds up to some 230,859 km^2.It extends to the southeast as far as the Chechian-Austrian border. Among the rivers are the Elbe and the IJssel, a

---

[1]   Deputy secretary, Common Wadden Sea Secretariat, Virchowstrasse 1, D2940 Wilhelmshaven, Germany

tributary of the Rhine. In addition a substantial part of the Rhine water enters the Wadden Sea via the North Sea through a coastal flow along the Dutch coast;

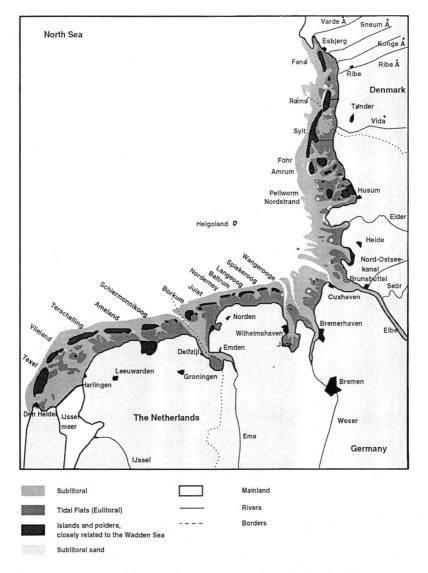

Figure 1:   Some topographic features of the Wadden Sea

2.   The Wadden Sea is a net sediment importing system. The sediments originate almost completely from the North Sea and are carriers of heavy metals and other contaminants. Due to the net North Sea current (fig. 2), a substantial part of North Sea sediments -and consequently polluting substances- is deposited into the Wadden Sea;

3.   The Wadden Sea lies at the rim of northwest Europe. An important part of its contamination is caused by the wet and dry deposition of airborne particles which originate from the highly industrialized northwest and east European countries.

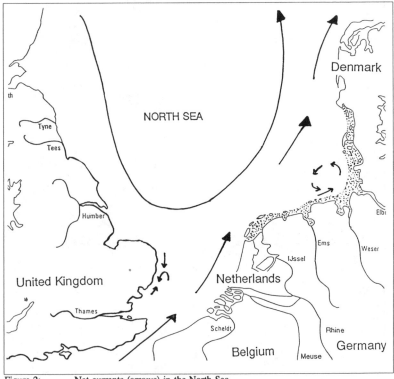

Figure 2:   Net currents (arrows) in the North Sea

During the last five years the water quality policies, both national and international, of the Wadden Sea littoral states have to an important degree been determined by the international conferences on the protection of the North Sea. At the 2nd and 3rd Conference (London, 1987; The Hague, 1990) important decisions regarding the reduction of inputs of toxic substances and nutrients were agreed upon:

-     between 1985 and 1995 inputs to the North Sea via rivers and estuaries of 36 priority substances which are toxic, bioaccumulating and persistent must be reduced by at least 50% (for dioxins, cadmium, mercury and lead 70%);

-    between 1985 and 1995 atmospheric emissions of 17 priority substances must
     be reduced by at least 50%;
-    between 1985 and 1995 inputs of nutrients to areas where they cause
     problems must be reduced by 50%.

The three Wadden Sea states are full parties to the North Sea Conferences. At the 3rd
North Sea Conference, the trilateral Wadden Sea cooperation submitted a joint
statement to the Conference in which additional measures for environmental
protection of the Wadden Sea were announced. These include the reduction of
pesticide emissions, the prohibition of discharges of offshore installations and the
development of a joint monitoring program and an early warning system for incidents
with hazardous substances.
At the 6th Trilateral Governmental Conference (Esbjerg, 1991) it was furthermore
decided to reduce inputs of polycyclic aromatic hydrocarbons (PAHs) by 50%
between 1985 and 1995 and to examine the possibilities of designating the Wadden
Sea catchment area a sensitive area under the EC Nitrate and Municipal Waste
Directive.

In the framework of the evaluation of the effectiveness of these decisions an
assessment of trends in inputs and concentrations of micropollutants and nutrients
over the period 1980-1991 was carried out (de Jong *et al*, 1993). In the following a
brief summary of this evaluation is presented after which possible implications for
policy and management are discussed.

Inputs

Input figures for the year 1990 have been summarized in table 1. Rivers are by far
the most important source of pollution of the Wadden Sea. Of these the river Elbe
is the major source of riverine inputs of contaminants and nutrients. The Treaty for
the sanitation of the Elbe, which was made possible through the German unification,
and which was signed in 1990 by Germany, Chechoslowakia and the European Community, will most probably have a positive effect on the reduction of contaminant and
nutrient loads from this river.

Data on direct run-off of freshwater to the Wadden Sea is very incomplete.
Freshwater run-off is probably an important source of pesticide input since the area
directly bordering the Wadden Sea is an agricultural area. Recent investigations in
the Netherlands make clear that inputs of lindane are substantial and that many other
pesticides are present which are not routinely monitored (Steenwijk *et al*. 1992).

The dumping of dredged materials is a considerable source of contamination with
heavy metals and PAHs (table 2).

In table 2 only figures for dredged material originating from harbors are given. The
total amount of sludge dumped in 1990 was almost 10 million tonnes. The figures
on dumping have been listed in a separate table because, depending on the dredging
site, the amounts may already be included in riverine inputs.

	Flow (10⁹m³)	N (t)	P (t)	Cd (t)	Hg (t)	Pb (t)	Zn (t)	Cu (t)	PCB (kg)	gHCH (kg)
Netherlands										
Rivers/run-off										
IJsselmeer	13.48	48700	1970	2.56	0.54	68.99	387.09	66.19	15.00	137[1]
Industrial/communal	N.I.	N.I.	N.I.	0.17	0.04	2.60	17.13	4.00	4.10	1.01
Total NL	13.48	48700	1970	2.73	0.58	71.59	404.22	70.19	19.10	138.01
Germany										
Rivers/run-off										
Ems	2.32	17000	350	0.06	0.16	9.00	32.00	5.60	20.00	6.00
Weser	8.20	45000	2400	2.01	0.71	26.12	256.80	90.39	60.40	120.00
Elbe	16.10	125600	8150	6.00	9.60	177.00	1472.00	181.40	68.00	212.00
Eider	0.83	4333	208	0.03	0.01	0.26	4.60	1.07	N.I.	3.68
Industrial/communal[2]	0.06	750	80	0.07	0.05	1.00	5.70	1.93	1.00	N.I.
Total D	27.45	192333	11158	8.69	10.52	213.26	1764.30	280.00	149.00	341.86
Denmark										
Rivers/run-off										
County Ribe	1.32	7758	223	N.I.	N.I.	N.I.	N.I.	N.I.	N.I.	N.I.
County Sonderjylland	0.84	3373	125	N.I.	N.I.	N.I.	N.I.	N.I.	N.I.	N.I.
Industrial/communal[3]	N.I.	711	13	N.I.	N.I.	N.I.	N.I.	N.I.	N.I.	N.I.
Total DK	2.16	11842	361							
Atmosphere[4]		7633		1.84	0.01	28.88	76.84	5.63	N.I.	25.00
Total Wadden Sea	43.15	260858	13519	13.27	11.12	313.85	2252.16	355.82	168.50	504.87
	+?	+?	+?	+?	+?	+?	+?	+?	+?	+?

NI= No Information;
1. According to Van Steenwijk et al. (1992), Figure includes IJsselmeer discharge (120 kg) and land run off (17 kg);
2. Heavy metal and organochlorine data only federal state of Niedersachsen;
3. Only Esbjerg fishmeal industry
4. Sea also Table 3.2.3.

	Amount	Cd (t)	Hg (t)	Pb (t)	Zn (t)	CuB (t)	PCB (t)	PAH (t)
Netherlands	5506.7	1.00 (0.57)	0.55 (0.24)	69.2 (32.3)	226.2 (78.9)	28.7 (0.3)	0.02 (0.02)	2.17 (2.17)
Germany	3960.5	2.78	1.11	117.0	304.8	40.6	0.02	NI
Denmark	514.6	0.29	0.02	22.6	92.1	14.0	NI	NI
Total	9981.8	4.07	1.68	208.8	623.1	83.3		

Table 2:   Dredge spoils from harbors in the Wadden Sea in 1990. Figures between brackets: antropogenic load. Source: Oslo Commission.

The atmosphere accounts for at least 3% of total nitrogen inputs to the Wadden Sea. Based on samples taken at Norderney in the period 1988-1990 atmospheric inputs of cadmium are as high as 14% of total cadmium inputs to the area.

Little is known about the inputs of contaminants and nutrients from and into the North Sea although it is generally accepted that the North Sea is one of the major sources of Wadden Sea contamination.

Inputs from industrial and communal sources in the Wadden Sea area itself are low compared to above sources. Data are however incomplete.

Temporal trends

It is difficult to estimate contaminant input from rivers and land run-off because of the problem of the computation of loads from concentrations and water flow. The monitoring location is of utmost importance because salinity and turbidity at the monitoring site influence the measurement. Also the sampling frequency is important. It should be high enough to be able to reveal any temporal trends (Hupkes, 1990; 1992). Any evaluation of contaminant and nutrient input should take water flow into account. Since in 1989, 1990 and 1991 the fresh water flow was relatively low, specific care should be taken in establishing temporal trends over the last years.

Nutrients

As can be concluded from the time series of freshwater inputs (tables 3 and 4) the annual water flow strongly influences the yearly loads of nutrients. The relation between flow and load is not always linear: according to an analysis of nutrients in the river Elbe (Gaumert, 1991) there is a positive correlation between water flow and nitrate concentration. This is explained by the fact that through high precipitation large amounts of nitrate enter the river through land run-off.

The Dutch Tidal Water Division has analyzed nutrient inputs and concentrations in the period 1972-1990 (table 3). From the analysis it is concluded that nitrogen concentrations show no significant reduction in the period 1981-1990 whereas phosphorus concentrations have decreased by 40%. Nitrogen loads have decreased by some 15% which cannot have been caused by lower concentrations. Phosphorus *loads* have decreased by half in the period 1981-1990. The 1991 data are in accordance with the above assessment.

	load N 1000 ton	load P 1000 ton	conc.N mg/l	conc.P mg/l	flow $10^6 m^3$
1972	24.96	1.27	2.41	0.12	10377
1973	41.48	2.11	3.19	0.16	13007
1974	60.20	2.70	3.97	0.18	15165
1975	53.64	2.70	3.75	0.19	14311
1976	30.75	1.72	3.63	0.20	8465
1977	46.16	2.66	3.62	0.21	12760
1978	49.27	3.02	3.70	0.23	13308
1979	60.93	3.80	3.86	0.23	16541
1980	59.28	3.86	3.87	0.25	15318
1981	74.71	5.09	4.02	0.27	18590
1982	49.87	3.61	3.31	0.24	15046
1983	72.04	4.33	3.85	0.23	18708
1984	81.94	4.42	4.38	0.24	18703
1985	53.30	3.66	3.61	0.25	14769
1986	61.29	3.45	3.95	0.22	15523
1987	79.73	3.57	3.82	0.17	20865
1988	77.30	3.63	3.68	0.17	21026
1989	48.31	2.34	3.73	0.18	12939
1990	48.70	1.97	3.61	0.15	13477
1991	43.58	1.43	N.I.	N.I.	11771

Table 3:   N and P inputs to the Wadden Sea via the IJsselmeer drainage
                and annual average concentrations of N and P at discharge site.
                Period 1972-1991. Source: Rijkswaterstaat, DGW

Nutrient loads of the Elbe have reduced tremendously since 1987. The extent to which this has been caused by the reductions in flow can not yet be calculated. According to Gaumert (1991) the reductions in the last two years must be attributed to the closing of factories in the former DDR. Additional reductions are expected as a result of the introduction of sewage treatment plants.

Micro-pollutants

With regard to heavy metal inputs, the studies of Hupkes (1990, 1992) clearly indicate that a thorough analysis over a period of at least ten years is necessary to be able to reveal significant trends.

Table 4 shows riverine loads of the Elbe. There seems to be a decrease in riverine loads of the heavy metals Hg, Cd and Pb and a number of chlorinated hydrocarbons

from 1988 to 1991. The very low flows in 1989, 1990 and 1991 are however at least partially responsible for this reduction.
Heavy metal loads of the IJsselmeer show no clear increasing or decreasing trends over the last 10 years.

	1983	1984	1985	1986	1987	1988	1989	1990	1991
Flow $(10^9 m^3)^1$	19.6	18.4	17.6	22.6	35.6	27.6	16.4	14.1	12.1
Cadmium	13.51	9.50	8.40	10.00	10.00	10.40	6.00	6.00	5.00
Copper		98.55	182.50	250.00	400.00	310.00	200.00	190.00	180.00
Mercury	5.48	7.30	7.30	15.00	25.00	15.30	10.00	10.00	10.00
Lead		36.50	219.00	200.00	300.00	220.00	160.00	180.00	270.00
Zinc		244.55	1825.00	2000.00	3000.00	2600.00	1700.00	1500.00	1500.00
gHCH	0.70	0.60	.70	.50	.80	.50	.44	.21	0.20
Sum-PCB				.50	.20	.20	.14	.02	<0.03

Table 4:   Contaminants loads Elbe 1983-1991 (t/y). Data from JMG. gHCH data 1983-1985 from Gaul (1991). 1991 data from Wassergütestelle Elbe.

Of the organic micropollutants only for *lindane* (g-HCH) and *PCBs* time series on riverine inputs are available. Lindane loads of the Elbe have been decreasing since 1983. PCB loads of the Elbe show a gradual decrease as of 1986 (Table 4).
IJsselmeer loads of both PCB and lindane since 1986 show no decreasing or increasing trend.

Concentrations
An analysis of concentrations of nutrients, heavy metals and some organochlorines over the period 1980-1990 was carried out.

From the analysis it is concluded that especially in the south-western Wadden Sea no changes are present in concentrations of *nitrogen* compounds, while the concentrations of *phosphorus* compounds are decreasing. In the Elbe estuary and the eastern Wadden Sea both nitrogen and, to a lesser extent, phosphorus concentrations have been decreasing in recent years.
These observations are in agreement with trends in nutrients loads from the rivers.

Lindane concentrations in water have decreased in the eastern Wadden Sea which can be attributed to the reductions in Elbe inputs.

Cadmium concentrations in sediment generally are decreasing (fig. 3). This is not the case for concentrations of cadmium in blue mussel which where either stable or increasing.

Concentrations of *mercury* in sediment in some areas slowly decrease and are rather constant in other areas. In area NS1 levels have increased since 1986.

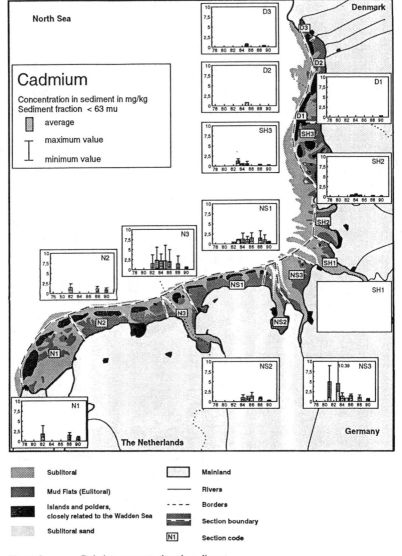

Figure 3:   Cadmium concentrations in sediment

Copper concentrations in sediment are rather constant or decreasing. Copper concentrations in mussel only show a decreasing trend in area SH3. In most of the other areas concentrations are stable or increasing.

Lead concentrations in sediment are decreasing or stable. Lead concentrations in mussel show no decreasing or increasing trends.

Zinc sediment concentrations are only decreasing in area N3. In other areas they are rather stable. The zinc concentrations in blue mussel show a highly variable pattern.

It can be concluded that, with the exception of cadmium, no clear links between riverine inputs of metals and their environmental concentrations could be established. This is due to a number of factors. Firstly there is a time lag between a change in input and a change in sediment concentration. Because there is a clear relation between sediment contamination and mussel contamination -mussels eat phytoplankton which adsorp metals from the sediment- the time lag with changes in concentration in mussel will even be longer. Secondly the high variability in input and concentration data, the lacking of data, both in time and space and the relative short evaluation period will certainly obscure possible trends. It is furthermore important to note that only two input sources have been used for the assessment. Inputs from the North Sea and the atmosphere are considerable but not enough reliable data are available.

Effects on the ecosystem
In the last decade many changes in the ecosystem have been documented (CWSS, 1991) of which only a few can be clearly attributed to contamination. In the 60s the Wadden Sea sandwich tern (*Sterna sandvicencis*) and eider duck (*Somateria mollisima*) populations were decimated as a result of contamination with drins. The pesticides originated from factories discharging into the southern North Sea. After a stringent reduction of inputs the populations recovered (Koeman and van Genderen, 1972). Another well-documented example is the negative effect of PCBs on the reproductive success of the harbor seal (*Phoca vitulina*) in the Dutch Wadden Sea (Reijnders, 1986).

Fish diseases, especially flatfish, may be related to pollution factors, probably in combination with other stress factors (Vethaak, 1991; Wahl *et al.*, 1992). The causes for the seal epidemic, which reduced the Wadden Sea harbor seal population by 60% in 1988 are still unclear. There are indications that the immune system is negatively affected by pollution (Brouwer *et al*, 1989).

In general it is however difficult to determine ecosystem effects of contamination. Concentrations of most substances are below levels of acute toxicity. Furthermore the Wadden Sea is a very dynamic area and exposed to much natural stress. These two factors make it difficult to differentiate between natural and man-induced effects.

De Jong *et al.* (1993) have compared concentration levels with so-called natural or background values. The concentrations of mercury, cadmium, lead and zinc in sediment are two to five times as high as reference levels. It is not clear what the effects of these enhanced concentrations on the ecosystem will be.

There is general consensus among experts that eutrophication has a very clear impact on the ecosystem which is generally judged as negative. Changes observed which can -at least partly- be attributed to increased nutrient input are, amongst others, an increase in primary production, phytoplankton biomass and duration of *Phaeocystis* blooms, a shift in phytoplankton species composition, an increase in toxic and nuisance species, a shift in macrobenthos species composition towards more oppor-tunistic short-lived species like worms and mass development of macro algae. Another problem that could be related to excess primary production and high sediment oxygen consumption rates by mineralizing bacteria are the so-called "black spot", areas. Black spots were observed for the first time in 1984 and have increased in frequency and size from 1987 onward. Within the spots, which can have sizes of up to several square meters, the oxidized surface layer has disappeared. In the black reduced layer that remains macro- and meiofauna are either killed or have moved to other areas (Kolbe, 1991; Michaelis et al., 1992).

Effects of reductions in nutrient inputs are however difficult to predict. Important in this respects is amongst others the N/P ratio of concentrations in water. According to Riegman et al. (1992) novel nuisance algal blooms may be the result of major shifts in N/P and NH_4^+/NO_3^- ratios rather than a general N+P enrichment. It is furthermore unclear in how far eutrophication boosts the energy flow through the bacterial foodweb and which the effects of changes in nutrient inputs and ratios will be. It is with regard to the above important to note that the N/P ratio is increasing in the western Wadden Sea and decreasing in the eastern Wadden Sea.

Water quality policies: Immission approach

Marine water quality policies in northwest Europe are generally governed by the 'emission' approach. In this approach pollution of the marine environment is to be reduced by the setting of Emission Standards (ES) for discharges of a number of dangerous substances.

Important fora where international ESs have been agreed upon are the Paris Convention and the European Commission. In these fora there has been a long-lasting discussion between the United Kingdom and other contracting parties about the application of Uniform Emission Standards (UESs), which were favored by the latter and Environmental Quality Objectives (EQOs) which were applied in UK water management (Boehmer-Christiansen, 1990). The UK use of EQOs for marine waters involves the setting of limit concentration values in its coastal waters for the substances for which other countries apply UESs.

At the International Conferences on the Protection of the North Sea (Bremen, 1984; London, 1987; The Hague, 1990) the use of both the UES and the EQO approaches was underlined. At all three Conferences it was furthermore stated that the two approaches should be integrated. The decisions of the 2nd North Sea Conference (London, 1987) and in particular the decisions to reduce inputs of dangerous substances and nutrients to the North Sea by 50% between 1985 and 1995 make clear that the role of UESs in the international policy has become the dominant one. This leading position was consolidated at the 3rd North Sea Conference (The Hague, 1990) and has been strengthened by the general acceptance of the Precautionary Principle.

Quite independent from the above described controversy there has been a development in the past years of elaborating and applying Ecological Quality Objectives for the marine ecosystem. This development follows that in fresh water and terrestrial ecosystem policies. In the past decade a general need for ecological objectives and ecological standards has become apparent. In the foregoing some problems with regard to the evaluation of emission-reduction goals have been described. Immission goals may serve as instruments in an environmental policy that not only relies on the control of *emissions* but also measures the effects of policy measures by way of monitoring the ecosystem (*immission approach*). Also the increasing use of environmental impact assessments has increased the need for ecological standards in order to be able to assess the possible impact of present and planned uses relative to the standards. *Ecological* Quality Objectives should not be confused with *Environmental* Quality Objectives. Although there does not yet exist a general definition of Ecological Quality Objective, it is possible to give some basic differences from EQOs. Ecological Quality Objectives are not limited to chemical or physical ecosystem parameters. They can be species numbers, species diversity, biological process parameters or a qualitative description of a specific feature of the ecosystem under consideration. Ecological Quality Objectives furthermore differ from EQOs in that they are to be assessed as a group of parameters which together are considered to be representative for the quality of the ecosystem and not as single parameters.

In the three Wadden Sea countries the development and application of quality objectives for marine waters is in different stages of progress (de Jong, 1992b).
At the 6th Trilateral Governmental Wadden Sea Conference (Esbjerg, 1991) the concepts of wise-use and sustainable development were adopted as a common basis for the conservation policy of the Wadden Sea (CWSS, 1992a; de Jong, 1992a). According to the above concepts such a policy requires that the natural potentials of the ecosystem be defined and maintained. A description of the natural potentials of the ecosystem is called the Reference Situation. The participants furthermore agreed to address at the 7th Trilateral Wadden Sea Conference (1994) a set of common ecological targets and a comprehensive set of measures to reach those targets.
Guidance with regard to the nature of the Reference is provided by the Guiding Principle which was adopted at the same Conference. The Guiding Principle of the trilateral Wadden Sea policy is to achieve, as far as possible, a natural and sustainable ecosystem in which natural processes proceed in an undisturbed way.
This principle aims at:
-    maintaining the water movements and the attendant geomorphological and pedological processes;
-    improving the quality of water, sediment and air to levels that are not harmful for the ecosystem;
-    safeguarding and optimizing the conditions for flora and fauna including
     =   preservation of the Wadden Sea as a nursery area for North Sea fish;
     =   conservation of the feeding, breeding, moulting and roosting areas for birds, and the birth and resting areas of seals as well as the prevention of disturbance in these areas;
     =   conservation of the salt marshes and dunes;
-    maintaining the scenic qualities of the landscape.

A trilateral expert working group is presently working on the development of an ecological Reference for the Wadden Sea ecosystem and the setting of Ecological Targets for the year 2010.

The development of an Ecological Reference requires first of all the selection of parameters which are considered indicative for the quality of the ecosystem. The second step is to assign reference values to each of the parameters.

Reference parameters may be validated by one or more of the following methods:
- application of historical data;
- application of ecosystem models;
- laboratory models (i.e. mesocosms studies);
- comparison with areas outside the Wadden Sea;
- comparison with Reference areas within the Wadden Sea.

On the basis of expert judgement it must be decided which method (or combination of methods) is most suitable for a certain parameter. The resulting list of Ecological References is in fact a *list of agreed facts* about the natural potentials of the Wadden Sea ecosystem. This automatically implies that on the basis of new findings amendments to the list are made. In some cases it may be necessary to wait for results of developments in undisturbed reference areas if these do not yet exist. Another possibility is that specific research is initiated in order to come to a better validation.

The value given to each of the parameters should ideally reflect an ecosystem which is sustainable and in which 'natural processes proceed in an undisturbed way'. It may be assumed that an ecosystem which is not affected by human activities, possesses these qualities.

As to the *physical* situation it will be possible to give a description of a situation in which the Wadden Sea was not influenced by human action, i.e. a Wadden Sea without dikes. A complicating factor in this case is: what is known about natural (physical) dynamics of such a system, in other words, is it possible to give more or less fixed values for relevant parameters for an evolving system like the Wadden Sea.

For a number of *chemical* substances it will in most cases be possible to derive "natural" or "background" values for concentrations in water, sediment and/or biota. Various methods exist on how to determine background concentrations (e.g. Laane *et al.*, 1992). Sediment background values can for example be determined through the analysis of cores taken from undisturbed sediments. According to Höck & Runte (1992) sediment concentration profiles from the Wadden Sea near Sylt of the trace metals Cu, Pb, Zn and Cd show that at a core depth of about 80 cm, which is roughly 100 years ago, concentrations of all four metals started to increase. The vertical distribution of seven PCB congeners in sediments from the Dutch Wadden Sea (Kramer *et al*, 1989) makes clear that PCB contamination started around the year 1933. PCB production started in the 1920s. The geochemical background for PCBs is of course zero.
Background values can also be derived from comparable areas which are not or hardly influenced by pollution. Such a method can for example be used to compare contaminants in biota.

Finally background values can be computed with dispersion models which take natural levels of inputs as starting parameters.

With regard to *biological* parameters there are two main problems: the first one is to find values for a situation without human influence. The second one is how to cope with the natural development of the system. The use of reference areas which are excluded from human activities is a possible option. It is however unavoidable that such reference areas are being influenced by other human effects, for example contamination.

On the basis of Reference values Ecological Targets for the year 2010 must be developed. In the setting of these targets a certain level of human use of the area -fishery, coastal industry, recreation- is to be included.
One of the major problems will be to relate certain targets to specific measures to be taken. As has been indicated above there are only few cases in which a clear relationship between contamination and effect in the ecosystem has been proven.

Summary and conclusions

Inputs and concentrations of nutrients and organic micropollutants have been analyzed over the period 1980-1991.
In the western Wadden Sea inputs of phosphorus are decreasing whereas nitrogen inputs show a stable pattern. This has resulted in an increasing N/P ratio. In the eastern Wadden Sea which is mainly under the influence of the river Elbe, both N and P inputs are decreasing since 1987. Because N inputs show a stronger decrease, the N/P ratio in this part of the Wadden is decreasing. The implications for policy and management are not clear because not enough is known about the ecosystem effects of changes in N/P ratios.
Inputs of trace metals of the Elbe have decreased. IJsselmeer inputs (the second largest source) show no clear trends although levels are generally lower in the second half of the past decade compared to the first half. Metal concentrations in the ecosystem do not (yet) reflect changes in inputs. This is due to a combination of factors: a high yearly variability in inputs caused by variable flows; a time-lag between change in input and change in environmental concentration; a lack of data, both in time and space.

The problems in assessing effects of reduction measures may -at least partly- be overcome with the aid of ecological objectives. This implies that, in addition to the setting of reduction goals specific ecosystem goals are agreed upon. The quality of the ecosystem is assessed with the aid of an Ecological Reference. In the trilateral Wadden Sea policy and management it has been agreed to develop an Ecological Reference for the Wadden Sea and specific problems encountered in the development of an Ecological Reference are the selection of parameters which are indicative for the quality of the ecosystem and the validation of these parameters. Moreover there is a general lack of knowledge about cause and effect relationships which makes it difficult to establish links between ecological targets and measures to reach the targets.

References

Boehmer-Christiansen. 1990. Environmental Quality Objectives versus Uniform Emission Standards. *In*: David Freestone and Ton IJlstra (eds.), *The North Sea: perspectives on regional environmental cooperation*, pp. 139-149. Graham & Trotman/Martinus Nijhoff.

Brouwer, A., P.J.H. Reijnders & J.H. Koeman. 1989. Polychlorinated biphenyl (PCB)-contaminated fish induces vitamin A and thyroid hormone deficiency in the common seal (Phoca vitulina). *Aquatic Toxicology*, 15:99-106.

CWSS. 1991. The Wadden Sea; Status and Developments in an international perspective. Common Wadden Sea Secretariat, Wilhelmshaven. 200 pp.

CWSS. 1992. Wise use and conservation of the Wadden Sea. Common Wadden Sea Secretariat, Wilhelmshaven, FRG.

Enemark, J.A. 1993. The protection of the Wadden Sea in an international perspective. Planning, protection and management of the Wadden Sea. *This volume*.

Gaumert, T. 1991. Trend-Entwicklung der Nährstoffe im Elbwasser von 1980 bis 1989. ARGE Elbe.

Höck, M. & K-H. Runte. 1992. Sedimentologisch-geochemische Untersuchungen zur zeitlichen Entwicklung der Schwermetallbelastung im Wattgebiet vor dem Morsum-Kliff/Sylt. *Meyniana* 44: 129-137.

Hupkes, R. 1990. Pollution of the North Sea imposed by west European rivers. International Center of Water Studies, Amsterdam, Netherlands. 102 pp.

Hupkes, R. 1992. North Sea Pollution: River Input (1984. 1990). ICWS-report 92.07. pp. 68 + Annexes. International Centre of Water Studies, Amsterdam.

Jong, F., de., J.F. Bakker, K. Dahl, N. Dankers, H. Farke, W. Jäppelt, P.B. Madsen & K. Koßmagk-Stephan. (*eds*.4) 1993. Quality Status Report of the North Sea. Subregion 10, the Wadden Sea. In preparation. Common Wadden Sea Secretariat, Wilhelmshaven, Germany.

Jong, F., de. 1992b. Ecological Quality Objectives for Marine Coastal Waters: the Wadden Sea Experience. *International Journal of Estuarine and Coastal Law*, Vol 7, No. 4: 255-276.

Jong, F., de. 1992a. The wise-use concept as a basis for the conservation and management of the Wadden Sea. *International Journal of Estuarine and Coastal Law*, Vol 7, No. 3: 175-194.

Koeman, J.H. & H. van Genderen. 1972. Tissue levels in animals and effects caused by chlorinated hydrocarbon insecticides, chlorinated biphenyls and mercury in the marine environment along the Netherlands coast. pp. 428-435 *in: Marine pollution and sea life*. M. Ruiveo (ed). FAO Fishing New Books Ltd. England.

Kolbe, K. 1990. Zum Auftreten 'schwarzer Flecken' oberflächlich anstehender, reduzierter Sedimente, im ostfriesischen Wattenmeer. Forschungsstelle Küste Norderney. 23 pp.

Kramer, C.J.M., Misdorp R. and R. Duijts. 1989. Contaminants in sediments of North Sea and Wadden Sea. Report GWWS-90.001 Tidal Waters Division, Ministry of Transport, Public Works and Water Management and Laboratory for Applied Research, MT-TNO

Laane, R.W.P.M. (editor). 1992. Background concentrations of natural compounds in rivers, sea water, atmosphere and mussels. Tidal Waters Division (Directorate-General of Public Works and Water Management), Report no. DGW-92.003. Summary of the group reports written during the international Workshop on Background Concentrations of Natural Compounds, The Hague 6-10 April 1992. pp. 84.

Michaelis, H., K. Kolbe & A. Thiessen. 1992. The 'black spot disease (anaerobic sediments) of the Wadden Sea. ICES paper 1992 E:36.

Reijnders, P.J.H. 1986. Reproductive failure in common seals feeding on fish from polluted coastal waters. Nature 324:456-457.

Reineking, B. 1993. The ecosystem of the Wadden Sea. This Volume.

Riegman, R., A.A.M. Noordeloos & G.C. Cadée. 1992. Phaeocystis blooms and eutrophication of the continental coastal zones of the North Sea. Marine biology 112: 479-484.

Smayda, T.J. 1990. Novel and nuisance phytoplankton blooms in the sea: evidence for a global epidemic. pp. 29-40 in: E. Graneli, B, Sundström, L. Edler & D.M. Anderson (eds), Toxic Marine Phytoplankton. Elsevier, New York.

Steenwijk, J.M., J.M. Lourens, J.H. van Meerendonk, A.J.W. Phernambucq & H.J. Barreveld. 1992. Speuren naar sporen I. Nota RIZA 92.057; Rapport DGW 92.040. pp 110 + bijlagen.

Directie Noordzee. 1991. Verontreinigingsrapportage 1991.

Vethaak, D. 1991. Diseases of flounder in the Dutch Wadden Sea and their relation to stress factors. Netherlands Journal of Sea Research (accepted).

Wahl. E., H. Möller, K. Anders, A. Köhler-Günther, H.J. Pluta, P. Cameron, U. Harms, H. Büther & K. Söffker. 1992. Fish diseases in the Wadden Sea. ICES paper C.M. 1992/E:27.

The Ecosystem of the Wadden Sea

B. Reineking[1]

Abstract

The Wadden Sea is a unique landscape and one of the last large and relatively undisturbed ecosystems in northwest Europe and therefore a wetland of international importance. At the same time the Wadden Sea, which extends over parts of the Netherlands, Germany and Denmark, is an area where people live, work and recreate and a lot of human uses take place.

The Wadden Sea is a natural ecosystem with a wide range of ecosystems and habitats for flora and fauna, in which natural processes take place in a more or less undisturbed way. The importance of the Wadden Sea, considering the high ecological value, the functions for the North Sea and its vulnerability, are presented in this paper by means of the following:

The area is of great value for the species diversity in general, for a number of endemic species in particular and for the fauna in large parts of the northern biosphere. The Wadden Sea serves as a cleansing site for the North Sea, it is an important nursery ground for numerous fish species and provides feeding grounds for nearly all Palaearctic species of waders and waterfowl.

Due to the important ecological function of the Wadden Sea for the whole North Sea, both are considered to be as sensitive as its most sensitive part, the Wadden Sea. It is described that besides the Wadden Sea as a vulnerable system, simultaneously the Wadden Sea has a large potential for survival due to its dynamics and chance of restorations.

By means of changes in the Wadden Sea ecosystems in the past and their causes, the natural changes and the still occurring impacts and effects of human uses and activities will be stated with examples.

Introduction

The Wadden Sea is one of the last large and most natural areas in northwestern Europe, in which natural processes have lead to a highly esteemed ecosystem of outstanding scenic beauty. It is a shallow coastal area, not deeper than 10 m with extensive tidal flats, which form a very rare habitat type not only in Europe but in the whole world. Due to regular tides, the Wadden Sea area is a periodically submerging coastal region. It consits of a network of tidal channels, sandbars, mud flats, salt marshes, dunes and islands. The area extends for about 500 km along the

---

[1] Deputy secretary, Common Wadden Sea Secretariat, Virchowstrasse 1, D 2940 Wilhelmshaven, Germany

coast from Den Helder (the Netherlands) in the southwest along the German coast to the peninsula Skallingen near Esbjerg (Denmark) in the north.

The exceptional ecological importance of the Wadden Sea is reflected by the variety and interaction of flora and fauna, in particular birds and fish. A number of North Sea fish is dependent on the Wadden Sea and requires it as the main nursery ground. Without the Wadden Sea as feeding and resting ground, the survival of migrating wader populations would not be ensured. The importance of this coastal area is not only based upon the vital necessity for birds and fish-stocks, but it has more and more been recognized in its entirety as a nature area of biological and scientific international importance, which has been manifested in the development of nature conservation activities for more than two decades.

Even today, this unique landscape is indeed a relatively undisturbed ecosystem, but at the same time, it is an area where people live, work and recreate, and thus a number of human uses and activities take place. The Wadden Sea is situated near the most developed and populated countries in Europe and, therefore, exposed to a number of threats.

In this chapter, the outstanding importance of the Wadden Sea, considering the high ecological value and the function for the North Sea, will be presented. The vulnerability of the Wadden Sea system will be described, but at the same time it is an open system, which is provided with a large potential of natural regeneration abilities due to its dynamics.

The changes of the Wadden Sea ecosystem in the past and their causes will be stated as well as the still ongoing threats caused by the impact and effects of human activities.

The Wadden Sea ecosystem

The Wadden Sea covers an area of about 9,000 km², including about 1,000 km² islands and about 350 km² salt marshes and summer polders. The remaining area consists of large tidal flats. Most parts of the Wadden Sea, in particular in the Netherlands and Lower Saxony, are sheltered by barrier islands and contain smaller or wider areas of intertidal flats. Between the Weser estuary and the island Amrum, the area is relatively broad and open to the North Sea. Because of embankments only four large sheltered bays have remained in the total area; the Ho Bugt in Denmark, the Jadebusen and the Leybucht in Lower Saxony and the Dollard in the Dutch-German border area. Twenty three islands with sand dunes, as well as fourteen high sands without dunes form a barrier to the North Sea. In the past, the natural processes of accretion and erosion caused the islands to slowly change their position. Since the last century, the majority have been kept in place by dikes, groynes and beach nourishment (fig. 1 in: de Jong, 1993, this volume).

The present type of coast is a consequence of the low altitude of the mainland, as well as the high tidal amplitude, which is between 2 and 4 m. Besides the Danish-

German-Dutch Wadden Sea, this type of coast only exists in the Delta area (the Netherlands) and estuaries of the United Kingdom.

The Wadden Sea hydrology is mainly determined by the daily tides. With each high tide an average of 15 km^3 of North Sea water enters the Wadden Sea, thereby doubling the volume from 15 km^3 to about 30 km^3. With the North Sea water several substances, which are dissolved in the water, such as nutrients, suspended particular matter, reach the Wadden Sea, as well as floating material.

Ecological aspects and biological processes
The Wadden Sea represents a complex ecosystem with many interacting processes. This ecosystem is an open system importing organic matter from land and seaside, accompanied by a high turnover on the flats and an export back on a higher trophic level. For marine organisms, the Wadden Sea is an extreme environment. Nevertheless, it is characterized by a very high biomass production. The high primary production of microalgae is the basis for both, an outstanding secondary production and a high remineralization activity. Eutrophication, the situation in which the import of nutrients and/or organic matter is larger than the export, causes an increase in primary production and consequently stimulates other biological processes. In the Wadden Sea, this has resulted in a substantial increase of mineralization.

Primary production
The major primary producers are unicellular microalgae, both at the surface of the tidal flats, called microphythobenthos, and suspended in water as phythoplankton. Other plants, like macroalgae and seagrass, only play a negligible role in primary production.

All primary producers need light, carbon dioxide and nutrients to produce organic matter. Nutrients in form of phosphorus and nitrogen compounds are the limiting factors for phythoplankton growth. Primary production of phythoplankton is generally larger than that of microphythobenthos. However, the Wadden Sea is dominated by diatoms -benthic microalgae- which require, besides nutrients, also silicon for building their silica shell.

Consumption
Various zooplankton groups are grazing on phythoplankton, and benthic microalgae are consumed by benthic grazers. On the next trophic level, the zooplankton stocks are exploited by jelly-fish and some fish species. However, as secondary producer, benthic macrofauna, such as filter feeders, here mainly mussels and cockles, which filter suspended algae, are much more important concerning biomass. High food supply results in high standing stocks of shellfish, as well as deposit-feeders like worms, which feed on algae living in the sediment. The biomass of benthic fauna in the Wadden Sea amounts to the highest values reached anywhere in the North Sea and are exploited by many invertebrate carnivores, such as starfish, ragworms and whelks, as well as fish and birds, the latter mainly during low tide.

Mineralization
The Wadden Sea is a place of aerobic and anaerobic mineralization of organic material (detritus), which is not or only very slowly degradable in the water column. Detritus enters the area by tidal currents. Mineralization products are carbon dioxide, inorganic nitrogen and phosphorous compounds which run again into the cycles. The only way, to eliminate nitrogen from the marine system is denitrification, which occurs in the sediment of the tidal flats. These processes are mainly biological functions of microorganisms, but also of the zoobenthos. The importance of this process can be seen, in particular, in the high contribution of the Wadden Sea to the biological self-purification of the North Sea.

Extreme biotope
The Wadden Sea can be described as a vulnerable ecosystem. On the other hand, this ecosystem is a very dynamic environment with regular and unexpected changes from one extreme situation to another, such as temperature with the possibility of ice, salinity, storms, waves and currents. Only species, which are able to survive these extreme conditions exist in the Wadden Sea ecosystem, and, therefore, the Wadden Sea species, and consequently the ecosystem itself, have a large potential for survival. Until now, the large capacity for survival offered the chance for and ensured the restoration of the Wadden Sea system. Nevertheless, the extinction of a number of species from the Wadden Sea ecosystem has also shown the possibility and threat of limitation for such a survival capacity.

Main communities and subsystems of species groups
The Wadden Sea provides a multitude of marine habitats and transitional zones to the land and freshwater environment, which supports an outstanding biological diversity. The extremely high habitat diversity is followed by an exceptional species richness. This includes 2,000 species of invertebrates in the salt marshes and 1,800 in the marine and brackish areas. Among these organisms, there is a high degree of ecological specialization.

Contrary to the salt marshes with their high diversity of species, only some species of flora and fauna have adopted to the extreme environment of the tidal flats. However, here an exceptional high number of individuals per species can be stated. The biological productivity of the tidal flats is comparable with the one of tropical rain forests. This high productivity results from the tidal stream of the North Sea which brings the essential nutrients to fuel the entire system twice a day.

Tidal flats and their biota
Areas which periodically submerge, due to regular tides, are called tidal flats. The extent of tidal current velocities results in a spatial differentiation of the sediment grain size. Generally, a positive correlation exists between the maximum current velocity and the size of sediment grains. Muddy sediments (grain size < 63 µm) are deposited in low energy areas, mixed sediments (grain size 63 - 100 µm) on tidal flats with transitional sediments. In total about 75 % of the tidal flats in the Wadden Sea are sandy, 18 % mixed and 7 % muddy sediments.

The sediment type reflects a great number of abiotic factors and, therefore, the occurrence and density of a number of species is correlated to the type of sediment. The highest biomass and the greatest number of species occurs in areas with an average time of submergence and a average sediment composition. Filter feeders, such as mussels, cockles and *Mya arenaria* only occur in great densities up to more than 1,000 g fresh weight per m², in areas with more than 50% of the time submergence. The sediment feeders, such as *Arenicola, Corophium, Scoloplos, Heteromastus* and other species occur throughout the whole intertidal area. Some have a specific preference for a particular type of sediment. The highest parts of the intertidal area are important for the juveniles of a number of species like *Arenicola, Macoma balthica, Cerastoderma,* probably because in these areas, the predation pressure of fish, shrimp etc. is strongly reduced.

Because of the dynamical character of the Wadden Sea, large fluctuations occur in the biota of the tidal flats. Fluctuations are partly determined by natural processes, for example storms and ice-winters, but human influences, especially mussel and cockle fishery with dredges, can also have serious negative effects.

Mussel beds
Mussels (*Mytilus edulis*) can be found everywhere in the Wadden Sea except for areas close to tidal inlets and in brackish parts of the estuaries of the rivers. Mussel beds are ecologically of great importance for the biomass production in the Wadden Sea. They filter water, recycle nutrients and provide habitat for a rich assemblage of associated organisms. Changes in distribution and coverage of mussel beds can therefore have far reaching effects on the entire ecosystem.

The ecological, as well as economical importance of the Wadden Sea is also documented by the high catches of mussels, cockles and shrimp. Mussel fishing in the Wadden Sea dates back to the last century, whereas mussel cultivation was introduced in 1949.

Eelgrass beds
Eelgrass beds of the two species common eelgrass (*Zostera marina)* and dwarf eelgrass (*Zostera noltii),* which belong to higher plants, provide a breeding area for a number of rare fish species and a feeding area for herbivorous birds. Eelgrass leafs are an important food resource for brent geese (*Branta bernicla*) and widgeon (*Anas penelope*) in autumn. In the past, eelgrass occurred on an extensive scale in the Wadden Sea, common eelgrass covered 6,000 - 15,000 m² in the sublittoral zone of the western Dutch Wadden Sea.
In the 1930s a dramatic decline of common eelgrass occurred throughout the North Atlantic, caused by an epidemic. The subtidal eelgrass beds never recovered, the role of increased turbidity linked to pollution and fishery is discussed as possible causes.

The dwarf eelgrass was not effected by the epidemic. However, since 1965, the surface covered by this species has greatly decreased in the Dutch Wadden Sea. In the 1970s, it covered smaller areas than in previous years, this has also been true for the area near Sylt and in the Jadebusen.

Macroalgae
Green algae like *Enteromorpha* and *Ulva* grow in the entire tidal zone. These macroalgae are either anchored in the sediment, attached to stones, mussels or shells, or they build floating associations. Since the mid 80s, mass developments of macroalgae, in form of algae mats on the flats, have been reported in the Wadden Sea. An excess of nitrogen combined with high temperatures is thought to be the cause. These masses of algae constitute a considerable burden for the ecosystem. During calm weather periods, the algae mats remain in the Wadden Sea for a longer period of time. Underneath the mats, the bottom fauna dies from lack of oxygen.

Salt marshes
The salt marshes of the Wadden Sea are the largest coherent salt marsh areas of Europe and constitute an essential element of the Wadden Sea ecosystem. They are the transition zone between sea and land, and form a habitat for a specialized terrestrial flora and fauna. The salt marshes are the habitat of a limited number of highly specialized plant species most of which are salt-tolerant. Also many invertebrate species, especially insects and spiders, depend on the salt marshes, both for ecological and geographical reasons; more than 250 species, sub-species and eco-types are endemic in the Wadden Sea.

Wetland and coastal birds use mainly salt marshes as breeding area. Breeding birds that settle on the salt marshes, some of which occur in large numbers, use the food resources on the tidal flats, in the water channels and in the salt marshes themselves. For many migrating birds, the salt marsh is an important feeding area and an important resting ground during high tide.

Marine mammals
Four species of marine mammals, the harbor seal (*Phoca vitulina*), the grey seal (*Halichoerus grypus*), the harbor porpoise (*Phocoena phocoena*) and the bottlenose dolphin (*Tursiops truncatus*), can be regarded indigenous Wadden Sea species. Starting in the sixties, there have hardly been any sightings of bottlenose dolphins and harbor porpoises. Since 1989, sightings of harbor porpoises have increased and it is evident, that there has been an increase in the presence of this species in some areas very close to the coastline.

After 1500, the grey seal virtually disappeared in the Wadden Sea as a result of intensive hunting. Small populations of grey seals have occurred again in this area since the mid 1960s. Today, two breeding sites can be located, near the Dutch island of Terschelling with about 90 animals and a second small reproductive colony of about 30 - 40 seals on a sandbar near Sylt and Föhr.

The harbor seal is the most numerous native marine mammal species in the Wadden Sea. Today more than 5,700 individuals of seals are living in the entire Wadden Sea, founded on counts during aerial surveys. According to calculations based on hunting data, the seal population in the whole area must have consisted of some 37,000 animals in 1900. After that, the population size is assumed to have declined continuously. At the beginning of the aerial counts in 1960, it had diminished to about 5,000 individuals. In the period 1960 to 1974, the seal population further decreased to a minimum of 3,600 animals. Since then, probably promoted by the ban of seal hunting in the 60s and 70s, the number of seals stabilized or increased slightly in the different parts of the Wadden Sea, with the exception of the Dutch seal stock. The total Wadden Sea seal population increased in size until the epidemic in 1988, during which the population was reduced by 60%, based on estimations. Now, five years after the epidemic, it can be stated, that the population is recovering very well after the virus disease.

The importance of the Wadden Sea for birds and fish
The various habitats of the Wadden Sea have their specific significance for breeding and migrating birds. The Wadden Sea is a breeding, moulting, feeding and roosting area for several million individuals of more than 50 species throughout the year. This function of the Wadden Sea for birds is commonly known and has led to the establishment of special nature reserves, which are of international importance. These areas have, for example, also been designated as Ramsar sites under that convention. (Enemark, 1993, this volume).

Breeding birds
For more than 30 species of birds, the Wadden Sea is an important reproduction area. For some endangered species, like Little Tern (*Sterna albifrons*) and Kentish Plover (*Charadrius alexandrinus*), the Wadden Sea has special significance. The most important breeding areas are the salt marshes, and, to a lesser extent, the dunes and beach plains of the islands. Typical Wadden Sea birds are Redshank (*Tringa totanus*), Black-tailed Godwit (*Limosa limosa*), Oystercatcher (*Haematopus ostralegus*), Ringed Plover (*Charadrius hiaticula*), Avocet (*Recurvirosta avosetta*) and a number of species of ducks, geese, gulls and terns.

Migrating and staging birds
The Wadden Sea plays a vital role for about 50 bird species originating from a large part of the northern hemisphere. Among these are many rare and threatened species. Every year, several millions of birds use the Wadden Sea area regularly as stepping stone between their arctic breeding and their wintering areas in the south. An average of 6 - 12 million birds pass through this area on their migration route from the breeding grounds in Siberia, Iceland Greenland and northeast Canada to the wintering grounds in Europe and Africa. They feed on the tidal flats, which are the most nutritious areas of the Wadden Sea. These tidal areas account for 60% of all tidal areas in Europe and North Africa, and they are extremely important for the West Palearctic Flyway.

The migratory routes of the birds are relatively fixed, and the staging areas along the flyway are visited around the same time every year. The Wadden Sea is by far the most important spring staging area for arctic waders in the West Palearctic and form the link between other tidal areas elsewhere, like for example The Wash in Europe, and the Banc d'Arguin in West-Africa. The strong connection between these tidal areas, however, means, that the size of the populations, using the Wadden Sea, is determined by factors affecting all breeding, staging and wintering areas along the flyway.

Fish
102 species of fish have been recorded in the Wadden Sea until now. 22 species of these are common, 26 fairly common, 16 scarce, 12 rare and 22 extremely rare. Only a few species are continuously present in the Wadden Sea and most species migrate and therefore are only present in some periods of the year.

Nevertheless, for a number of species, the Wadden Sea is of vital necessity. This area has a decisive nursery function for economically important fish species like herring, plaice and sole. 80 % of the plaice and 50 % of the sole, and in some years a large part of the North Sea herring, grow up in the Wadden Sea. The nursery function is based on the existing high food supply, the protection from predators and the higher temperature in this area.

Changes in the Wadden Sea ecosystem and their causes

The Wadden Sea borders northwest Europe, one of the world's most developed and industrialized areas. As a result of a number of developments, both from within, as well as from outside the area, this region is under strong pressure. A number of human activities, which take place within the Wadden Sea and in the surrounding areas, had and may have an adverse environmental impact on the Wadden Sea system.

Habitat destruction

A number of human activities cause loss, damage, destruction and alteration of biotopes. Some of these activities are land reclamation, embankment, coastal protection measures, port and dam constructions, as well as constructions of infrastructure for tourism, for example, hotels, marinas, artificial beaches etc. In the past, the most serious and permanent destruction of natural habitats in this shallow estuarine area were caused by embankment and reclamation of the coastal marshes. Since 1963, more than 40,000 ha have been embanked. Most of the embanked salt marshes are used for intensive agriculture which makes them worthless for breeding birds.

Pollution

Biotic changes, which means structural devaluation of environmental quality can be stated as effects of a number of human activities. In particular, the Wadden Sea is subject to biotic changes resulting from pollution by an overload of nutrients, heavy metals and organic micro-pollutants, which reach the Wadden Sea via a number of rivers, the mainland, the North Sea, the atmosphere and as a result of shipping activities in the North Sea. Pollution, as one of the most serious threats to the Wadden Sea with the most adverse impacts on its ecosystem, is described in more detail by de Jong (1993, this volume).

Exploitation of natural resources

Exploitation of natural resources in the Wadden Sea, which entails, among others, fishing of mussels, cockles and shrimp, may cause biotic alterations. Two habitats, the oyster beds and the reefs of *Sabellaria spinulosa* , were exploited and destroyed by fishing. Mussel and cockle fishing have considerable impact on benthic communities and sometimes even on the sediment stability.

The operation of oil and gas exploration and exploitation sites in the Wadden Sea and adjacent marine areas constitutes an environmental risk. The construction of platforms and other construction works have an impact on the ecosystem because they interfere with natural processes and devaluate the scenic qualities.

Disturbances

Disturbance can be defined as any activity which by means of acoustical, mechanical or visual action interferes with or influences the normal biological behavior or processes. Disturbance of animals results in an unnecessary loss of energy which can lead to lower breeding success and lower survival rates. Due to recurrent disturbances, specific areas may become unsuitable for breeding, hauling-out or feeding, which means in fact a loss of habitat. Disturbance of wildlife, in particular of birds and seals, comprises all human activities like, for example, shipping movements, air traffic, military activities, hunting, as well as a number of recreational activities like pleasure boating, mud flat walking, flying a kite etc. Most human activities in the Wadden Sea cause visual disturbance, mainly of breeding birds and nursing seals.

In general, it can be stated, that developments caused, in particular, by pollution have increased during the last 10-20 years and have reached international dimensions, whereas the threats concerning loss of biotopes resulting from reclamation, embankments and coastal protection measures have decreased and are more of a local and regional character. However, destruction and alteration of habitats by exploitation of mussel and cockle fishing, as well as the category of disturbance of wildlife have a serious adverse impact on the Wadden Sea ecosystem.

Due to a number of human activities in line with the increasing development and industrialization of the area, the Wadden Sea is under strong pressure.The uniqueness of the Wadden Sea bears rare features and requires extensive conservation and protection measures and management to achieve, as far as possible, a natural and sustainable ecosystem (Enemark, 1993, this volume).

References

Common Wadden Sea Secretariat (CWSS, 1991): The Wadden Sea. Status and developments in an international perspective. 1991. 200 pp.

Dankers, N., K.S. Dijkema, P.J.H. Reijnders, C.J. Smit. 1991. The Wadden Sea in the future - why and how to reach? Research Institute for Nature Management 1991 - 1, 108 pp.

Enemark, J.A. 1993. The protection of the Wadden Sea in an international perspective. Planning, protection and management of the Wadden Sea. *This volume.*

Jong, F. de .1993. Water Quality of the Wadden Sea. *This volume.*

Jong, F. de; J.F. Bakker; K. Dahl; N. Dankers; H. Farke; W. Jäppelt; P.B. Madsen & K. Koßmagk-Stephan. (*eds.*) 1993. Quality Status Report of the North Sea. Subregion 10, the Wadden Sea. In preparation. Common Wadden Sea Secretariat, Wilhelmshaven, Germany

Prokosch, P. 1988. Arktische Watvögel im Wattenmeer. (Dissertation) Corax Band 12, Heft 4, 442 pp.

Prokosch, P.; S. Mielke; D. Fleet 1991. 1991. The Common Future of the Wadden Sea. A report by the World Wide Fund for Nature, WWF. Flensburg 1991. 454 pp.

Wolff, W. J. 1990b. Ecological developments in the Wadden Sea until 1990. *in:* Dankers, N.; Smit, C. J. & Scholl, M. 1992. *Proceedings of the 7th International Wadden Sea Symposium, Ameland 1990.* 1992, pp. 23-32.

Wolff, W.J. (ed). 1983. The Ecology of the Wadden Sea. Volume 1-3, Rotterdam, 108 pp.

Wolff, W.J. 1991. Ecology of the Wadden Sea. *in:* Prokosch, P.; S. Mielke; D. Fleet 1991. (eds.) *The Common Future of the Wadden Sea. A report by the World Wide Fund for Nature,* WWF. Flensburg 1991, 13 - 21.

Development and Implementation of the Coastal Defence Policy for the Netherlands.

R.Hillen[1], Tj. de Haan[1]

Abstract

In 1990 the Government and Parliament of the Netherlands have decided on a new national coastal defence policy. To ensure the enduring safety of the polders and the sustainable preservation of the dune area, it was decided to preserve the 1990 coastline.

The main method to preserve the 1990 coastline (the "basal coastline") is beach nourishment. On a yearly basis 5 to 7 million m^3 of sand is added to the Dutch beaches. The new coastal defence policy appears to be a powerful tool to maintain the coastline in its present position, even with an increased sea level rise. In the coming years innovative coastal defence methods will be applied as well: a shoreface nourishment project will be carried out and for coastal stretches with very severe erosion "seaward" coastal defence options are in study.

Coastal defence in the Netherlands is not only a State affair. Provincial author-ities as well as local waterboards have specific tasks as outlined in the Water Defence Bill.

In the Netherlands, the stress in the coastal areas is increasing rapidly. On the one hand there are more and more initiatives to develop the dune and beach areas by housing, industries and recreational facilities. On the other hand, as a consequence of the Nature Policy Plan, steps are taken to safeguard the scenic and ecological beauty of the Dutch coast.

Introduction

The entire western and northern half of the Netherlands can be considered as a coastal zone. It is among the most densely populated areas in the world, protected from the sea by natural sand-dunes and high dikes. Large rivers -Rhine, Meuse and Scheldt- are flowing through this low lying area into the North Sea. In fact, the Netherlands is part of a large delta system which have been occupied by man ever since about 5000 years ago. At first, people lived on the higher elevated grounds such as beach barriers and levees. Later they colonised the marshes be-hind the dunes. The new inhabitants drained the marshes and started a process of subsidence which is still going on since 20 centuries (De Haan 1991a). For that reason they had to built mounds and (since 1000 years) dikes to protect them-

---

[1] Ministry of Transport, Public Works and Water Management, P.O. Box 20907, 2500 EX Den Haag, the Netherlands

selves against flooding by the sea and the rivers. At present, 8 million people (more than 60% of the Dutch) are living in their polders up to 6 meters below mean sea level. Nevertheless, the country is considered safe from flooding by storm surges.

The Dutch have always been fighting the sea, often winning this struggle, sometimes losing. The last flooding disaster occured in 1953, with more than 1800 death casualties and a damage of appr. 14% of the GNP. After this event the national parliament adopted new safety standards against flooding. These standards are defined in the Water Defence Bill which provides a basic legal framework for all coastal defence measures in the Netherlands. For the coast of central Holland, for example, the sea defences (dunes and dikes) are able to withstand a storm surge level which is exceeded only once in ten thousand years on average (i.e.: the probability of exeeding this level is one tenthousandth per year). For other parts of the coast other safety standards apply, basically depending on the economic value (real estate, infrastructure, etc) of the polderland.

The coastline of the Netherlands is appr. 350 km long; 254 km consist of dunes, 34 km of sea dikes, 38 km of beach flats and 27 km of boulevards, beach walls and the like. The width of the coastal dunes varies between less than 200 metres and more than 6 km.

The dune coast and the beach flats (occurring at the extremes of the Wadden Islands) are dynamic in character. At some locations there is sand accretion, at other locations erosion prevails. Erosion and accretion patterns also vary in time. Since the middle of the 19th century the position of the dune-foot and the high- and low-water lines are measured every year. For this purpose fixed reference poles (beach posts) has been set up on the beach at intervals of 200 to 250 metres. Since the middle of the 1960's the annual coastline measurements are performed through a combination of remote sensing (onland) and sounding techniques (offshore). At every beach post a coastal profile is measured, extending from appr. 200 metres landward of the beach post to appr. 800 metres seaward. The result of this annual coastal monitoring is a unique data-set available for all types of coastal research and evaluation.

A typical example of the application of the monitoring data is the sand balance of the Dutch coastal system between the 8 m depth contour and the top of the first dune row (fig. 1; Stive et al, 1990). From this figure the following general conclusions can be drawn:
1.   In the North there is a structural loss of sand to the Wadden Sea, several stretches of the North Sea beaches are eroding;
2.   On the central part of the coastline (the "closed" Holland coast), sand is being transported from the deeper part of the foreshore to the shallower part resulting in a steepening of the foreshore;
3.   In the Delta area in the Southwest, sand is deposited in front of the closure dams. As a result of shifting gullies close to the coast, many beaches in the southwest of the Netherlands are subject to erosion.

Comparable sand balance studies have also been performed on different time and space scales. Based on that information, shoreline predictions are made, indicating locations where accretion and erosion can be expected in the coming decades.

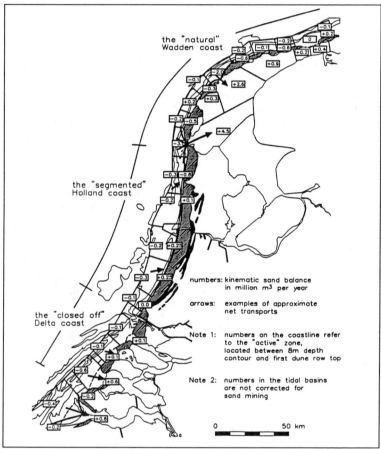

Figure 1:   Large-scale sand balance of the Dutch coastal system (Stive *et al.*, 1990)

Coastal defence policy

Public discussion on a new national policy for coastal defence started in the 1980's (De Haan 1991a). Until 1991 an ad-hoc policy was followed: measures were only taken when the safety of the polderland was at stake or when special values in the dune area (e.g. drinking water supply areas, nature reserves) were threatened. After the 1953 floods, the dikes and dunes along the North Sea were strenghtened to meet the required safety level, thereby ensuring the safety of the polders. However, if no measures are taken against ongoing coastal erosion, tens of kilometres of coast will become unsafe and hundreds of hectares of valuable dune area will be lost every decade. An accelerated rise in sea level will enhance this problem even further.

In 1989, the so-called Discussion Document was presented including four policy alternatives (Min. Transp. & Public Works, 1990):

1. *Retreat*: coastal recession will only be counteracted at those locations where erosion threatens the safety of the polders;
2. *Selective Preservation*: intervention would not only be pertinent to those locations where the safety of the polders is threatened, but also where major interests in the dunes or on the beach may be lost;
3. *Preservation*: the entire coastline would be maintained at its 1990-location;
4. *Expansion Seaward*: at locations of concentrated erosion, artificial defences extending into the sea would be built, bringign coastal recession to a standstill. Elsewhere along the coast, the 1990-coastline would be preserved.

Benefits and costs for all policy alternatives were calculated for the period 1990-2090 (Louisse & Kuik, 1990; Min. Transp. & Public Works, 1990). In 1989 and early 1990 an extensive public discussion was initiated among national, provincial and local authorities, scientists, environmentalists and other people concerned with the dune and beach areas. (De Haan 1991b). Out of the four policy alternatives the preservation alternative was almost unanimously preferred by all parties. In November 1990, the national Parliament decided for the Preservation alternative. This policy choice is primarily aimed at enduring safety against flooding and sustainable preservation of the values and interests in the dunes and on the beaches. To emphasize the wish for the preservation of the natural dynamics and character of the dune coast, the chosen alternative was specified and called "Dynamic Preservation".

The policy choice in 1990 marks a new era in coastal defence policy in the Netherlands. The most important aspect of this choice is that for the first time in history the coastline is to be maintained at a fixed position. Until 1991, large sections of the Dutch coast were eroding, at some locations resulting in a retreat of 5 kilometers in 4 centuries. From 1990 onward, all structural erosion is to be couteracted. For this purpose the concept of the "basal coastline" has been developed.

Other important aspects of the policy choice include a yearly budget for coastline maintenance (DFL 60 million, 1990 price level), administrative measures such as the definition of tasks assigned to different authorities, and the choice for sand nourishment as the main method to combat erosion.

The policy choice for "dynamic preservation" in 1990 was in a way facilitated by the severe storm surges of January and February of that year. During those storm surges, extreme water levels and severe dune erosion occurred and the Dutch once again realised the strenght of the sea. Directly after those events public awareness and support for coastal defence were at its maximum. But to safeguard the yearly budget for coastline maintenance after several relatively calm winters is obviously more difficult.

Public opinion c.q. public support plays a key role in the succesfulness of a new policy. Therefore much attention is paid to inform the public through brochures, video's and the press. Some examples of public discussions on "hot issues" during the last two years are:

- The technique of beach nourishment has been questioned by many. During the last 40 years the Dutch are used to "hard" Deltaworks to be constructed for centuries. What people see now is that money is spent on sand which is for a part out of sight after the first storm surge. It is apparently not fully understood that sand which has been replaced from the beach to the fore-shore, is not lost for coastal defence. In 1991 a brochure and video on sand nourishment have been produced which are distributed and shown at the nourishment sites. Moreover, a project to evaluate nourishment projects is presently carried out. Today we see the first results: ecologists and local ad-ministrators are explaining the benefits of beach nourishment to the press;

- The public does not understand the difference between structural erosion and incidental storm damage. The new coastal defence policy involves the coun-teraction of structural erosion, but it is still impossible to avoid all damage from individual storm surges on the dunes. On this subject a press informa-tion bulletin has recently been prepared;

- In the new coastal defence policy beach flats are permitted to develop more or less without restriction. These beach flats are almost exclusively found on the Wadden Islands, more particularly at their extremes. Stopping all coastal defence measures at these locations implies more dynamics and optimal chances for nature development. For the island of Rottumeroog (the eastern-most Wadden Island) the long-term consequence would be that the island would disappear into a large tidal gully. Although nobody actually lives on the island and in this area and history has proven that new islands develop in the course of time, a spontaneous public discussion arose and showed that a large majority of the Dutch did not agree to stop all defence measures. A pressure-group "Friends of Rottummeroog" was born and a bank donated money for the maintenance of the island. Eventually the Minister agreed to continue a limited maintenance on the island to lengthen its life. Now the "friends" do help Rijkswaterstaat with their own hands to maintain the island.

Preservation of the coastline

Over the last 2½ years several important steps forward have been made to imple-ment the new coastal defence policy: the basal coastline has been calculated for the entire Dutch coastline; for each coastal province a so-called Provincial Con-sultative Body has been formed, every year a number of sand nourishments (totalling 5 to 7 million m^3 sand per year) is carried out.

But other problems arise, such as the increasing pressure on the dunes (e.g. hou-sing, recreation, land reclamation) in spite of the decision of the central government that the majority of the dunes will be nature reserve areas. Moreover, adverse effects of an increased rise in sea level are expected.

The basal coastline concept

The new coastal defence policy implies "preservation of the 1990-coastline" and "counteracting structural erosion". For hard coastal structures, such as dikes, there is no discussion on the position of the coastline. But *where* is the 1990-coastline for a dune coast? And what is *structural* erosion? For these questions the concept of the "basal coastline" has been developed. The "basal coastline" is in fact the coastline-to-be-preserved. Every year the position of the coastline will be compared with the basal coastline to control if the basal coastline has not been crossed.

The position of the basal coastline for a dune coast is measured for each fixed reference point along the Dutch coast. First the so-called transient coastline is determined from the results of the yearly coastal measurements. The transient coastline for a certain location and for a certain point in time is the result of a volumetric integration of the most dynamic part of the coastal profile (fig. 2). The amount of sand on the beach and on the shallow shoreface in fact determine the position of the transient coastline.

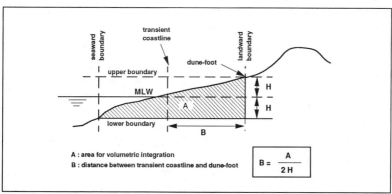

Figure 2:   Method to calculate the transient coastline

To calculate the position of the basal coastline for a certain reference point, the position of the transient coastlines over the period 1980-1989 are plotted against time (fig. 3). The position of the trend-line on the 1st of January 1990 is the position of the basal coastline for that particular reference point.

Thus the basal coastline for the entire coastline of the Netherlands has been calculated. The results of these calculations have been discussed among coastal morphologists and within the Provincial Consultative Bodies for the coast (see section 4). Within a few months (i.e. early 1993) the Minister will officially establish the position of the basal coastline for the entire country. Then the standard for the preservation of the coastline is fixed.

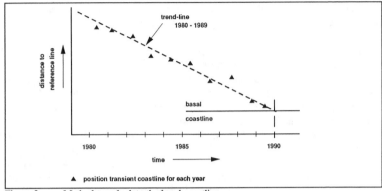

Figure 3:   Method to calculate the basal coastline

Every year, the trend in coastline development is calculated from the transient coastlines of the past 10 years. If this trendline will cross the basal coastline in the next few years, preventive action will be taken (fig. 4). In practice this means that a timely sand nourishment is carried out. Since 1992, yearly calculations are performed and nourishment works are planned according to this method.

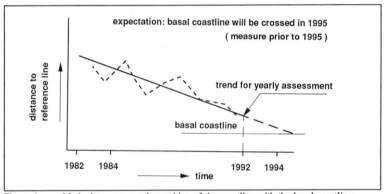

Figure 4:   Method to compare the position of the coastline with the basal coastline

It is interesting to note that the results of individual storm surges do not really affect the position of the basal coastline. The concept is aimed to identify locations with structural erosion in the first place. The effects of dune erosion as a result of storm surges are "filtered out" by using a volumetric approach for the calculation of the transient coastline, by calculating the basal coastline over a period of 10 years and by comparing a 10 years trend line of transient coastlines with the basal coastline. This implies that preservation of the basal coastline does not mean that all dune damage from storm surges will be prevented in the future.

Sand nourishment

Sand nourishment is a common measure to combat coastal erosion in the Netherlands since the end of the 1970's. Over the years sand nourishment has proven to be an effective, flexible and financially sound method (Roelse, 1990). Prior to the policy choice of 1990, sand nourishments were mainly carried out to repair the damaged coastline at selected locations. Since 1990 the nourishments are no more repair works, but they are meant as a buffer: preventing crossing of the "basal coastline". The sand is placed on the beach, thus creating an transient coastline in a more seaward position. The nourished sand forms a buffer against the ongoing erosion and will be placed on the eroding beach *before* the basal coastline is exceeded.

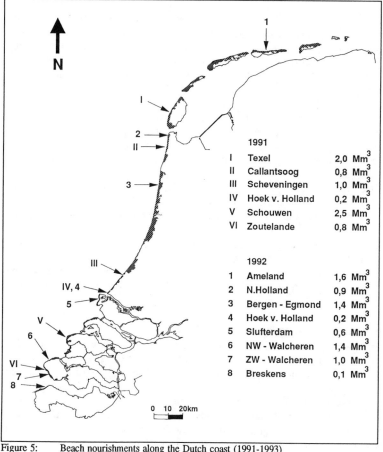

	1991	
I	Texel	2,0 Mm^3
II	Callantsoog	0,8 Mm^3
III	Scheveningen	1,0 Mm^3
IV	Hoek v. Holland	0,2 Mm^3
V	Schouwen	2,5 Mm^3
VI	Zoutelande	0,8 Mm^3

	1992	
1	Ameland	1,6 Mm^3
2	N.Holland	0,9 Mm^3
3	Bergen - Egmond	1,4 Mm^3
4	Hoek v. Holland	0,2 Mm^3
5	Slufterdam	0,6 Mm^3
6	NW - Walcheren	1,4 Mm^3
7	ZW - Walcheren	1,0 Mm^3
8	Breskens	0,1 Mm^3

Figure 5:   Beach nourishments along the Dutch coast (1991-1993)

Every year some 5 to 7 million m^3 of sand is added to the Dutch beaches (fig. 5). For this purpose a yearly budget of appr. 60 million Dutch guiders (35-40 M US$; January, 1993) is available. In fact these costs can be considered the maintenance costs for the coastline. Just for comparison: the average costs for the maintenance of one km of sandy coastline is less than the average maintenance costs of a km motorway.

Preservation of the dynamics of the Dutch coast

The dune coast of the Netherlands is of great scenic beauty and represents international biotic and abiotic values. Nature conservation organisations and ecologists fully support the policy choice of preservation of the coastline and the choice for "soft" coastal defence methods. At their request also the term "dynamic" was added to ensure the dynamic character of the Dutch coastline. Several nature conservation organisations now plead for a less strict policy with regard to the maintenance of the foredunes. Suggestions for the formation of so-called slufters (wet dune valley influenced by the tides) and dune areas with more aeolian dynamics (e.g. sand drifts, blow-outs, mobile dunes) have been presented recently (Stichting Duinbehoud, 1992).

From the viewpoint of coastal defence, there are possibilities for natural development of coastal areas, but not everywhere and unconditioned.
On the beach plains at the extremes of the Wadden islands no active coastal defences measures are carried out. Basal coastlines will not be established for these areas.
On the other hand, several dune areas (especially in the Delta area) are too narrow to allow nature development experiments.
For the remaining dune areas a less strict stabilization policy could be considered as long as the safety of the polderland is not endangered. This might imply a different management of the dune area. Presently, certain zones of the dunes are set aside to realize the coastal defence requirements, other zones are nature conservation areas or drinking water supply areas. In the future a more integrated management of the dune areas could be considered (Van der Meulen & Van der Maarel, 1989).

Institutional and legal framework

One of the lessons from of the Dutch history is that landowners could only get good protection against the water when they cooperated. They joined forces and founded waterboards. The costs of the tasks of the waterboards -except some grants of the central government- are met by the landowners inside the territory of the waterboard. They pay their waterboard taxes proportional to the area they own. Nowadays buildings are taxed too and there are so much buildings, that the owners of buildings pay together more than the landowners in some waterboards, but all proportional to the value of their properties.
The law provides the power for the waterboards to execute their tasks. This way the waterboards are governing bodies operating on the same level as the municipalities but with a specialised task.

Soon there was a need to supervise the waterboards in the tasks they execute, for instance because a lot of the waterboards were small bodies. In the 17th and 18th centuries the landlord was the supervisor. Later the provinces -then the almost independent federal states of the Republic of the United Netherlands- took over the supervision. Around 1800 the centralistic French occupiers gave the central government also a task in supervision. This lead to the following system:
* Waterboards manage dikes and dunes. To manage means:
    - to maintain in the technical sense and to strengthen the dikes and dunes if necessary;
    - to protect the dikes and dunes from damage by acts of men. Nobody is allowed to do anything in, on or near a dike or dune without a licence from the waterboard;
* The provincial governments supervise the waterboards in the execution of their technical duties, but also in their administrative and financial powers. The provincial governments have the power to give the waterboards instructions;
* The central government -in particular the Minister of Transport, Public Works and Water Management- supervises the provincial governments in the way they supervise the waterboards. In particular this is important when the territory of a waterboard is within the territories of two provinces. The minister has the power to give the provincial government instructions. If the province and the waterboard do not execute these instructions, the minister can execute measures at the expenses of the waterboard.
This system also applies for the water defences along the North Sea coast.

From 1991 the central government protects the coast against structural erosion by maintaining the basal coastline. The waterboards still have the task to maintain the strength of the sea defences (dikes and dunes) to meet the safety standard. They have to repair the damage of stormsurges on the dunes and to prevent narrow dunes from damage by aeolian transport of sand.
These tasks will be founded in the Water Defence Bill. This Bills orders the Minister of Transport, Public Works and Water Management to underttake the works to counteract structrural erosion. Before the works are carried out, the minister has to consult the Provincial Consultative Bodies. In these bodies at least the provincial government, the waterboards and rijkswaterstaat (Dept. of Public Works and Water Management) are participating. In some provinces, also municipalities and/or nature conservation organisations are represented in the Provincial Consultative Bodies (fig. 6). However, the Water Defence Bill does not specify any specific task for these authorities and organisations. The provincial governments (chairing the bodies) have the freedom to invite municipalities and organisations.
Although the Water Defence Bill is still in discussion in Parliament now, the authorities cooperate as it is ordered in the Bill because there is consensus among all parties and the Parliament.

The presence of the members of the bodies reflect the wish of all parties to approach coastal defence matters in an integrated way. In practice, the Provincial Consultative Bodies deal with all matters relevant to the preservation of the coast, including the struggle against erosion. Concepts on integrated coastal zone management could well be realized through the Provincial Consultative Bodies.

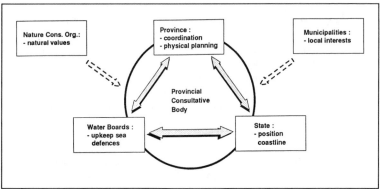

Figure 6:   Collaboration within the provincial consultative bodies

Looking into the future

The new coastal defence policy of "dynamic preservation" is in fact only at its infancy. A thorough evaluation cannot be given at this stage, but developments since 1990 are encouraging. A set of instruments has been developed (institutional framework, yearly budget, etc) and new developments are investigated (both fundamental coastal research and innovative coastal defence measures). At the request of the national Parliament a thorough evaluation of the policy will be presented in 1995.

Coastal research and monitoring

Out of the yearly budget for coastline preservation, about 3 million Dutch guiders are spent on coastal research and monitoring. The coastal monitoring system, as briefly described in section 1, has proven to be necessary for both research and evaluation purposes. Without these data no accretion/erosion patterns can be made and no yearly calculation of the transient coastline can be carried out.

The coastal research in the Netherlands is mainly concentrated in the Coastal Genesis project, a multidisciplinary project aimed to understand coastal processes and to predict future coastal developments. The information and knowledge gained through this project has been of great value to evaluate the various policy alternatives.

Together with the Discussion Document, 20 Technical Reports were published in 1989. These reports can be considered as the state-of-the-art in the field of coastal defence of that moment. Based on these technical reports also the "white spots" in knowledge could be identified: for example, the response of the coastal system to a rise in sea level, the limited knowledge of cross shore transport processes, the "sand wave" features along the coastline that are not yet completely understood, and the processes governing transport of water and sediment in tidal inlets. Present Coastal Genesis studies are primarily directed towards these "white spots".

Innovative coastal defence techniques

With the choice for "dynamic preservation", the government and Parliament have chosen sand nourishment as the main approach towards protection of the dune coast. This does not mean, however, that other forms of coastal protection are excluded. Sand nourishment may not offer the best or cheapest solution at every point along the Dutch coast. At some locations, the added sand is rapidly carried away via deep tidal gullies; this occurs for example, near large tidal inlets in between the barrier islands in the north of the country. To carry out sand nourishments more frequently and in larger volumes could become a very expensive solution here.

In such situations, alternative, more suitable measures are also considered, such as the construction of dams perpendicular to the coast, or the protection of the shoreface with stones (both examples of "hard" coastal protection measures), possibly in combination with sand nourishment. For two locations along the Dutch coast feasibility studies are worked out for alternative defence measures (De Ruig, 1992; Rakhorst, 1992).

Innovative sand nourishment techniques are also considered. In 1993 a shoreface nourishment project will be carried out on the island of Terschelling at a water-depth of 5 to 7 meters. A desk study has indicated that foreshore nourishment, under certain conditions, will be less expensive than beach nourishment (Hillen et al, 1991). Moreover, during the execution of a foreshore nourishment project, recreational activities on the beach are not interrupted. For the extensive modelling and monitoring aspects of such an innovative nourishment project, co-operation with Danish and German coastal research institutes has been established and financial support from the MAST-programme of the European Community, through its programme on Marine Science and Technology MAST, will be given.

Characteristic of all studies into alternative coastal defence methods is that it aims to preserve the coastline with a minimum of means and effort. Management of the coastline means nothing more nor less than regular maintenance of the coastline. This is done with the most efficient and effective means available, taking into account the other interests in the coastal zone (for example: nature and recreation) as much as possible.

Increased rise in sea level

An increase in sea level rise is widely considered one of the most serious threats for low-lying countries such as the Netherlands. Studies show that if the most likely IPCC sea level rise scenario of appr. 60 cm in the next 100 years is adopted, the total costs for the Netherlands would amount about 12,000 million Dutch guilders (appr. 7,500 million US$, price level Jan. '93). For the preservation of the coastline, the additional costs would be 10 million Dutch guilders per year, i.e. an increase of the present nourishment budget by 15% (Rijkswaterstaat, 1991). Both in terms of finances and know-how the Netherlands will be able to cope with an increased sea level. As compared to many of the low-lying developing countries and several small island states, the Netherlands are in fact not very vulnerable to an increase in sea level rise (IPCC/CZMS, 1992).

Developments in the coastal zone

Through the preservation of the coastline on its position of 1990 at the minimum, the State Government in fact realizes a basic condition for other functional uses in the dunes and on the beach. The new policy also offers opportunities for new developments in the coastal zone, e.g. opportunities for the restoration and development of nature, for the application of new coastal defence techniques, and for a more integrated management approach in the coastal zone. Such opportunities can probably best be realized through initiatives on a regional or local level.

Developments that are considered harmfull for nature development in the dune areas can possibly be prevented through legislative action. As a follow-up of the national Nature Policy Plan (Min. Agriculture, 1990), steps are taken to bring the entire coastal dune area of the Netherlands under the Nature Conservancy Act prior to 1998. Thus the ecological and scenic beauty of the Dutch coast can, for a large part, be safeguarded.

However, in the meantime the stress on the coastal zone is increasing rapidly. Especially the pressure for additional housing and recreation facilities in the dunes is growing step by step. Over the past few years also plans for land reclamation along some coastal stretches have been developed. Such plans could best be considered in an integrated way by the authorities concerned with physical planning on a regional level. The discussion on consequences of such plans for coastal protection could very well be channeled via the Provincial Consultative Bodies for the coast.

References

De Haan, Tj. (1991a) A new coastal defence policy for the Netherlands. *MAFF-Conference for River and Coastal Engineers*, Loughborough July 1991.

De Haan, Tj. (1991b) Public support, keep it awake. *IABSE-colloquium*, Nyborg May 1991.

De Ruig, J.M.H. & P. Roelse (1992) A feasibility study of a perched beach concept in the Netherlands. *Proceedings ICCE Conference 1992*, Venice.

Hillen, R., Van Vessem, P. & Van Der Gouwe J. (1991) Suppletie op de Onder wateroever (Foreshore Nourishment). *Rijkswaterstaat/Tidal Waters Division report* (in Dutch).

IPCC/Coastal Zone Management Subgroup (1992) Global Climate Change and the Rising Challenge of the Sea. *Supporting document for the IPCC-update report 1992.*

Louisse, C.J. & Kuik, A.J. (1990) Coastal Defence Alternatives in the Netherlands. *Paper # 1 in: The Dutch Coast, report of a session on the 22nd International Conference on Coastal Engineering 1990.*

Ministry of Agriculture, Nature Management & Fisheries (1990) Nature Policy Plan of the Netherlands. The Hague.

Ministry of Transport & Public Works (1990) A new Coastal Defence Policy for the Netherlands. The Hague.

Rakhorst, H.D. (1992) Onderzoek naar kusterosie op Texel (Coastal erosion on Texel island). *Land + Water*, Aug./Sept. '92 (in Dutch).

Rijkswaterstaat (1991) Rising Waters, Impacts of the Greenhouse Effect for the Netherlands. *Rijkswaterstaat, Tidal Waters Division.*

Roelse, P. (1990) Beach and Dune Nourishment in the Netherlands. *Paper # 11 in: The Dutch Coast, report of a session on the 22nd International Conference on Coastal Engineering 1990.*

Stichting Duinbehoud (1992) Duinen voor de Wind, een Toekomstvisie op het Gebruik en Beheer van de Nederlandse Duinen (Dunes for the Wind). *St. Duinbehoud*, Leiden (in Dutch).

Stive, M.J.F., Roelvink J.A. & De Vriend H.J. (1990) Large-scale Coastal Evolution Concept. *Paper # 9 in: The Dutch Coast, report of a session on the 22nd International Conference on Coastal Engineering 1990.*

Van der Meulen, F. & Van der Maarel E. (1989) Coastal Defence Alternatives and Nature Development Perspectives. *In: Perspectives in coastal dune management*; pp. 183-195.

The Protection of the Wadden Sea in an International Perspective

Planning, protection and management of the Wadden Sea

J.A. Enemark[1]

Abstract

This paper describes the state of the protection, conservation and management of the Wadden Sea by the three countries, Denmark, Germany and the Netherlands in the framework of their joint cooperation to protect the Wadden Sea as a coherent nature area. It will encompass an analysis of the international cooperation on the protection of the Wadden Sea based upon the "wise use" principle as an example of international cooperation in the field of environmental and nature protection. It will further include an overall analysis and comparison of the coastal planning system with respect to the protection of the Wadden sea, the application of nature protection instruments and the management of the area.

Introduction

Nature conservation in the Dutch-German-Danish Wadden Sea has a long tradition. In the beginning of this century, smaller uninhabited islands were protected for the purpose of, in particular, breeding bird colonies of coastal birds. Later, this was extended to salt marshes and, on a limited scale, to the sea-territory of the Wadden Sea itself (Wolff, 1990, Rudfeld, 1990a). Some 25 years ago, it became evident, in conjunction with an increasing awareness of the areas' outstanding national and international importance, that the traditional terrestrial and species conservation was inadequate to preserve the Wadden Sea ecosystem as such. Large scale embankments, made possible by advances in technical possibilities, rapid increase in tourism in the area, harbor and industrial developments and pollution from adjacent areas endangered, or, in some cases, turned over the more or less existing balance of traditional use of the area and the conservation of the system (Wolff, 1976; Reineking, this volume).

About 25 years ago, the initiatives to protect the Wadden Sea as an ecological entity commenced by scientists, nature conservation interest groups and policy makers, e.g. by establishing a Wadden Sea Working Group of several Dutch scientists in collaboration with nature conservation organizations in 1965. Since 1970 Danish and German scientists joined the common efforts which resulted in the 80s in both the establishment of a national based Wadden Sea conservation policy, encompassing

---

[1] Secretary, Common Wadden Sea Secretariat, Virchowstrasse 1, D2949 Wilhelmshaven, Germany

extensive nature protection schemes, and the establishment of a trilateral Dutch-German-Danish cooperation on the protection of the Wadden Sea.

The protection of the Wadden Sea requires the application of a wide scale of legal instruments and an integrated management to ensure its conservation and wise use. This chapter will focus in particular on two aspects of the protection of the Wadden Sea: - the protection on the national level of the three states, so as to examine differences in approaches identifying issues of consideration with respect to the cooperation on the international level, and
- the cooperation between the three Wadden Sea states with the goal of examining the approach and systems with regard to the protection of a shared coastal system.

Whilst both levels are analyzed separately, it should be stressed that the national and the trilateral level are closely interrelated, being an example of coastal management of a marine ecosystem and its conservation and wise use, based on many years of experience in attempting to resolve issues that such coastal areas are confronted with.

Trilateral cooperation: establishing political cooperation
In the 70s, it was recognized by the governments of the three adjacent states that the Wadden Sea as such needed to be protected, and in order to protect the Wadden Sea as an ecological entity, arrangements in the field of nature protection policy and management between the three states were necessary. A problem related to such common arrangements between the states is the fact that there are differences between the countries in terms of legal and administrative systems (Zwiep, 1990).

In order to overcome such differences, a draft Convention on the Conservation of the Wadden Sea Region was prepared by the International Union for Conservation of Nature and Natural Resources (IUCN) in 1974 and submitted to the governments of the three Wadden Sea states (Wolff, 1975). The proposed convention was a framework for the protection of the Wadden Sea as a whole, establishing an arrangement for intergovernmental cooperation. It was, however, not accepted by the concerned states in essence, it can be assumed, at the time, because of the lack of legal protection measures on the national level in conjunction with the binding character of the proposed convention according to international law. Judged by modern standards of environmental protection, the convention was by no means a "chocking" proposal, but it introduced some new elements in conservation like the landward buffer-zone concept, and the introduction of a joint commission for the management of the area. Developing a model for the protection of the Wadden Sea, based on extensive research and examination of the legal and administrative possibilities, was necessary (Mörzer Bruyns & Wolff, 1983).

Therefore, both the national and the trilateral line of protection and cooperation continued. The Wadden Sea was protected according to a series of national initiatives during the 80s starting with the Wadden Sea Memorandum in The Netherlands in 1980 and later the establishment of a Nature Reserve in the Danish part in 1982, and National Parks in the German part from 1985 on. The whole Wadden Sea from

Esbjerg in Denmark in the north to Den Helder in the Netherlands in the west is now under protection.

Parallel hereto talks between the three governments were initiated resulting in the first Trilateral Governmental Conference on the Protection of the Wadden Sea in 1978. At the Third Governmental Conference in Copenhagen in 1982, the three governments agreed to a Joint Declaration on the Protection of the Wadden Sea. According to the Joint Declaration, the governments declared their intention to consult each other in order to coordinate their activities and measures to implement a number of international legal instruments with regard to the comprehensive protection of the Wadden Sea region as a whole. The international legal instruments mentioned are the Ramsar Convention on Wetlands, the Bonn Convention on conservation of migratory species, the Bern Convention of conservation of European wildlife and natural habitats, and the relevant EC Directives in particular the EC-Bird Directive (Zwiep 1990).

The Joint Declaration resolved a dilemma. It is in essence a declaration of intent, stating a political commitment to work towards a common goal, but it includes a number of legally binding international instruments. It can be contested that it is the intention of the parties that counts rather than the legal character of the instrument. (van den Mensbrugghe 1990).

The Joint Declaration served as a catalyst. The cooperation between the Wadden Sea countries was extended, and, in order to further structure collaboration, a common secretariat was established. The Wadden Sea was designated as a Wetland of International Importance (Ramsar Site) and larger parts as Special Protection Areas (SPA) according to the EC-Bird Directive. In addition, the German and the Dutch part have been declared a Man and Biosphere area (MAB) of the UNESCO (table 1).

	Ramsar	SPA	MAB
Denmark	1987	1983	-
Schleswig-Holstein	1991	-	1990
Hamburg	1991	1983[1]	1992
Niedersachsen	1978	1982	1992
Netherlands	1984	1991	1987

Table 1: Status of International Designations. Source: CWSS, 1991c (modified).
[1] Only the islands Neuwerk and Scharhörn

National Wadden Sea policy: conservation and integration

The protection measures implemented in the Wadden Sea area in the 80s, according to national legislation as indicated above, share the common objective that they aim at conserving and protecting the Wadden Sea as a nature area of international importance (fig. 1). This objective has been pursued according to basically two "models" of protection, namely the Dutch model of planning and conservation and the German model of establishing national parks in the Wadden Sea. The Danish model takes a somewhat intermediate position between the two. The measures entailed in both "models" aim at resolving basically common problems. It is therefore

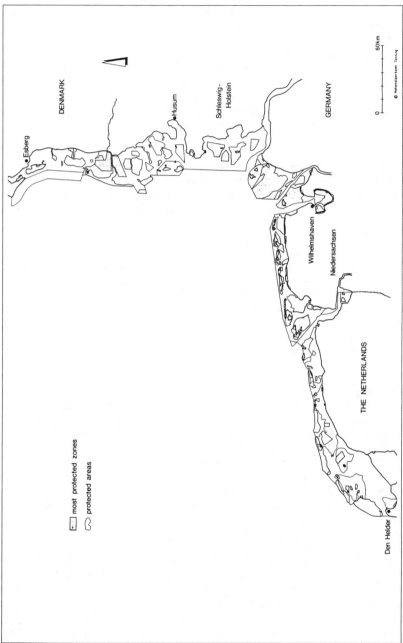

Figure 1: The Wadden Sea and the protection regimes

worthwhile considering these models in order to examine their resolving capabilities for some of the most important common issues.

National parks as an instrument of nature protection

The German coastal states of Schleswig-Holstein, Lower Saxony and Hamburg have designated their parts of the Wadden Sea as national parks in 1985, 1986 and 1990 respectively. The federal states are responsible for nature conservation in the framework of the Federal Nature Conservation Act, in which provisions for nature protected areas and national parks are laid down.

Whilst there are some differences between the national parks, e.g. in terms of delimitation of the area, regulation of activities and utilization, they share some common basic features of which the most important are:

a.   the national parks are divided into zones (two to three) in which different activities and exploitation are allowed; the ecologically most important areas are encompassed in zone 1, the core zones, extending in the different parks from 30 to 50% of the territory, where admittance is prohibited and planned to declare them as non-use zones;

b.   national park authorities have been established in the three parks which are responsible for the implementation of the provisions of the National Park Orders and Acts, in order to ensure a unified protection and management regime within the boundaries of the Wadden Sea national parks; the jurisdiction of regional and local authorities in the framework of nature conservation has been limited or ended in the national parks (CWSS, 1991; Peet & Gubbay, 1991, Burbridge, 1991).

Since the establishment of the national parks progress has been made with respect to improving the management of the national parks, e.g. salt marshes and the resolving of some economic activities by implementing the zoning system e.g. for mussel and cockle fishery, which have a negative impact on the Wadden Sea. The approach ensures, to a large extent, that nature conservation interests in the national parks are pursued, and a unified management according to the guidelines of the national park authorities will be implemented (Andresen, 1992).

There are however, a number of developments and activities, which have an impact on the Wadden Sea, that cannot, or can only indirectly, be solved by the establishment of national parks. This concerns the complex ecological, management and economic situation in the coastal area. Firstly, many of the developments which have a significant negative impact on the Wadden Sea ecosystem originate from outside the boundaries of the parks. This is in particular true for pollution, shipping in the North Sea and recreation, only to mention some of the most important ones. Such developments can only be controlled by taking measures in adjacent areas of the Wadden Sea national parks.

Secondly, the national parks cannot operate effectively in coastal areas unless they form a part of a broader regional planning system. The national parks must be part of the physical planning schemes of the national, regional and local levels. This also applies to developments within the parks. Whereas the national park authorities can pursue a wide range of conservation interests, as regulated by the national park orders and acts, other federal and state authorities have basically maintained their responsibilities. This requires a well developed sectoral planning system and close cooperation between the sectoral agencies to promote the objectives of the national park (Burbridge, 1990; Andresen, 1992).

In summary, the policies and management of the national parks should be seen in the framework of a broader coastal system, and an integration of the planning and the activities of sector agencies.

Dutch - Danish Wadden Sea policy: integrated planning

The outset of the Dutch Wadden Sea policy differs in some essential features from the German approach. The key governmental decision on the protection of the Wadden Sea, the Wadden Sea Memorandum, adopted in 1980 and amended in 1993, is a national physical planning document for the Dutch Wadden Sea as a basis for all further planning, conservation and management for the area of all state, regional and local authorities. The Memorandum states the overall objective for the Wadden Sea policy, the policies with respect to activities and utilization of the area and general arrangements with respect to coordination of policy and management.

The Memorandum is basically implemented along two lines. Firstly, as opposed to the German policy where the regional and local jurisdiction of the area has been curtailed, the Dutch Wadden Sea has been deliberately brought under the jurisdiction of the adjacent provinces and municipalities with the aim of ensuring an integrated physical planning of the area down to the lowest level and with public participation in the planning phase. This has resulted in the adoption of regional planning schemes and local development plans, which are binding for individuals, for the Wadden Sea.

The second line of implementation is the designation of the major part of the Dutch Wadden Sea as a State Nature Monument under the Nature Conservation Act. The Nature Monument determines that, without permission, it is prohibited to undertake activities which destroy and damage the protected area including its flora and fauna. An overall management strategy and a system of management plans have been adopted to ensure the implementation of policies in management and the necessary coordination between sector interests (CWSS, 1991).

The Danish conservation and planning scheme for the Wadden Sea is a mixture of the German and Dutch approach. The Danish part of the Wadden Sea was designated as a Wildlife and Nature Reserve in 1979/1982, encompassing a zoning system comparable in certain aspects with the German system. In addition, the Wadden Sea, major parts of the Wadden Sea islands and the adjacent marsh land have been designated as Ramsar Site and a Special Protection Area according to the EC Bird Directive, which appoints the area with a priority position for regional and sectoral planning and which furthermore has special implications according to national legislation and administrative regulations. In general, according to Danish law, this ensures the Wadden Sea the highest protection status (Koester, 1989).

From the outset, the Dutch Wadden Sea policy aims at an integrated approach to the protection and management of the Wadden Sea as a part of a larger coastal area. The Memorandum states that a number of activities outside of the area of the Wadden Sea shall be taken into account according to the degree in which such activities have a negative impact. An issue related to the integrated planning and management system is the overall coordination of sectoral policies. In contrast to the German approach, in which e.g. the responsibilities of the regional authorities regarding nature conservation in the area has been curtailed or ended, there are basically several authorities responsible which may not always ensure unified management   and, therefore, complex coordination mechanisms are necessary.

A further essential issue is that the Dutch approach is related to a multi-purpose-concept use of the Wadden Sea (CWSS, 1992b). In the framework of the sustainable protection of the Wadden Sea, human activities are basically possible according to the Memorandum. It may however not always be possible to pursue conservation objectives and ensure a wise use of the area without restricting certain activities in time and/or space. A zoning of a number of activities, in particular of recreation and fisheries, has therefore been introduced progressively as a management instrument and is currently part of the amended Memorandum.

Trilateral Wadden Sea policy: sustainable utilization of a shared coastal system

As can be noted on the basis of the examination of the national protection measure, there is a convergence in terms of principles and objectives. Put in simplified terms: the German approach works from within to the outside whereas the Dutch-Danish approach works from the outside to within. It was noted at the 6th Trilateral Governmental Conference on the Protection of the Wadden Sea in 1991, at which the three governments took stock of the cooperation between the three countries almost 10 years after the adoption of the Joint Declaration, that there were indeed a number of major similarities in protection measures in terms of objectives of protection and in terms of management based on a zoning system.

It was noted though that there also existed a number of differences between the protection regimes of which the major ones are:
-       differences with respect to implementing the concept of the development of natural processes including the weighing of interests;
-       differences with respect to the delimitation of the Wadden Sea, both landward and seaward;
-       differences with respect to the regulation of human activities and management of the area (CWSS, 1992a; Ministerial Declaration).

Beyond these, there are also the existing differences in the legal and administrative systems, as indicated above, e.g. with respect to responsibilities in the field of nature conservation. The German federal states are responsible for nature conservation according to the Constitution within the federal framework act, whereas the federal government is responsible for foreign relations. In Denmark and the Netherlands, both are the responsibility of the central government. Such differences explain, to a certain extent the differences in protection regimes.

Further, it was concluded that the assessment of the current state of the Wadden Sea leads to the conclusion that the quality of the ecosystem needs to be significantly improved in order to restore and maintain its natural potentials. The sustainable utilization, in a way compatible with the maintenance of the natural properties of the ecosystem, is the wise use of wetlands as defined by the Ramsar Convention, which is one of the Conventions of the Joint Declaration.

It was therefore decided by the three governments to define the wise use principle for the Wadden Sea as a shared wetland system for its conservation and sustainable utilization by adopting a common guiding principle, management principles and common objectives for the human utilization of the area. These principles and objectives shall further assist in bridging the differences in legal and administrative systems, so that it is in principle aimed at solving common problems on a common basis and thereby increasing the mutual effectiveness of the measures.

Guiding principle and management principles
The guiding principle of the trilateral Wadden Sea policy is to achieve, as far as possible, a natural and sustainable ecosystem in which natural processes proceed in an undisturbed way.

This principle aims at:
- maintaining the water movements and the attendant geomorphological and pedological processes;
- improving the quality of water, sediment and air to levels that are not harmful to the ecosystem;
- safeguarding and optimizing the conditions for flora and fauna including
  * preservation of the Wadden Sea as a nursery area for North Sea fish;
  * conservation of the feeding, breeding, moulting and roosting areas of birds, and the birth and resting areas for seals as well as preventing disturbance in these areas:
  * conservation of the salt marshes and dunes;
- maintaining the scenic qualities of the landscape, in particular the variety of landscape types and the specific features of the wide, open scenery including the perception of nature and landscape (CWSS, 1992a; Ministerial Declaration)

These elements are the basic conditions for the Wadden Sea ecosystem aiming at maintaining the ecological integrity of the system as a whole (de Jong, 1992).

In order to further narrow the differences in legal and administrative systems, seven common management principles have been adopted for the Wadden Sea (CWSS, 1992a; Ministerial Declaration, CWSS, 1992b). An important common management principle is the precautionary principle, namely to take action to avoid activities which are assumed to have significant damaging impact on the environment even where there is no sufficient scientific evidence to prove a causal link between activities and their impact. The definition has, in the framework of Wadden Sea cooperation, been extended from including substances to also include activities.

A further management principle is the principle of careful decision making on the basis of the best available information. This means, in a number of cases environmental impact assessment studies are requested. Regarding the Dutch part of the Wadden Sea, a special Executive Order on Environmental Impact Assessment Procedure is being issued with more activities and lower thresholds for individual activities than for the Netherlands in general. It has been further agreed between the Wadden Sea countries to aim at harmonizing their environmental impact assessment procedures for the Wadden Sea.

Other principles concern the principle of avoidance of activities which are potentially damaging to the Wadden Sea, the principle of translocation of harmful activities and the principles of best available technology and best environmental practice.

Common objectives

The Wadden Sea area is an area where people live, work and recreate. Activities in the Wadden Sea and utilization of its resources are possible within the framework of the principle of conservation and sustainable utilization. The three governments have agreed to a set of common objectives which defines the principle of conservation and wise use with respect to human activities and utilization of the Wadden Sea as a whole.

The common objectives cover all the main common activities and utilization in the Wadden Sea and in adjacent areas for those activities that may have an adverse impact on the Wadden Sea. It is essential to emphasize that the Wadden Sea area is dealt with as a coherent system with a level of regulation accordingly. It would be too extensive to go into detail with all the objectives and therefore only examples of agreements that concern the Wadden Sea itself, adjacent areas of the Wadden Sea and a larger area are stated.

Concerning the Wadden Sea itself, including the salt marshes outside of the seawall and the coasts, objectives have been set for a large number of activities on different levels of specification, e.g.:
-    it is in principle agreed to prohibit embankment and to minimize unavoidable loss of biotopes by sea defence measures;
-    the negative impact of mussel fishery and cockle fishery, in areas where this is allowed, shall be limited by closing considerable parts of the Wadden Sea permanently including intertidal and subtidal areas;
-    hunting of migratory species will be progressively phased out.
It appears from the agreements that it is left up to the responsible national authorities to decide on the relevant time-schedule and the spatial scale, which may depend on the conditions in terms of e.g. morphology and differences in intensities of utilization in the different subregions of the Wadden Sea.

Further hereto, there are a number of activities in adjacent areas which have an adverse impact on the Wadden Sea. These activities concern shipping in the North Sea, harbor and industrial facilities immediately adjacent to the Wadden Sea, the use of wind energy in a zone adjacent to the Wadden Sea and civil air traffic. The level of agreed objectives is in particular directed at maintaining the present level of activities so as to minimize the impact on the Wadden Sea itself from outside.

A further level of common activities is the reduction of the input of surplus nutrients and pollutants. The improvement of the chemical situation is a basic condition for the sustainable utilization of the Wadden Sea (de Jong, this volume). Effective measures can only be taken in the catchment area of the North Sea in conjunction with other adjacent states at, in particular, the North Sea Conferences but also in the framework of the European Community.

Concerning the wide international importance of the Wadden Sea for migratory birds, the Wadden Sea states support the initiatives towards establishing a Western Palearctic Waterfowl Agreement in the framework of the Bonn Convention, which will set up a mechanism for the protection and management of migratory birds along the flyways from the arctic breeding areas through the European staging areas, of which the Wadden Sea is a key site, to the wintering areas in Africa (CWSS, 1992a; Ministerial Declaration).

Implementation of the principles and objectives: Wadden Sea coastal management

These principles and objectives are now in the process of being implemented on the national level through the authorities which are responsible for the protection and management of the Wadden Sea, and by making use of the mechanisms existing for that purpose. The Wadden Sea states have further agreed, at the last Governmental Conference, to undertake the necessary steps to establish a coherent special conservation area covered by a coordinated management plan. Also, this agreement is in the process of being implemented in the framework of cooperation. The aim of protecting and managing the Wadden Sea area as an ecological entity is therefore an ongoing process.

The formulation of a policy of sustainable utilization for the Wadden Sea as a whole is, in summary, directed towards maintaining and, where necessary, restoring its natural potentials as part of a larger coastal system, and complementary bridging differences in legal and administrative systems between the countries and states. The wise use objectives must therefore be formulated in relation to the area and the level and intensities of the developments and utilization. The basic conditions for protecting, maintaining and, where necessary, restoring the integrity of the Wadden Sea as an ecological entity is the conservation and wise use of the system. Such objectives have also been formulated by non-governmental interest organizations previous to the Wadden Sea governmental Conference in 1991 (WWF, 1991).

The Wadden Sea states have been in the process of managing the Wadden Sea as a unique world wide important coastal system for about 25 years. Three aspects of protection and management should be emphasized in particular:
- the ecological integrity of the Wadden Sea can only be maintained and restored by the conservation and the wise use of the area;
- the Wadden Sea, including the adjacent land and sea territory must be protected and managed in an integrated way in order to maintain and restore its integrity by applying integrated planning, by applying conservation measures and by integrated management; the principle of sustainable utilization should therefore become an integrated part of all relevant sector activities;
- the Wadden Sea is a shared system which must be protected and managed on the basis of common principles and objectives cutting across differences in legal and administrative systems; these common principles and objectives of sustainable utilization can be applied to (shared) coastal systems for an integrated coastal zone management.

References

Agger, P. (1992): Aims and goals for the future management of the Danish Wadden Sea. *In* Dankers, N., Smit, C. J. & Scholl, M. (1992): *Proceedings of the 7th International Wadden Sea Symposium,* Ameland 1990, pp. 69-72.

Andresen, F.H. (1992); Future management of the Schleswig-Holstein National Park. *In*: Dankers, N., Smit, C. J. & Scholl, M. (1992): *Proceedings of the 7th International Wadden Sea Symposium,* Ameland 1990, pp. 73-75.

Burbridge, P.R. (1990): Protection and management of coastal areas: Integrated coastal zone management concepts as a contribution to the conservation of natural resources. Unpublished.

Burbridge, P. R. (1991): The potential designation of the Wadden Sea as a World Heritage Site. A report for UNESCO. Unpublished.

Common Wadden Sea Secretariat (CWSS, 1991): The Wadden Sea. Status end developments in an international perspective.

Common Wadden Sea Secretariat (CWSS, 1992a): Sixth trilateral Governmental Wadden Sea Conference, 1991. Ministerial Declaration-Seal Conservation and Management Plan- Memorandum of Intent-Assessment Report 1992.

Common Wadden Sea Secretariat (CWSS, 1992b): Wise use and conservation of the Wadden Sea.

Helbing, C. (1992): Future management in the Niedersachsen part of the Wadden Sea. *In*: Dankers, N., Smit, C. J. & Scholl, M. (1992): *Proceedings of the 7th International Wadden Sea Symposium,* Ameland 1990, pp. 83-86.

Jong, F. de (1992): Ecological Quality Objectives for marine coastal waters: The Wadden Sea Experience. *International Journal of Estuarine and Coastal Law,* Vol. 7, no. 4., pp 255-276.

Jong, F. de (1993): Water quality of the Wadden Sea (*this volume*).

Kleinmeulman, A.M.W. (1992): Future management of the Wadden Sea, a dynamic task. *In* Dankers, N., Smit, C. J. & Scholl, M. (1992): *Proceedings of the 7th International Wadden Sea Symposium,* Ameland 1990, pp. 87-89.

Koester, V. (1989): The Ramsar Convention. On the Conservation of wetlands. A legal analysis of the adoption and implementation of the Convention in Denmark.

Mensbrugghe, Y. v. d. (1990): Legal status of International North Sea Conferences Declarations. *In* Freestone, D. and IJlstra, T. (1990): The North Sea: perspectives on regional environmental cooperation: special issue of the *International journal of estuarine and coastal law*. London, pp. 15-22.

Mörzer Bruyns, M.F. & Wolff, W.J. *ed.* (1983): Nature conservation, nature management and physical planning in the Wadden Sea area. The ecology of the Wadden Sea. Report 11.

Peet, G. and Gubbay, S. (1990): Marine protected areas in the North Sea. *In* Freestone, D. and IJlstra, T. (1990): The North Sea: perspectives on regional environmental cooperation:special issue of the *International journal of estuarine and coastal law*. London, pp. 241-251.

Reineking, B. (1993): The ecosystem of the Wadden Sea *(this volume)*.

Rudfeld, L (1990a): 25 års beskyttelse af Vadehavet. Miljøministeriet, Skov- og Naturstyrelsen.

Rudfeld, L. (1990b): The Danish Wadden Sea - 25 years of protection. *In* Dankers, N., Smit, C. J. & Scholl, M. (1992): *Proceedings of the 7th International Wadden Sea Symposium,* Ameland 1990, pp.199-213.

Wolff, W.J. *ed.* (1975): Proceedings of the Conference of Wadden Sea experts held at the island of Schiermonnikoog. The Netherlands, 26-28 November 1975. Contribution nr. 3. of the Wadden Sea Working Group.

Wolff, W. J. (1990b): Ecological developments in the Wadden Sea until 1990. *In* Dankers, N., Smit, C. J. & Scholl, M. (1992): *Proceedings of the 7th International Wadden Sea Symposium,* Ameland 1990, pp. 23-32.

World Wide Fund for Nature (WWF, 1991): The Common Future of the Wadden Sea. A report by the World Wide Fund for Nature, Flensburg.

Zwiep, K. v.d. (1990): The Wadden Sea: A yardstick for a Clean North Sea. *In* Freestone, D. and IJlstra, T. (1990): The North Sea: perspectives on regional environmental cooperation: special issue of the *International journal of estuarine and coastal law*. London, pp. 201-212.

Fisheries in the southern North Sea

K. Popp Madsen[1], K. Richardson[2]

Abstract

Because of its hydrographic and bathymetric characteristics, the southern North Sea supports one of the most productive fisheries in the world. In this chapter, these fisheries are reviewed and the processes leading to the production of this region are described and discussed. The coastal regions of the southern North Sea are also subject to anthropogenic influences especially in the form of river-borne nutrients and contaminants. The potential threat of these anthropogenic inputs to fish and fisheries in the southern North Sea is discussed.

Introduction

The North Sea is one of the most productive fishing areas in the world and within its 220,000 square nautical miles, 2.5 - 3 million tonnes of finfish are harvested per year. This catch represents close to 30% of the total annual landings from the northeast Atlantic. Landing statistics for these fisheries are collected by the International Council for the Exploration of the Sea (ICES) and, for that purpose, the North Sea is divided into three regions (ICES Divisions IVa, IVb and IVc) which correspond to the northern, central and southern North Sea. These Divisions largely reflect the bathymetry of the area - IVb and IVc are shallow areas with depths of less than 100 m while Division IVa includes the deeper plateau of the North Sea as well as parts of the Norwegian Deep and the Continental Shelf (fig. 1).

There are 224 different fish species reported from the North Sea. Comparatively few of these are of economic importance and, in fact, about 90% (by weight) of the total landings from the North Sea can be accounted for by about 10 species. About half of the total landings are used for the production of fishmeal and oil (the so-called "industrial" fishery). Species such as sandeel, sprat and Norway pout are target species for this fishery. However, because of the low market price (approximately 0.08 ECU per kilo in 1992), industrial fisheries represent considerably less than half of the economic value of the fisheries in the North Sea.

Nearly all of the economically more important species and, especially, the flatfishes are dependant upon the sandy shallow inshore waters of the central and southern

---

[1] Senior Principal Scientific Officer, Danish Institute for Fisheries and Marine Research, Charlottenlund Castle, DK-2920 Charlottenlund, Denmark

[2] Head, Dept of Marine Ecology, Danish Institute for Fisheries and Marine Research

North Sea. The Wadden Sea, along the coasts of Denmark, Germany and the Netherlands, is of particular importance to the early life stages of several species and it is also here that the major part of the fisheries on invertebrates such as shrimps and mussels are conducted.

Because of its hydrographic and bathymetric characteristics, the combined effects of the major water inflows occuring to the north of Scotland and (much less importantly) through the English Channel, Coriolis force and water outflow via the Norwegian Coastal Current give rise to a general anti-clockwise circulation pattern in the North Sea (e.g. Backhaus, 1989). Riverine input of freshwater to the North Sea amounts to approximately 200 km^3 annually (Portmann, 1989). As a result of

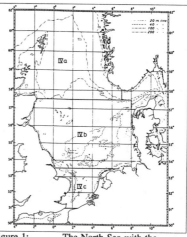

Figure 1:   The North Sea with the divisions used by ICES for statistical purposes.

the circulation pattern described above and the density differences between waters of differing salinity, the freshwater entering via rivers does not mix with the entire volume of the North Sea. This means that the nutrients and contaminants which are carried to the North Sea in river water are not diluted by the entire North Sea but, instead, remain at relatively high concentrations in shallow coastal waters and, in some cases, have been documented to affect the ecosystem in these regions.

Thus, the fish and the fisheries of the North Sea are directly dependant upon the very region of the North Sea which is most directly impacted by anthropogenic influences and man's activities in the coastal region (pollution, eutrophication, land reclamation, recreation, etc). Human activities have the potential of influencing the conditions for fish and the state of the fisheries in the entire North Sea. In this paper, the biology of the economically most important fish species in the North Sea and the recent history of the fisheries on these species are briefly described. In the final section of the paper, the potential interaction between anthropogenic activities and these fisheries is discussed.

Plaice (*Pleuronectes platessa*)

Adult plaice occur throughout most of the southern and central North Sea at depths of less than 80 m. The most important spawning grounds for this species are found south of the Dogger Bank, in the southern Bight, and in the eastern part of the English Channel. Plaice eggs and larvae are found in a continuous belt from the English Channel and into the southern and central North Sea (fig. 2).

The spawning season for plaice lasts from December to April usually with a peak occurring in January. After hatching, the larvae drift in the direction of the nurseries, of which the Wadden Sea is, by far, the most important. The young larvae enter the inlets of the Wadden Sea by means of selective tidal stream transport (Rijnsdorp and van Beek, 1991). Once in the Wadden Sea, the pelagic plaice larvae appear to be transported passively until they settle to the bottom, mainly during April - May.

By the end of the first year of life, the plaice begins its migration into deeper waters. Figure 3 shows the distribution of 1 to 3 group plaice in the North Sea. The migration is, undoubtedly, triggered off by size and faster growing individuals begin their migration first. The distribution of older plaice resembles that of the 3-group, but for a marked decline of fish in the coastal waters and increasing concentrations towards the deeper water north of the Dogger Bank.

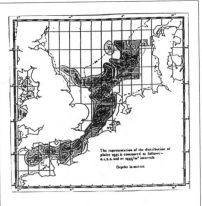

Figure 2:   The distribution of plaice eggs in the English Channel and the Southern Bight and the Central North Sea (from Harding et al. 1978).

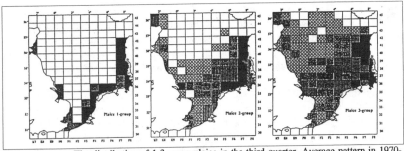

Figure 3:   The distribution of 1-3 group plaice in the third quarter. Average pattern in 1970-1986 (from Rijnsdorp and Van Beek 1991).

The North Sea plaice fishery is conducted primarily using beam trawl, otter trawl, Danish seine and gill nets. The total nominal landings in 1988 amounted to 138,412 metric tonnes. The distributions[3] by area and country are given in table 1.

The historical development of the plaice landings is quite interesting. From 1909 until World War II (with the exception of during World War I), plaice landings remained steady between 50 and 60,000 tonnes. Immediately following WWII (1946), plaice landings shot up to about 110,000 tonnes and then fell gradually 51 to 63,000 tonnes in 1955. Since 1955, there has been a steady increase in landings to approximately

---

[3]   According to ICES.

170,000 tonnes (fig. 4). The observed increase 1733 is partly due to an increased fishing effort but there has also been an increase in the frequency of good year classes in this period which has raised the 116 mean level of recruitment to the fishery. The cause of this increased frequency of good yearclasses is unknown.

	IVa	IVb	IVc
Belgium	+	7044	4483
Denmark	273	19978	51
France	11	29	1733
Germany	828	1622	116
Netherlands	982	44583	31159
United Kingdom	6411	16500	1586

Table 1: Plaice landings in tonnes per area and country (1988).

Figure 4: Plaice in the North Sea. Nominal landings in thousand tonnes (1960-1990).

Sole *(Solea solea)*

The sole is a lusitanian-boreal species with 58° N as the northern boundary of its distribution. The North Sea sole stock consists of a number of separate spawning populations which spawn close to the coasts of Denmark, Germany, the Netherlands and Belgium. Spawning takes place in the spring with April and May being the peak months. After spawning, the adult sole migrate into deeper water where overwintering also takes place. The return to the spawning areas the following spring appears to be quite fast.

Sole eggs hatch quickly (in about 8 days at 10° C) and the pelagic phase of the larvae is also relatively short. As a result, the dispersal of eggs and larvae is much more limited for sole than for plaice. The 0 and 1-group sole undertake a migration to deeper waters in order to overwinter at higher temperatures than those found in the littoral waters. They return to shallow coastal waters for another season of feeding and growth in the spring. With increasing age, sole disperse more evenly over the southern North Sea (fig. 5). Older sole are distributed rather similarly to the 3-group but with a tendency to concentrate more in the western part of the North Sea.

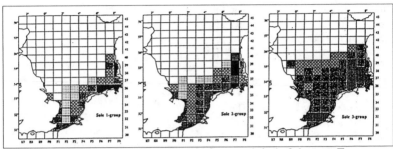

Figure 5:   Average distribution pattern of 1-3 group sole in the 3rd quarter (From Rijnsdorp and van Beek, 1991).

The North Sea sole fishery is carried out using beam trawl, otter trawl and gill nets. The total official landings in 1988 were 13,363 tonnes (table 2).

It should be noted, however, that the official figures are, in some cases and perhaps especially for sole, likely to be underestimates of the actual landings.

Sole landings since 1960 are shown in figure 6. Note that the values shown are somewhat higher than the official landing figures due to the inclusion of unallocated catches.

	IVa	IVb	IVc
Belgium	-	187	1012
Denmark	1	450	165
France	-	-	487
Germany	5	432	20
Netherlands	9	4336	5500
United Kingdom		189	570

Table 2:   Sole landings in tonnes per area and country (1988).

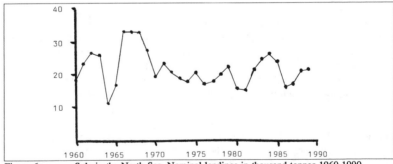

Figure 6:   Sole in the North Sea. Nominal landings in thousand tonnes 1960-1990.

With the exception of a few years in the 60s, landings have remained at a rather steady level of about 20 to 25,000 tonnes. The marked drop in 1964 was the result of the very cold winter in 1962/63 which not only increased the catchability of the cold weakened sole but also also increased natural mortality of the sole by increasing their vulnerability to predation (especially by cod).

An interesting feature of sole biology in the North Sea that cannot be deduced from the landing figures is that there has been a significant increase in growth rate during recent years. This change in growth rate appears to have begun in the mid 1960s and, during the course of the next decade, the weight at age of female sole almost doubled. For males, the increase in weight at age was about 50%.

Cod *(Gadus morhua)*
The cod is an arctic-boreal species and the English Channel is close to the southern boundary of its distribution. It is found over almost the entire North Sea. Cod eggs are also found over nearly the entire North Sea although elevated concentrations occur in several well defined areas. The spawning season extends from January to April and begins first in the southern part of the North Sea.

International trawl surveys indicate that juveniles (0-group; i.e. less than 1 year) and even more strikingly, the 1-group cod are found in the highest concentrations along the coastal areas of Denmark, Germany and the Netherlands. Juvenile cod do enter the Wadden Sea. However, this region is not the indispensible nursery ground for cod that it is for plaice and sole. The open part of the German Bight is, on the other hand, an important assembly area for young cod as demonstrated by the establishement of the so-called "Cod box" around Helgoland and off the Dutch west coast for the purpose of easing the fishing pressure on the young yearclasses. Inside this box, it is not permitted to use fishing gear with mesh sizes smaller than 100 mm during some periods of the year.

Tagging experiments have shown that the adult cod do not undertake extensive migrations. Fish tagged in the central North Sea and Southern Bight showed the most movement but, even here, 95% of the recaptured fish were still within 140 miles of their release point after 3 years (ICES, 1971).

	IVa	IVb	IVc
Belgium	34	425	5049
Denmark	2601	29243	3104
France	1661	131	6531
Germany	1421	5493	793
Netherlands	48	3962	12958
United Kingdom	32836	24798	7244
Other	3949	22	

Table 3:   Cod landings in tonnes per area and country (1988).

The main fisheries for cod are carried out using trawl, Danish seine and gill nets. In 1988, the total North Sea landings amounted to 142,306 tonnes (table 3)

Figure 7 illustrates the development in cod landings from the North Sea since 1960. From the beginning of the century and up to 1965, the total annual catch has fluctuated between 60 and 100,000 tonnes. After 1965, landings increased sharply to a peak of 350,000 tonnes in 1972 and, after a period of decline, another peak of close to 300,000 tonnes occurred in 1981. Since then, the landings have declined steadily towards the level of 25 years earlier. A similar pattern in the development of landings is found for other North Sea round fish such as haddock and Norway pout. This peak in landings in the 70s and 80s is known as the "gadoid explosion" and was the result of a series of good to very good yearclasses. However, the underlying causes of these series of good yearclasses is unknown.

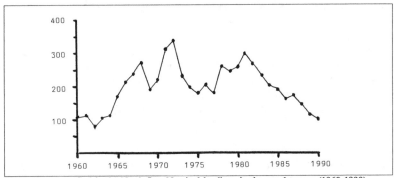

Figure 7:   Cod in the North Sea. Nominal landings in thousand tonnes (1960-1990).

Herring *(Clupea harengus)*

As late as the beginning of this century, herring were one of the more important sources of animal protein in northern Europe. The fish was inexpensive and could be preserved in various ways for long periods of time. The herring stocks in the central and southern North Sea were of strategic importance and have even been the underlying cause of a number of wars at sea! The most important of the herring stocks in the North Sea is the autumn spawning herring which is usually considered to consist of three spawning populations - the Shetland, the Dogger and the Downs herring. They spawn in the western parts of the North Sea beginning in the Shetland area in August and ending up in the eastern English Channel and the southern Bight in December-January.

Larvae from the Shetland-Orkney grounds are transported south and southeast across the North Sea towards the Skagerrak and the Danish coasts. Some of the larvae hatched on the eastern coast of the UK grow up here while others are transported eastwards towards the German Bight. Larvae from the southernmost grounds will drift northwards along the Dutch coast and spend their earliest stages in the Wadden Sea.

The nursery areas for the 0-group stages are widely distributed along the UK coast and, especially, along the Dutch, German and Danish North Sea coasts. Varying but significant numbers of larvae are transported into the Skagerrak and even into the Kattegat and spend about two years in these waters before they return to the North Sea. The 0-group larvae in the North Sea leave the inshore waters well before winter cooling has set in (i.e. when they reach a length of between 10 and 13 cm). During the following spring, the juvenile herring (1-group) concentrate outside of the 40 m depth contour south of the Dogger Bank and spread gradually over the North Sea as they increase in size. At the age of three, the young herring reach maturity and join the spawning and feeding migrations of the adult herring.

The central and southern North Sea have small stocks of spring spawning herring in both the western and eastern regions. Both stocks spawn in very shallow water of low salinity and are of only local importance. In the western stock, the Zuiderzee herring comprised the largest spawning element. However, with the closure of the Zuiderzee in the Netherlands in 1932, the herring lost their access to this spawning ground and the stock is now represented only by the Bayswater herring which spawn in the Thames estuary. The eastern stock has also traditionally consisted of a number of different spawning communities. The River Elbe herring have now disappeared but other communities are still found in the Danish Ho Bay, Ringkøbing Fjord, Nissum Fjord, and the Limfjord. The Ho Bay may be considered as the northernmost part of the Wadden Sea while the three fjords are all landlocked. In two of these fjords, the herring have to force sluices in order to reach and to leave their spawning grounds. Clearly, the survival of these small herring stocks is completely dependant upon human activities.

Fishing for herring is carried out with bottom trawls, pelagic trawls and purse seines. Drift nets are only used locally in inshore areas despite the fact that this was the only fishing method used prior to WWI and still of significant importance in the first decade after WWII. The official landing figures for 1988 are presented in table 4. The total landings of 488,400 tonnes given by ICES are undoubtedly an underestimate of the actual figures.

From 1906 and into the 1950s, herring landings from the North Sea fluctuated around 500 - 600,000 tonnes with the exception of the two war periods. Figure 8 shows the development in landings since 1960. The high catches in the beginning of the 60s were the result of an exceptionally rich year class in 1956. This year class was followed by an even stronger yearclass in 1960 which spawned for the first time in autumn 1963.

	IVa	IVb	IVc
Belgium	72395	22150	+
Denmark	542	8595	467
France	10337	4884	-
Germany	181611	37192	2669
Netherlands	53611	39635	-
United Kingdom	4024	17318	328
Others	-	-	-

Table 4:   Herring landings in tonnes per area and country (1988).

The spent herring from the Shetland and the Dogger spawnings overwinter along the Norwegian Deep and the huge shoals in this area attracted a lot of extra fishing effort. As a result, the herring landings reached an historic high in the mid-60s. There followed a period of steady rapid decline in recruitment, spawning stocks and landings until 1977 when the EC and Norway agreed to a total ban on herring fishing in order to save the stock from extinction. During the 1980s, the North Sea herring regained its strength and landings are now back to the former level of 500 - 600,000 tonnes.

Overfishing has been blamed for the collapse of the herring industry in the North Sea. However, it has also been suggested that the successive recruitment failures during the late 60s and 70s may have been caused by adverse environmental conditions (an unusual circulation pattern in the North Sea during this period) rather than by low spawning stock sizes (Corten, 1986). Larval surveys suggest that herring larvae may not have been transported from the spawning to the nursery grounds and that large numbers of larvae may have perished in deep water due to unfavourable current regimes.

Figure 8:   Herring in the North Sea. Nominal catch in thousand tonnes (1960-1990).

Other fish species

The four species described above are the economically most important species which are dependant upon the shallow coastal areas in the central and southern North Sea. Other commercially important species that use the coastal areas and, especially, the Wadden Sea as nursery areas are sprat, whiting, dab and flounder. The latter spawns off the coast in winter - early spring and the eggs and larvae are transported into the Wadden Sea by the residual currents. The juvenile flounder can penetrate into freshwater in river estuaries such as the Elbe, the Weser, the Ems and semi-enclosed regions such as the IJsselmeer and the Danish fjords.

A special representative of the "nursery" group of fish is the eel which spawns in the Sargasso Sea and drifts as larvae back to the European coasts. Some of the young "glass eel" remain in the Wadden Sea but most of them ascend into the rivers where they stay for 7 to 12 years until they return to the saltwater as "silver eels". The eel is the economically most important finfish species in the IJsselmeer and in the Danish fjords. The eel fisheries in these areas are heavily dependant upon free access for the fish past sluices and dams and these fisheries are probably more vulnerable to pollution and other anthropogenic influences on the environment than other fisheries.

Invertebrates

The major non-finfish fisheries in the North Sea are targeted on mussels, cockles and shrimps. These fisheries are typically connected to inshore areas and, especially, to the Wadden Sea. Some of these fisheries have a short history while others have themselves become history! The endemic oyster, *Ostrea edulis*, has, for example, disappeared from the North Sea. Oyster culture still exists, however, based on recruits (primarily of *Crasstostrea gigas*) from other areas.

A German fishery for the soft shell clam, *Mya arenaria*, has only been conducted during short periods (i.e. during WWI and in the period from 1945 - 1949). It is recognised, however, that the shellfish resources in the shallow coastal regions of the North Sea may not be fully exploited and several pilot projects are underway in order to investigate the potential of these resources.

Shrimp fisheries

The brown shrimp, *Crangon crangon*, is a typical coastal species thriving where the bottom is soft, silty or sandy. Large populations of commercial interest are found in and off the Wadden Sea. The high primary production supported by riverborne nutrients and the transport of larval shrimp by tidal currents to the Wadden Sea makes this region the ideal shrimp nursery.

Originally, shrimp were landed only for the production of meal which was used as an additive to the fodder for poultry. However, the average price of shrimps for human consumption increased rapidly from about the middle of the 1950s and so did consumption landings. The landings of undersized shrimp for industrial purposes declined as it became obvious that the previous exploitation pattern was far from optimal. The introduction of selective trawls and of sieves for sorting catches on deck have reduced the catch of undersized shrimp and the unwanted by-catch of juvenile finfish such as plaice, sole, cod and whiting.

The shrimp is a short-lived species and only a few reach the age of 2 years. The breeding period is between February and June and some of the juveniles have already by July reached a length that makes them vulnerable to the trawl fishery. In accordance with the growth rate, the catch of undersized shrimp culminates between July and September and that of consumption shrimps between September and November. The shrimp move into deeper waters during the winter and are followed by the shrimp trawlers to as far as 30 miles off the coast.

Both the consumption and small shrimp landings are about 10 - 20,000 tonnes annually. Low catches of shrimp seem to coincide with good yearclasses of cod which may invade the German Bight and the Wadden Sea. Invasion of the area by good yearclasses of whiting is also known to occur and the very poor shrimp fishery in 1990 is believed to have been the consequence of such an invasion.

Blue mussel

Exploitation of the blue mussel was originally a Dutch undertaking and the culturing of mussels was also first practiced in the Dutch Wadden Sea. Later, Germany has begun mussel culture in its parts of the Wadden Sea. Mussel culturing started in earnest in about 1950. The earlier fisheries for "wild" mussels were of minor significance except for during World War I, when a maximum of 120,000 tonnes was reached in 1918. Juvenile mussels are caught outside the culture banks in spring and summer and then transplanted to the culture banks. Culturing sites are usually placed in deeper water where the mussels are always covered by water. In this way, food uptake and growth are not interrupted as they are for mussels in the tidal zone.

In both the Netherlands and Germany, the culture banks are supervised by the government. The fishing is regulated by a licence system and the culture lots are rented to the fishermen. In the period 1981 - 90, the total industrial landings have shown a yearly average of 133,000 tonnes with 95,000 tonnes (in 1988) being the lowest and 176,000 tonnes (in 1982) the highest catches.

The influence of human activities on fisheries in the North Sea

By comparison with many freshwater and more enclosed marine systems, the North Sea is still a relatively clean environment. Nevertheless, there are approximately 21 million people living on or near the shores of the North Sea. Pollution and refuse generated by this community as well as active oil and shipping industries all contribute to the degradation of the North Sea environment. The highest concentrations of contaminants are generally found in organisms taken from inshore regions near river outflows. However, to date, the observed contamination levels for various pollutants in fish and shellfish from most of the North Sea have generally not given any cause for alarm with respect to human health standards (Portmann, 1989).

However, there are a number of indications that at least some members of the marine biota in the coastal regions of the North Sea respond negatively to the higher contamination levels observed in coastal as opposed to more offshore regions. Reijnders (1986) has presented evidence that the reproductive capabilities of the seals in the Wadden Sea are diminished by PCB contamination. Contamination levels of heavy metals and organic pollutants in fish and shellfish have been shown to be higher in coastal than in more offshore waters of the North Sea (Portmann, 1989) and, although no direct cause and effect mechanism has been established, the frequency of fish diseases is also greatest in the more contaminated (i.e., the coastal areas) of the North Sea (Dethlefsen, 1989).

Laboratory studies indicate that high concentrations of various organic contaminants in the gonads of fish decreases the viability of the eggs they produce (von Westernhagen et al., 1989) and there is some suggestion that there may be a greater occurrence of malformed fish embryos produced in the coastal regions of the North Sea than farther offshore (Dethlefsen, 1989). Thus, although there is little cause for alarm for the human consumer of fish caught in the North Sea, it seems clear that pollution (just as any other environmental interaction) does affect the marine ecosystem - especially in the coastal regions of the North Sea! Nevertheless, there is some ground for optimism with respect to the pollution of the North Sea in that good biochemical techniques are being developed which will allow the scientist to quantify the ecosystem effects of pollution and (best of all) the concentrations of most metal and organic contaminants appear to be falling in many parts of the North Sea (Portmann, 1989).

The situation with respect to the effects of eutrophication (enrichment with the nutrients that stimulate plant growth) on fish and fisheries in the North Sea is, perhaps, somewhat more complicated than that of pollution. Nutrients are, in themselves, not damaging to fish. However, larval and some juvenile fish are dependant upon the plankton food chain for their survival. Eutrophication alters the planktonic food chain. In cases where a simple stimulation of plankton production occurs, one could imagine that eutrophication could increase the survival of young fish and give rise to better yearclasses through better recruitment. Indeed, it has been hypothesized that this mechanism may have played a role in the development of the plaice fishery as well as the change in growth patterns observed for sole in recent years.

However, eutrophication can lead to more than a simple stimulation of plankton production. The plankton composition can, for example, be altered so that the dominant species cannot be easily eaten by young fish. There can also be an increase in the number of annual peaks in plankton production ("blooms") where the food chain is unable to consume all of the organic material being produced. This organic material sinks to the bottom and its degradation can lead to oxygen depletion ("hypoxia" or "anoxia") which can be lethal for those fish who are unable to escape from the affected region. Oxygen depletion events have been recorded in the eutrophied eastern North Sea in 1981, 1982, 1983 and 1988 (fig. 9).

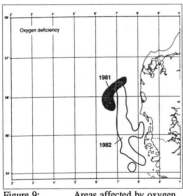

This area has been identified as feeding ground for a number of fish species. If an oxygen depletion event lasts long enough to smother some or all of the bottom fauna in an affected region, then the region's potential as a feeding ground for fish may be altered for some time to come. Identifying the eutrophication affects the marine ecosystem is another active research area for marine scientists currently working in the North Sea.

Figure 9:   Areas affected by oxygen
              depletion in 1981 and
              1982.

Finally, it should be mentioned that the fishing industry, itself, just as all other human activities carried out in the North Sea, affects the ecosystem there. In addition to altering the numbers of various fish species, some forms of fishing (i.e. some trawls) damage and kill bottom fauna. This not only may alter the composition of benthic communities but the presence of damaged organisms on the bottom may increase the food availability for a number of different scavengers (including fish). Fishing activities can also increase the food availability for some bird species and decrease it for others! In addition, the fishing fleets of the North Sea can generate litter in the same manner as other shipping activities.

As it seems unlikely that man's activities in the North Sea will be abolished, it may be a useful exercise to consider both eutrophication and fishing (as well as all other anthropogenic activities) as components of the North Sea ecosystem, anno 1993. The job of the scientist is to quantify the interactions occurring between all of the components within the ecosystem. A far more difficult job is to set limits as to the degree to which various components of this ecosystem should be allowed to affect others and this is the challenge facing us all in the years to come.

References

Anon. 1971. Report by the North Sea Roundfish Working Group on North Sea Cod. *ICES Doc. C.M.* 1971/F:5 (Mimeo)

Backhaus, J.O. 1989. The North Sea and the climate. *DANA, 8*: 69-82.

Corten, A. 1986. On the causes of the recruitment failure of herring in the central northern North Sea in the years 1972-1978. *J. Cons. int. Explor. Mer, 42*: 281-294.

Dethlefsen, V. 1989. Fish in the polluted North Sea. *DANA, 8*: 109-129.

Harding, D., Nichols, J.H., and Tungate, G.S. 1978. The spawning of plaice (*Pleuronectes platessa* L.) in the southern North Sea and English Channel. *Rapp. P.-v. Réun. Cons. int. Explor. Mer, 172*: 102-113.

Portmann, J.E. 1989. The chemical pollution status of the North Sea. *DANA, 8*: 95-108.

Reijnders, P.J.H. 1986. Reproductive failure in common seals feeding on fish from polluted coastal waters. *Nature, 324*: 456-457.

Rijnsdorp, A.D. and Van Beek, F.A. 1991. Changes in growth of plaice *Pleuronectes platessa* L. and sole *Solea solea* L. in the North Sea. *Netherlands J. Sea Res. 27* (3/4): 441-457.

Westernhagen, H. von, Cameron, P., Dethlefsen, V. and Janssen, D. 1989. Chlorinated hydrocarbons in North Sea whiting (*Merlangius merlangus* L.) and 3 effects on reproduction. I. Tissue burden and hatching success. *Helgoländer Meeresunters. 43*: 45-60.

Coastal Planning for Recreation in Denmark along the North Sea Coast

I. Vaaben[1]

Abstract

The aim of this article is to describe the background for and planning of the near-shore areas along the Danish North Sea coast, especially on the basis of the high recreational value of the areas. A description will be made of the funds which have been used to protect and at the same time utilize the coastal areas during the past centuries and the new Danish strategy for coastal zone management will be outlined.

Introduction

The Danish North Sea coast is partly a coast with sandy beaches and partly a coast of tidal flats. The overall length of the coast is approximately 450 km (fig.1).

Figure 1: The Danish North Sea coast.

[1] Head of division, Ministry of the Environment, Haraldsgade 53, DK2100 Copenhagen, Denmark

Of this area the northernmost 340 km of coastline consists of sandy beaches. This part is characterized by wide beaches, lagoon-like inlets and posterior dunes and is equally unique (fig. 2). Despite many summer cottage areas, some towns, technical installations, harbours, etc., there are only a few other places in western Europe, where original coastal nature can be experienced as here.

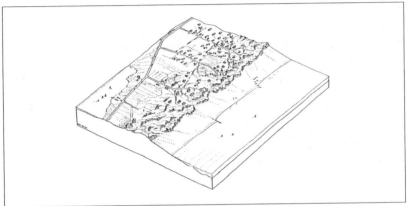

Figure 2:   Artist impression Danish west coast.

Coastal landscapes are important for the immediate experience of nature where land and sea meet, and are therefore of much importance in general for outdoor life, holiday homes and tourism, for activities, stays, sailing, fishing and hunting.

Due to its unique and unspoilt nature the west coast of Jutland is exposed to great pressure from tourism. Special measures by the authorities are necessary. It is the policy of the Danish Government that Danish tourism is to be developed on a sustainable basis. I.e.: consideration of nature and the environment has top priority. At the same time, however, this has the effect that the tourist trade is ensured optimum conditions in the long term. If nature and the environment are ruined, the tourists will stay away.

The protection and use of coastal landscapes
The Danish soil and surface structure has not changed significantly since the Stone Age, except for its coasts which are constantly changing character. A stretch of coast is not quite the same today as yesterday or as it will be tomorrow. In some places land is eroded, in other places land is deposited. The coast is formed by the sea and the wind.

On the Danish West Coast securing and protecting the coastline is carried out in accordance with the rules and regulations of the special Act on coastal protection (Laustrup, 1993).

The West Coast not only suffers coastal erosion but also has a continuous problem with sand drift in areas along the coast and often further inland. Through the ages sand drift has ruined forests and agricultural areas along the West Coast. It does not take many centimetres before an otherwise fertile field loses its agricultural value. Over 200 years the Danish State has therefore tried to implement measures which could limit the adverse effects of sand drift through various Acts - from the Sand Drift Decree of 1792 through the Sand Drift Act of 1867 to the Nature Conservation Act of 1992. Since 1935, for example, general coast conservation regulations have prohibited or governed construction on the beach and in all dunes, but with a maximum limit of 500 m from the inner-most edge of the beach. The reason for these restrictions is the wish to protect the vegetation which binds the dunes together. To this is added that the vegetation could be subject to wear and tear by people's coming and going in the terrain. Such traffic will naturally be more intense, the more dense the buildings are.

In turn, public access has been secured for walking and staying for short periods of time in the preserved stretches. This opportunity for stays and bathing can, however, be restricted in consideration of the danger of sand drift.

With its unique landscape values and natural resources the West Coast of Jutland is also protected through the general protection provisions of the nature conservation legislation and through the conservation and nature management provisions. In addition to the general protection, this has had the effect that in the course of the past 75-100 years the public authorities have preserved large areas and have bought up areas in order to ensure the continued existence of these values.

Furthermore, many areas along the coast have been designated as Ramsar areas and EC bird preservation areas (fig. 3).

As a general rule for the many protected areas, the public is allowed as much access as possible to come and go and stay in the natural areas, and emphasis must be put on the importance which an area can have for the general public due to its location.

Figure 3: A. Ramsar areas and EC bird preservation areas.
B. Stateowned and preserved areas.

The areas along the coast are relatively sparsely populated. The northernmost 340 km only has 5-6 ports with more than 1000 inha-bitants. In addition to this, there are 15-20 smaller towns. As generally the whole stretch of coast has good bathing beaches, the area has many holiday and leisure facilities: first and foremost expanding summer cottage areas with thousands of summer cottages (fig. 4).

Spatial planning.

In a vulnerable area such as the West Coast area physical planning is of great importance as the overall framework for the use of the areas. Through such spatial planning it is possible to establish an overview of the area's composition and to make decisions for the areas' use and protection on the basis of a broad perspective.

Through the past 20 years, physical planning in Denmark has been based on a simple planning system which has, among other things, resulted in a set of regional plans. Especially with regard to the open country, the regional plans have provided the public authorities with a unique management basis in connection with decisions on area application issues.

Figure 4:   * Towns.
              ▓ Summercottage areas.

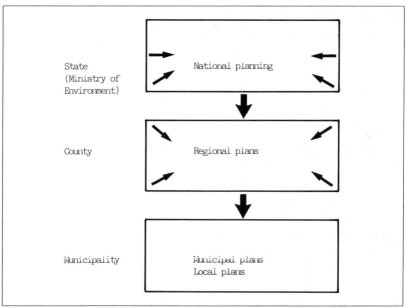

Figure 5:   The planning system

In brief, the planning system can be described as follows (fig.5).
There are three political planning levels: State, County and Municipality. The State lays down the overall planning framework but does not prepare the plans themselves. The counties and municipalities, which are elected bodies, prepare the physical plans. The counties do this for the whole region, which typically covers 20 municipalities. The municipalities prepare the plans for the municipality's area.

The counties' plans, the regional plans, must not be contrary to the overall state guidelines and directives. The municipal plans, have to be in accordance with the regional plans and thus also in accordance with the state guidelines. Within this framework the municipalities prepare the most detailed plans, the local plans, for individual areas in the municipality.

The area along the West Coast is located within the geographical area of 5 counties and 24 municipalities. Physical area planning for the area is carried out by these authorities for their individual geographical area, but in joint cooperation.

The state framework for coastal planning

Even though the West Coast is protected in pursuance of Danish nature conservation and sand drift legislation, a relatively large part of the coast is reserved for urban purposes, summer cottages, large holiday and leisure facilities, etc. of which holiday and leisure facilities make up the predominant part.

An attempt has been made to counter this development, which can be seen along all Danish coasts with changing intensity, through a Governmental stop to prevent the building of new summer cottage areas (1977) and by a national directive (1981) with binding guidelines for county and municipal planning of holiday and leisure construction, especially in the nearshore areas.

In 1991, the protection of nearshore areas in the whole country was extended to cover all activities in the nearshore zone which is defined as a 3 km wide belt along the coasts (fig. 6).
In the following an outline will be given of the background for and the further contents of the new coastal strategy on which the guidelines from 1991 are based.

The Danish coastal zone management strategy and planning

The basic point of departure is that the near-shore areas are an endangered and limited resource and there are many and often opposing wishes for functional uses in the coastal areas. Consequently, the aim of the strategy is to ensure through increasing requirements for coastal planning that building developments along the coast and near-shore areas are subordinated to the Danish coastal landscape.

Similar reflections have been made under the auspices of the European Community (EC). In February 1992, the EC's Ministers of the Environment decided to ask the Commission to prepare an overall strategy for general coastal zone management with the objective of preparing a coherent framework for the environment and for integral and sustainable development. In the EC's 5th environmental plan of action, such a strategy has been outlined together with a strategy for the development of tourism,

Figure 6:   The nearshore zone.

accompanied by deadlines for the implementation of the objectives set.

The requirements for the coming years' regional and municipal planning in Denmark are that planning must be carried out with a view to ensure that the coasts are only used for the functions which can only be placed in connection with the coast and for recreational purposes. New areas for housing, industries, etc. can only be made if documentation can be presented in support of planning or functional arguments for nearshore location. The planning of new summer cottage areas is still prohibited and holiday and leisure facilities can only be built in accordance with coherent political considerations regarding tourism for the whole area. Finally, a general requirement applies for ensuring and improving public access to and along the coast.

In addition, through their local planning, the municipalities are put under an obligation to ensure that concrete building and construction works in the nearshore areas are designed and adjusted to the coastal landscape in accordance with the coastal strategy. The local plan must provide information about the visual effect on the surroundings.

The counties will revise the regional plans in 1993 on the basis of the binding state guidelines on coastal planning. The Minister of the Environment can veto a regional plan if the plan is contrary to the guidelines. After the regional plans have been adopted in the course of 1994 and 1995, the municipalities will revise the municipal plans to bring them into accordance with the regional plans. Only then will the new coastal strategy have its full effect.

Since they were published at the end of 1991, the state guidelines have, however, imposed a duty on counties and municipalities to manage present planning and accompanying amendments in accordance with the intentions behind the coastal strategy. The most important fact is that the requirement for visualizing the effects of construction and building projects on the surroundings has entered into force.

The strategy is resulting in better planning, but the implementation of it will presumably create certain problems for a free development of the coastal areas. There is, however, no doubt that there is a high degree of political consensus regarding the objective of the coastal strategy. Only the future can tell whether the will to implement it in practice is present as well.

References

Laustrup, C. (1993) Coastal erosion management of sandy beaches in Denmark. *This volume*.

Shoreline Management of Eastern England

R. Runcie[1], K. Tyrrell[2]

Abstract

This paper gives a perspective on the length of coastline falling within the boundary of the National Rivers Authority, Anglian Region (NRA) in respect of the shoreline management carried out by the Authority and the Maritime District Councils (MDC).

Introduction

The coastline examined in this paper extends from the Humber to the Thames estuary. A geographical area covering almost a fifth of England, much of which is flat and below maximum recorded sea level (fig. 1). Low lying land, vulnerable to flooding, protected by sea defences is the responsibility of the NRA. Coastal protection to higher land prone to coastal erosion is the responsibility of the Maritime District Councils.

The area is protected by 1,300 km of sea defences and coastal protection works. These works have been constructed upon a wide range of coastal geomorphology and exposure to tide, wind and waves, and has resulted in a great variety of coastal engineering work. This background of geomorphological diversity, the forecast of climatic changes and sea level rise, will increase pressures to invest in new as well as maintain existing coastal structures. The NRA Anglian Region currently invests some £40 million annually on flood defence works. The Maritime District Councils spend £6 million which gives a total annual investment of £46 million. Funding for sea defence and coastal erosion is provided through central government by the Ministry of Agriculture, Fisheries and Food (MAFF).

It is against this requirement for an understanding of coastal processes and need for sound strategic plans, that the Anglian Sea Defence Management Study (SDMS) was conceived in 1987. The issues considered cross administrative boundaries and highlight the need for, and value of, effective Geographic Information Systems (GIS) for use in the shoreline management of eastern England.

---

[1] Project Manager, National Rivers Authority, Anglian Region, Kingfisher House, Goldhay Way, Orton Goldhay, Peterborough PE2 5ZR, England

[2] Principal engineer, Waveney Distric Council, Mariners Street, Lowestoft NR32 1JT, England

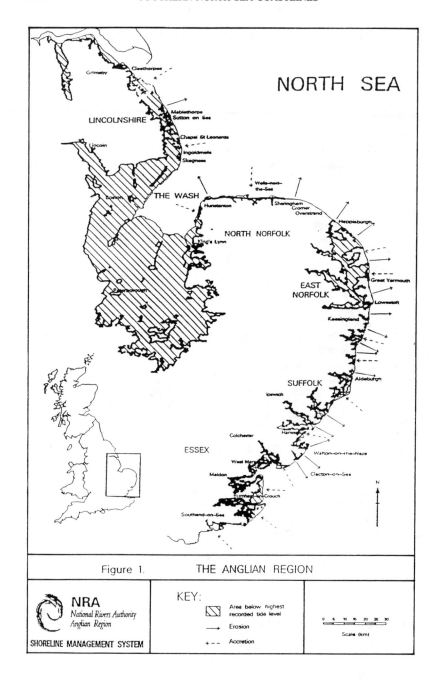

Figure 1.   THE ANGLIAN REGION

Responsibilities

Sea defences and coastal protection schemes in the Anglian Region have been built by a wide variety of public and private bodies each adopting standards that are considered to be appropriate and affordable.

The NRA was established in 1989. It is an independent body whose statutory duties include flood defence. The prime objective is to provide effective defences for people, the built and natural environments against flooding from rivers and the sea. The Coast Protection Act 1949 confers on the MDC's (Local Authorities) the powers to carry out works to protect land from coastal erosion.

Organisations nominally responsible for the maintenance of coastal defences fall into three categories and are given in table 1.

Organisation	NRA Anglian	Local Authority & Others (inc. Private O'ship	TOTAL
Length of Open Coast Defence	363	63	426
Length of Coastal Tidal Defence	812	68	880
Total Length	1175	131	1306

Table 1:   Distribution of Coastal Responsibilies

The Anglian coastline is an unsurpassed area of environmental resource (Runcie and Barham, 1992). Environmental assessment is therefore an integral part of coastal defence project development.

The sea defence management study

The vulnerability of the Anglian Region to both freshwater and tidal inundation is well recorded. The Anglo-Saxon Chronicle recorded such floods at "Martinmass" in 1099; Pepys noted in his diary for 7 December 1663 that the previous night "All Whitehall had been drowned". A surge that would have caused flooding in eastern England. The most notable present day event was on 31 January 1953 when over 200 people drowned in eastern England, 1,800 in Holland, as a consequence of an extreme North Sea Surge (ICE papers, 1953). 1978 saw higher levels than 1953 but fortunately loss of life and damage to property was not so great (Townsend, 1979).

North Sea surges are frequent events created by a combination of low atmospheric pressure and the funnelling effect of the southern North Sea. Surges in excess of 1 metre occur four or five times a year but unless coincidental with high tides, spring tides or onshore winds, do not cause undue problems.

The NRA has adopted a strategy for sea defence based on the principle of understanding and working with nature rather than against it.

The need to understand the coastal processes are paramount to the management of the shoreline. The flood defence needs against surge events in the recent past has dictated the location and often the type of defence. But it is the daily action of tides, currents, wind and waves that shape our coastline which ultimately dictates the management considerations for the future. To effectively manage the NRA flood defences requires a coastal process knowledge for the whole coastline irrespective of individual responsibilities or jurisdictions. The effects of the forcing processes resulting in erosion or accretion are summarised on figure 1. It was on this understanding that the SDM was developed and resulted in the Shoreline Management System.

The development of the Sea Defence Management Study database is well documented (NRA, 1991). The study's primary aim was to develop the tool necessary for the strategic management of sea and coastal defences in the Anglian Region. This was to be achieved through an understanding of all the forces acting upon the coastline. The Study (of the whole coastline of eastern England) was undertaken in four phases at a total cost of £1.65 million. The Government through the MAFF funding £1million of the Study cost. The development of the Geographic Information System (GIS) has proved to be highly complex. The system has been developed not only to display data but to permit detailed analysis (Townend and Leggett, 1992). The three year Study was completed in the Spring of 1991 and heralded the beginning of shoreline management in eastern England.

Anglian coastal authorities group (ACAG)
Another important initiative took place in 1987 with the setting up of the Anglian Coastal Authorities Group. This was intended to be a forum for engineers engaged in managing the East Anglia coastal defences.

Once established, the group expanded to cover the whole of the Anglia Region coastline and a set of objectives for the group were formalised:
(i)   to provide a forum for news about changes and trends, studies and schemes, exchange experience of successful (and unsuccessful) solutions to coastal problems;
(ii)   to create and maintain an inventory of past records, studies, schemes and problems;
(iii)   to identify the need for further studies and opportunities for liaison and collaboration;
(iv)   to work towards an overall coastal management strategy.

An important feature of the representation at ACAG meetings is the interchange of ideas and information between the NRA and MDC's. Although carrying out their functions under different legislation the areas of concern, techniques employed and many other issues are very similar.

The SDMS is a vital piece of work for those charged with the responsibility of managing any part of the coastline in the region. The study required the collection, collation and analysis of existing and new data sets. This immense task of data collection was assisted and supported by many organisations (Child, 1992). Many of whom will continue to actively participate in keeping the information up to date.

The demands made on the study findings and hence the amount of interaction will inevitably vary from Authority to Authority.

Shoreline management

Both the NRA and the MDC's are committed towards a strategic approach to the management of the shoreline under their care.

Although the coast protection and flood protection aspects of shoreline management are separately allocated the coastal processes being dealt with by both authorities are the same. This is reflected in the objectives behind the preparation of strategic plans. These are listed for the NRA and a typical MDC in Tables 2 and 3 respectively.

AIM
To provide effective defence for people and property against flooding from rivers and the sea.

OBJECTIVES

(i)	To implement a sea defence management strategy that ensures the integrity of sea and tidal flood defences.
(ii)	To ensure that such plans are based on sound economic and technical principles.
(iii)	To promote and make use of research and development into future flood defence needs.
(iv)	To protect and, where works are intended, identify opportunities for enhancement of landscape, amenity and conservation.
(v)	To liaise with and develop close working relationships with other responsible coastal authorities and organisations, to understand each others objectives and responsibilities, to resolve conflicts, to promote public awareness and understanding, and to work towards an integrated coastal management approach.

Table 2:   NRA Management Aims and Objectives

Monitoring

The data contained within the NRA GIS system is given in Appendix A which summarises the data type, format, distribution, source and frequency of update. The development of the Shoreline Management System has commenced with the completion of the SDMS.

The supporting studies conducted for the SDMS identified the requirements for the monitoring necessary to keep the system up to date. This updating is essential to conduct meaningful analysis, develop process understanding and performance monitoring of flood defence schemes. The Shoreline Management System therefore forms the basis for the development of the NRA Management Strategy for this coastline.

i)	To establish clear strategic guidelines and policies for effective management of the District's shoreline within which a cohesive programme of capital and revenue works can be developed.
(ii)	To assign the appropriate standards of protection to each section of the District's coastline commensurate with the nature and usage of the protected area. These standards are to be set after consideration of the effects of global warming and the associated rise in sea level.
(iii)	To identify opportunities for development and conservation which may arise from effective management of the shoreline.
(iv)	To reinforce our commitment to the concept of coastal planning contained in the objectives of the Anglian Coastal Authorities Group and to provide a clear statement of our concerns and intentions for the benefit of the group and its constituent members.
(v)	To obtain an assessment of the present condition of the natural and man-made coastal defences within the District and to establish the level and anticipated spending profile required to achieve the assigned standards of protection.
(vi)	To identify the requirement for continuing monitoring of coastal processes and the condition of the natural and man-made defences.
(vii)	To identify the requirement for further research and study to enable the other objectives of this strategy to be achieved.

Table 3:   Typical Objectives of MDC Strategic Plan

The coastal processes along the Anglian coastline are very complex. There are strong tidal and wave energy inputs. The mean spring tidal range varies from 2m - 3m along the East Norfolk coastline and increases to 6 m or more along the Lincolnshire coastline. Mean wave heights along the coastline exceed 0.4 m along the North Norfolk coastline decreasing to 0.1 m along the South Essex coastline and the Wash. The large tidal range along this coast results in a wide active beach frontage which is vulnerable to storm damage. This is the case for the sand veneer beaches overlying clays found along the Lincolnshire and Norfolk coastlines.

The complex sediment transport regime along this coastline is shown by the offshore bathymetry along the Anglian coastline (NRA, 1991). The extensive offshore banks from the Burnham Flats adjacent to the Wash to the series of banks off the Thames estuary clearly indicate that both near-shore and off-shore sediment transport paths are operative along the coastline. This is an area requiring much further investigation to develop our understanding of the sediment movements fundamental to the longer term shoreline decision making process.

The Regional topographic, bathymetric and aerial beach surveys carried out by the NRA Anglian Region, will help develop a holistic understanding of the complex coastal processes in operation. The survey philosophy is to take levels and soundings on cross-sections spaced at 1km intervals along the coastline. Topographic surveys are carried out bi-annually and bathymetric surveys on 5 yearly cycles.

The regional surveys form the most important ongoing input to the Anglian Shoreline Management System. The accuracy of information will determine the effectiveness of the shoreline management process. The current annual monitoring programme costs approximately £300,000/year.

Coastal units

To manage the coastline it is necessary to identify sections of coastline experiencing similar conditions. In the United Kingdom there has been much debate on the definition of coastal cells. For example, a coherent case can be made for considering Flamborough Head (Yorkshire) to the North Foreland (Kent) as one cell (Pethick and Leggett, 1993). This cell covers over 1500 km of coastline and has distinct boundaries to it. Under the Sea Defence Management Study, the GIS was utilised in order to define Coastal Process Units. These units lie within this large cell and each unit has particular characteristics.

The forces of waves and tides were combined to create an overlay of our dynamic coastline. This was then overlain with the geology and geomorphology of the coast, to produce a definition of process units. Each unit within the classification has coherent characteristics. However, processes are likely to change with time which requires re-classification and shifts in the boundaries of the process units. This is not considered to be problematic, in that the units provide a process framework from which to base management decisions. The management approach must be flexible to reflect the dynamic nature of the coastline and overview a sufficiently large section of coast such as a coastal cell. Clearly changes may also take place on the cell scale but these boundary changes are likely to be less dynamic in nature. The difficulty is therefore in the derivation and definition of the coastal cell.

The anglian region data model

Administrative boundaries may exist within coastal cells and units. In order to mitigate the effects of such boundaries an integrated approach to information gathering is important. This has been formalised through the preparation of a data model which documents the format of each variable and how one data set relates to others within the system.

GIS was the technology adopted for this application because of the ability to integrate, manage, analyse and query spatial data from a large number of sources. The NRA for the whole of England and Wales has adopted the data model for the Shoreline Management System. The data model provides the base line for data storage, collection and sharing of information. The model also facilitates national analysis of parameters within it to yield an overview of the English and Welsh coastline.

Coastal wildlife database

Shoreline management requires consideration beyond the brief for the development of the SDMS. An example of this is the Coastal Wildlife database, which provides the NRA with detailed, sound, ecological information for the Anglian Region. The real value of this database is exploited when species information is related, through the Shoreline Management System GIS, to coastal morphology and processes. Relational analysis between physical and biotic features of the coastline helping shoreline managers to plan strategically and in a fully integrated way. The process may be summarised as working with nature rather than against and means that, while considering ecological values, coasts can be managed as dynamic, not static systems. This outcome will result in coastal flood defence planning which properly seeks to ensure the maintenance and enhancement of conservation and ecology along the Anglian coast.

Coastal management planning

ACAG has developed proposals for establishing a Regional Shoreline Management Plan to resolve the conflicts between coastal users and to determine the best use of coastal resources with due recognition of the needs of neighbouring authorities (Oakes, 1992).

Following the Government's response to the House of Commons Environmental Select Committees report "Coastal Zone Protection and Planning (House of Commons, 1992) the group, with the agreement of MAFF and DOE, is promoting the development of a Regional Shoreline Management Plan (RSMP).

The preparation of the Regional Plan is through a "bottom up" approach. Each participating MDC would develop a Local Shoreline Management Plan (LSMP) and the NRA a Shoreline Management Strategy. This approach will deal with identification of the coastal processes affecting the shoreline, the hazards which those processes produce, standards of protection to be adopted and the development of an overall strategy for the engineering management of the coastline. An integrated strategic approach to shoreline management which will encompass engineering, environmental and economic criteria.

All such local plans would be compared and the major issues which they contained would, in combination with the NRA Shoreline Management System, form the basis of the RSMP.

Each MDC would also combine its LSMP with its other plans concerning the coastal zone (e.g. Tourism Strategy, Environmental Charter) to prepare a detailed strategy for the whole of its coastal zone. In turn the elements of this strategy would be incorporated into the District Development Plan. This would also include important statements concerning flood risk resulting from NRA inputs extending the Plan areas to the protected hinterland.

In this way, important aspects of the management by the MDC or NRA of the shoreline within a District would be enshrined in enforceable policy statements of the Local Development Plan.

Conclusions

Past practice in the UK has resulted in sea defence and coastal protection works being planned, implemented and managed on a piecemeal ad hoc basis.

The Shoreline Management System (GIS) is providing the operating authorities with the information necessary to develop technically sound, cost effective and environmentally sensitive shoreline management strategies for the Anglian Region of eastern England.

The production of a Regional Shoreline Management Plan under the aegis of ACAG is an important step forward in developing an integrated approach to the management of the national coastline.

The shoreline management approach is a first in the UK and sets a course for coastal zone management application in the foreseeable future.

References

Child M W 1992, The Anglian Management Study, *ICE Coastal Management 1992*, Blackpool, England.

House of Commons 1992 Environment Committee 2nd Report Coastal Zone Protection and Planning, House of Commons Paper 17-1.

ICE 1953, Conference on the North Sea Flood - *Conference Proceedings*, ICE, London, England.

NRA 1991, The Future of Shoreline Management, *Conference Proceedings*, NRA, Anglian Region.

Oakes T A 1992, Anglian Coastal Authorities Group, *ICE Coastal Management 1992*, Blackpool, England.

Pethick, J, Leggett D 1993, Morphology of the Anglian Coastline, *Proceedings Coastal Zone 1993*, New Orleans, USA.

Runcie R, Barham P, 1992, Project Development : An Anglian View, MAFF, Loughborough, England.

Townsend Lt Cdr J 1979, *Meteorological Magazine*, Vol 108, 1979 pp 147-153.

Townend I H, Leggett D, 1992, GIS For Shoreline Management, *Eurocoast 1992*, Kiel, Germany.

APPENDIX A

SHORELINE MANAGEMENT GEOGRAPHIC INFORMATION SYSTEM

TYPE	FORM	DISTRIBUTION	SOURCE	UPDATE
AGRICULTURE	Database, map		Land grade parcels from ADAS (1988)	Bi-annually
AMENITY	Map			Annually
COASTAL WORKS	Database, map		Herlihy report (1980)	Annually
ESTUARIES	Database, map		Field Survey, 1:10,000 and 1:50,000 OS maps. Oblique and	5-yearly
Estuary backshore	Database, map		vertical aerial photographs	
GEOLOGY			BGS maps or 1989 survey	None
1:10,000 survey	Map	12 sites	corridor extends 5km offshore by 2km wide	required
1:100,000 survey	Map	4 sites		
Foreshore - hinterland	Database, map	Whole coast		
INDUSTRY	Map			
Dredge site	Map		(Borough and County Council,	Bi-annually
Dump site	Database, map		Crown Estates and DoE)	
Port	Database, map		"Lloyds Ports of the World" 1988	Intermittently
Outfall	Database, map		Anglian, NRA	Annually
INFRASTRUCTURE	Map			Annually
MORPHOLOGY	Database, map		1:10,000, 1:50,000 OS maps and	
Backshore beach	Database, map		field survey (1989)	5-yearly
Backshore marsh	Database, map			5-yearly
Barrier beach	Database, map		and estuaries survey	5-yearly
Foreshore	Database, map			Bi-annually
Hinterland	Database, map			Intermittently
Hinterland/backshore	Database, map			Intermittently
Nearshore	Database, map		1:10,000, 1:50,000 maps	5-yearly
Offshore	Data model			
Landscape	Database, map			Intermittently
MILITARY	Map		OS Maps	Intermittently
PHOTOGRAPHS	Database, map	5 locations	ADAS, 1991, Scanned & aerial photos	Intermittently
SEDIMENTOLOGY	Data model			
HYDROGEOLOGY	Data model			

TYPE	FORM	DISTRIBUTION	SOURCE	UPDATE
TOPOGRAPHY Land Contour Beach survey	Data model Database, map	1km interval	BMT survey (1989) and subsequent NRA surveys along whole coast to LWM Summer 1989, Summer 1991, Winter 1992	Bi-annually
Bathymetric survey and contours	Database, map Map	1km intervals	NRA survey 2-4km offshore and Admiralty charts	5-yearly
CURRENTS Measured Modelled	Database, map Database, map	29 sites 1163 nodes	MIAS database current measurement records and modelling to cover Southern North Sea	
RAINFALL	Data model only			
RIVER DISCHARGE	Data model only			
TEMPERATURE	Data model only			
WATER LEVELS	Database, map	85 sites	Proudman long term records and shorter term NRA records	5-yearly
Constituents	Database, map			5-yearly
Extremes	Database, map			5-yearly
Historical tides	Database, map			Intermittently
Modelled	Database, map	1163 nodes		
Constituents	Database, map			Not required
Tide	Database, map			5-yearly
WAVES	Database, map	38 locations approx every 10-15km along the coastline		
Average	Data model only			
Energy	Data model only			
Extremes	Database, map			
			10 year hindcast studies	
WIND	Data model only			
BEACH PROFILES	Data model only			
LAND SUBSIDENCE	Data model only			
SHORELINE MOVEMENT	Database, map	250m spaced profiles	OS 1:10,000 maps from 1880-1970, 1567 calculation points along the open coast	1880, 1990 1950, 1970
SEA BED MOVEMENT	Data model only			
PROCESS CHARACTER	Database, map		SMS database information	

TYPE	FORM	DISTRIBUTION	SOURCE	UPDATE
PROCES FORCING	Database, map		SMS database information	
PROCESS POLICY	Database, map		SMS database information	
PROCESS RESPONSE	Database, map		SMS database information	
PROCES UNITS	Map		Hinterland geomorphology and landscape character data sets	

Morphological Trends of the Belgian Coast shown by 10 Years of Remote-Sensing based Surveying

P. De Wolf[1], D. Fransaer[2], J. van Sieleghem[2], R. Houthuys[3]

Abstract

The morphological evolution of foreshore, backshore, and adjoining dunes of the entire Belgian coast is surveyed by means of aerial remote sensing on a yearly basis since 1983. The evolution over this 10-year period is comprehensively illustrated by the land- and seaward shifts of the low-water line, high-water line and dune foot. It appears that nearly all major displacements are linked to human interactions, such as beach nourishments, groyne construction, etc. If these are left out of consideration, the main trend of the coastline evolution would be a landward retreat. Since 1986, the results of the bathymetric BEASAC® measurements of the nearshore are linked to the aerial remote sensing surveys. The combination of these two observation techniques has proved very effective for the monitoring and understanding of coastal morphodynamic processes.

Introduction

The Belgian coast is situated along the smooth, sandy southeastern shores of the southern bight of the North Sea. The morphological behaviour of this type of coast is the result of a dynamic equilibrium, determined by winds, waves, and tides. Changes in the boundary conditions, either natural or man-induced, will cause a response in this behaviour. Adequate monitoring is therefore very important for a rational management of the coast.

The morphological evolution of foreshore, backshore, and adjoining dunes of the Belgian coast is surveyed by means of aerial remote sensing on a yearly basis since 1983. For a profound understanding of the coastal morphological processes it is however necessary that the nearshore be included in the regular surveys. In Belgium, the nearshore of the coastal stretches subject to erosion is very efficiently and accurately surveyed by means of the hydrographic BEASAC® hovercraft[4] since August 1985. The results of these bathymetric measurements are routinely linked to

---

[1]   Senior Engineer, Service of Coastal Harbours, Vrijhavenstraat 3, B8400 Oostende, Belgium

[2]   Project Leader, Eurosense Belfotop N.V., Nerviërslaan 54, B1780 Wemmel, Belgium

[3]   Project Engineer, Eurosense Belfotop N.V.

[4]   BEASAC®: Belfotop Eurosense Acoustic-Sounding Air Cushion

the aerial remote sensing surveys to produce charts and volume data that cover the whole area from dunes to sea bottom.

This paper briefly describes the observation methods and the processing of the measurements. An account is given of the morphological changes that have been observed on foreshore and backshore since 1983. Results of the integrated beach and nearshore morphological monitoring are presented as a case study of the behaviour east of Zeebrugge harbour since 1986.

Remote-sensing based surveying of beach and nearshore morphology

The monitoring programme for the 65 km long Belgian coast combines two different observation techniques. One technique is based on aerial remote sensing; the other consists of bathymetric surveying by hovercraft.

Aerial remote sensing

Yearly, a survey flight is carried out in spring over the entire coast; additional recordings are performed in autumn for the most threatened parts of the coast (Kerckaert et al., 1989). The photo flight with a specially equipped aircraft takes place at low tide. The aircraft is a twin-prop offering optimal stability for data acquisition operations at low altitudes.

The entire beach and dune foot is stereoscopically covered using a Forward Motion Compensated (FMC) aerial photogrammetric camera. The FMC camera allows large aperture times, thus establishing sufficient detail even when modest light conditions occur during the flight. The low flight altitude (440 m) is important to measure heights with adequate precision. The planimetry and altimetry of the entire Belgian beach and dune foot area, together with some 5 km of the Dutch coast, are stereophotogrammetrically digitized after each flight.

The benefits of a survey method based on aerial remote sensing are obvious. The method is very fast, giving an instantaneous recording of the present situation. The precision in height and positioning is excellent over so vast an area characterized by a continuously changing morphology. Moreover, the method is not labour intensive and thus cost-efficient.

To facilitate plotting of height difference maps and the calculation of volume differences, etc., a digital terrain model (DTM) is established. A DTM is a numerical description of the terrain elevation, relating to a particular survey. It is essentially an enumeration of height values, geographically arranged in a rectangular or square grid. The unit distance between two adjacent DTM points determines the resolution of terrain coverage. The unit grid distance is selected in function of the detail required. In order to support the foreshore and dune DTM, a sufficiently large number of over 200 height points per ha is stereophotogrammetrically digitized.

BEASAC® acoustic depth soundings

The development of the BEASAC® hovercraft-based hydrographic system was undertaken by Eurosense in the early 1980s, with a special grant of the Belgian authorities (at present : the Ministry of the Flemish Community, Administration of Waterinfrastructure and Maritime Affairs). The system has been operational and cleared for the production of official nautical charts since 1985 (Maes, 1985; Kerckaert et al., 1989).

The advantages of employing a hovercraft platform in hydrography are numerous: fast operational speed, ability to measure in very shallow areas and high manoeuvrability and stability. The problems that had to be overcome during the development of the system mainly relate to the high rate of data acquisition, which is inherent to the high operational speed; the improvement and integration of positioning data; and the design of a system to compensate for the movements of the craft (De Putter *et al.*, 1992). The BEASAC® system integrates a sufficient number of redundant sensor data to improve positioning and to correct depth measurements. Some of the corrections, including positioning and filtering of the raw depth data, are carried out in real time. Other corrections are performed at Eurosense's Zeebrugge data processing centre. The most important of these post-survey corrections is the tidal reduction, using the M2 method. For soundings in the nearshore, use is made of simultaneous tide measurements performed on-purpose.

The tidal range at the Belgian coast is 4 to 5 metres, while the overall beach slope is gentle (1 - 2 %). As the aerial remote sensing recordings are performed at low tide, and the bathymetric BEASAC® soundings take place around the same day, but at high tide, an overlap in data is achieved in a strip of some 150 to 300 m wide. In this overlap zone, the deviations in terrain height between the two methods are less than 10 cm, and the height data obtained by the two survey methods can thus without difficulties be joined in an integrated data set.

From this combined data set, again a DTM is established, that covers the coastal relief in a 1.5 km wide strip stretching from the dune crest down to the sea floor.

Analysis of the coastal morphological evolution

The evolution between a given survey and an earlier reference survey is obtained by calculating the difference DTM. Its map representations, in which height differences are colour coded, make the study of erosion and accretion patterns very easy. The volume of accretion or erosion can also be derived from the difference DTM with very good precision. For the purpose of the monitoring programme, the Belgian coast has been divided into some 280 sections. The average section width is 250 m in a coast parallel direction. A section often coincides with the area between two groynes. Fig. 1 is a schematic representation of a typical section. It illustrates how dune, backshore, foreshore, nearshore, and sea bottom are defined along the Belgian coast. After each survey, the volumes of the different areas shown in the figure are calculated.

Analyses of the evolution in time of these areas are thus easy to perform. In several sections, it was found that volume changes (either due to accretion or erosion) are an almost linear function of time. In order to quantify this change, the mean annual rate of erosion or accretion per metre coast length is calculated for each section using a linear regression on the data of all the surveys. The regression fit, as expressed by the regression coefficient, allows to evaluate the reliability of the calculated accretion and erosion rates. A high regression coefficient means that there is a good fit between the calculated trend and the survey data.

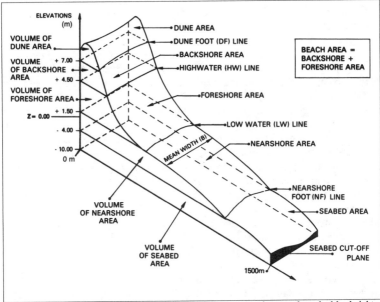

Figure 1:   Definition of dune, backshore, foreshore, nearshore, and sea bed by height
planes in a beach section. Heights in m above Z (= MLLWS).

Morphological evolution of the beaches and the dune foot between spring 1983 and spring 1992

The evolution of foreshore, backshore and adjacent dunes over the period from spring
1983 to spring 1992 can well be illustrated by showing the shift with time of the
"beach lines" : the low-water line (LW), the high-water line (HW) and the dune foot
(DF). These lines are height contour lines as defined in fig. 1. Their shifts are given
in the lower half of figs. 3, A-E (see fig. 2.1 for the situation of the different coast
stretches).

The lower half of fig. 3, A-E, is actually a map representation of the three beach
lines, showing both their 1983 and 1992 position. When the shift in position was
seaward, the area between the 1983 and 1992 lines has been shaded. A landward
shift, indicating erosion, has been left blank (see legend in fig. 2.2).

The upper half of fig. 3, A-E, is a graph representing the mean erosion rate (negative
value) or accretion rate (positive value) per year for each coastal section. The section
position and width has been plotted along the X axis using the same scale as in the
map representation below. The mean rate of erosion or accretion ("mean volume
difference per metre and per year") has been expressed by the height of the section
boxes. These boxes have been shaded according to the value of the regression
coefficient of the linear trend fit explained above (see shade legend in fig. 2.2).

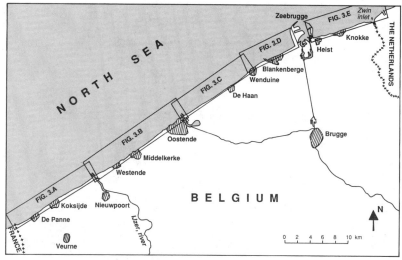

Figure 2.1: Overview map of the Belgian coast. Situation of the coast stretches represented in fig. 3A-3E.

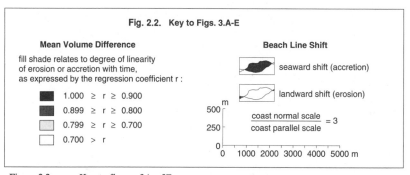

Figure 2.2: Key to figures 3A - 3E.

Between the French border and Nieuwpoort (fig. 3.A), significant accretion occurs in Koksijde and Nieuwpoort-Bad. There is a clear link between this morphological evolution and the construction of new, long groynes some years ago. The accretion in Koksijde is also related to yearly beach scrapings that include the supply of a few thousands of cubic metres of sand brought from outside that area. West of De Panne, there is a slight landward beach retreat.

Almost all of the shore between Nieuwpoort and Oostende (fig. 3.B) is protected by a continuous seawall and groynes. Again, a positive influence is seen near recently constructed groynes. A problem area exists east of the Nieuwpoort harbour entrance (left in fig. 3.B). The erosion here was not successfully countered by the Longard system, applied in 1981.

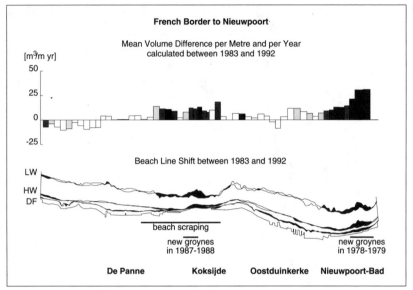

Figure 3A: French border to Nieuwpoort

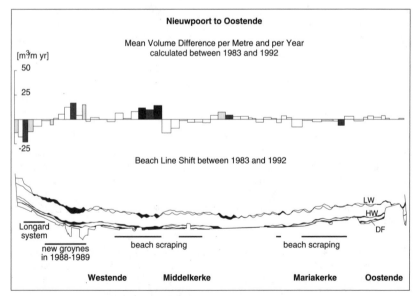

Figure 3B: Nieuwpoort to Oostende

Midway between Oostende and Wenduine (fig. 3.C), especially around De Haan, the dune foot line, high-water line, and particularly the low-water line, have been retreating significantly landward. This beach stretch without groynes west of De Haan suffers sand losses of almost 20 m³/m year. Here again, the erosion continues in spite of the Longard system that was applied along with a beach nourishment in 1978-1980. The erosion rates increase in the direction of De Haan. In the latter resort, however, no resultant volume losses are recorded. This is the result of a yearly effort to supply sand; over 150,000 m³ or 150 m³/m year have been applied since 1982. It is clear that in this area, a structural solution to halt the erosion was needed. As a "first" along the Belgian coast, an underwater berm was applied here in 1991-1992. Simultaneously, a beach nourishment was performed in De Haan in the late spring of 1992. The underwater berm is expected to act as a sand feeder berm for the De Haan beach (Helewaut & Malherbe, this volume). A specific monitoring programme is being performed to evaluate this experiment.

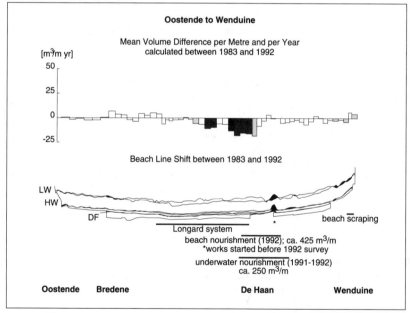

Figure 3C: Oostende to Wenduine.

In Wenduine and Blankenberge (fig. 3.D), the beaches have accreted since 1983. In Blankenberge again, human intervention is at the base of the accretion: construction of new groynes, beach scraping, and a yearly sand supply of some 25,000 m³ from dredging works in the Blankenberge harbour channel. Between Blankenberge and Zeebrugge is a 2 km long erosive beach stretch. The erosion here deserves special attention, the protective dune ridge being only a few tens of metres wide.

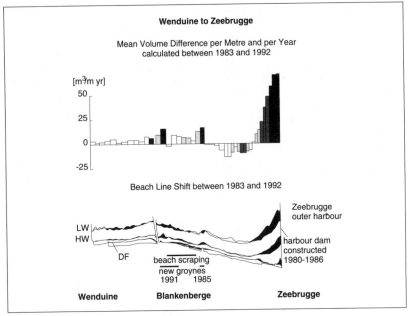

Wenduine to Zeebrugge

Mean Volume Difference per Metre and per Year
calculated between 1983 and 1992

Figure 3D: Wenduine to Zeebrugge.

The 1 km long beach immediately west of Zeebrugge harbour has accreted very much since the start in 1980 of the construction of the new outer harbour breakwaters. The breakwaters were completed by 1986, but the accretion is still going on today. Accretion is observed especially after stormy periods. The longshore sediment transport by wind and waves having a predominant west to east direction, much of the sand set free by storms is caught by the obstruction formed by Zeebrugge's western harbour breakwater.

Even more direct human interventions determine the morphological evolution of the beaches east of Zeebrugge (fig. 3.E). The longer-term resultant of the longshore sediment transport is directed from west to east. In the stretch of coast east of Zeebrugge, erosion was therefore expected after the construction of the harbour breakwaters, since the longshore transport would be intercepted by these breakwaters. A beach nourishment in 1977-1979 over 9 km of coast length, totalling over 8 millions m³, anticipated this negative evolution, and at the same time created wider and touristically more attractive beaches for the seaside resorts of Heist and Knokke. Like expected, most of the supplied beaches suffered gradual erosion afterwards. The most severe erosion was observed in a 2.6 km long stretch of beach at Knokke centre, at the most prominent part of the coast in that area (fig. 3.E). Here, in 1986 a second beach nourishment of 1 million m³ was carried out. The entire volume of sand supplied on foreshore and backshore was again gradually lost by erosion, over a 5 years' period. The apparent zero balance suggested by fig. 3.E has thus in reality been achieved by considerable efforts. The evolution of the beach at Knokke is discussed in more detail in the following section.

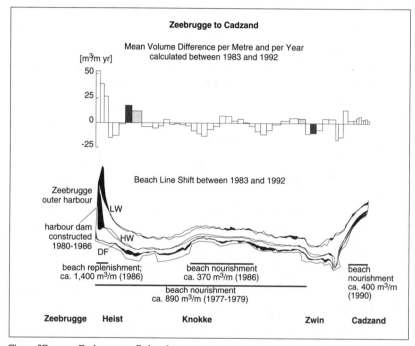

Figure 3E: Zeebrugge to Cadzand.

It is clear from this brief account of the morphological evolution of the Belgian coast during the past 10 years, that almost everywhere man-induced influences interact with natural evolution. According to their influence on coastal morphology, human interactions can be classified in 4 major groups :

-     beach scrapings: they constitute minor interactions and are carried out yearly in Koksijde, Westende, Middelkerke, Mariakerke, Wenduine, and Blankenberge. Sometimes they are accompanied with the supply of a few thousands m³ of sand from other areas. In the resorts mentioned this suffices to keep a stable beach. No harmful effects were noted at the adjacent beaches, much like Bruun's (1983) observations on US and Danish coasts;
-     beach nourishments involving the application of huge volumes of sand. They have been performed near De Haan and in Knokke-Heist and must be repeated after 5 to 7 years;
-     the construction, during the last decade, of 300 - 400 m long groynes in Koksijde, Nieuwpoort-Bad, Westende, and Blankenberge. They caused a marked local accretion. According to Dutch experience (Rijkswaterstaat, 1988;

Verhagen, 1989), groynes exert no marked long-term influence on beach development; however, when tidal channels are present, the groynes halt the migration of the channels. So far, two new groynes at Koksijde have successfully countered the migration of the tidal channel "Potje", which confirms the Dutch experience. In the other sections of the Belgian coast where new groynes have been built, no tidal channel was involved but beach accretion has nevertheless been observed in every case;

- the expansion of Zeebrugge outer harbour, with the construction in 1980-1986 of new, 3-4 km long breakwaters, continues to have a considerable influence on the evolution of a 1 km long beach west of the harbour, where accretion is observed, and over 5 km of beaches east of the harbour, where mostly erosion occurs.

Not apparent in fig. 3.A-E is the positive influence of osier hedges in the dune foot area. Osier hedges have been planted and kept along half of the Belgian coast length, mostly in the parts where no seawall is present. Thanks to this effort, hundred thousands of m^3 of sand have been trapped since 1983.

Human interventions make it difficult to single out the natural evolution. Longer-term beach observations however indicate that erosion and landward retreat is the dominant evolution at the Belgian coast.

At the Dutch coast observations over the last century have shown a cycle of beach erosion and accretion, propagating like a wave along the coast line ("sand waves", Verhagen, 1989). The existence of a similar morphological system at unprotected beaches of the Belgian coast has been suggested ("erosive megaprotuberances", De Moor, 1979, 1981). The monitoring programme of the Belgian coast as yet gives no formal evidence of the existence of such a system.

Evolution of the beach in relation with the nearshore : case study east of Zeebrugge

From 1986 on, systematic bathymetric surveys have been carried out in the nearshore and the adjacent part of the sea bottom, using the BEASAC system. The bathymetric data are integrated with the aerial remote sensing data into one terrain model. To illustrate this integrated approach, the evolution of the situation east of Zeebrugge harbour is presented here as a case study.

After the large-scale beach nourishment had been executed in 1977-1979 over a length of 9 km, the erosion was most intense at Knokke, from 4.1 to 6.7 km east of Zeebrugge. In a more localized, second beach nourishment, another 1 million m^3 were applied in the erosive stretch. This nourishment restored the 1979 situation at Knokke.

The monitoring programme based on integrated bathymetric and remote sensing data, shows a continuing erosion of the beach, together with a partial and temporary storage of eroded sand on the nearshore. The erosion rate on the Knokke beach after 1986 was higher than before (fig. 4). After the second, local nourishment, the total volume of sand, applied on fore- and backshore, was taken away by erosion in less than 5 years.

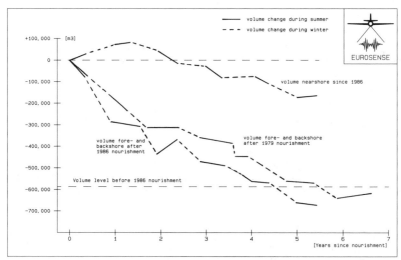

Figure 4:   Volume evolution of the 2.6 long Knokke beach after 1979 and 1986
                 nourishments.

Time series of beach profiles show a relatively steep ramp at the high-water line that
gradually retreats landwards. This morphological evolution was first noted in the
westernmost sections, and is spreading to the east, in the direction of the Dutch
border.
The nearshore at Knokke showed accretion in the first year after the nourishment (fig.
4). After that time, erosion set in. In the Knokke nearshore area, erosion was also
first observed in the westernmost sections, with the erosive zone later extending
eastward.
The beach profile evolution of the Knokke beach after the 1986 nourishment shows
typical characteristics of sand losses due to wave attack. The fore- and backshore
volume evolution graph of fig. 4 illustrates this: winter erosion rates systematically
exceed summer erosion rates. Part of the eroded sand is temporarily stored on the
nearshore. This fact is not only deduced from the general trend of the nearshore
evolution cui ve in fig. 4, but also from the observation that nearshore volume
reductions are milder in winter than in summer. The gradual eastward extension of
the erosion is in agreement with the long-term direction of the longshore sediment
transport at the Belgian coast.
Knokke's nearshore coincides with the landward flank of a tidal channel named
"Appelzak". Since observations began in 1986, this channel tends to become wider
and longer (fig. 5). This change might reflect an adaptation of tidal flows to the new
situation created by the expansion of Zeebrugge. Whatever may be at the origin of
the Appelzak's widening and prolongation, it is observed that beach-derived sand
deposited on the channel's south flank is in all conditions cleared away. This process
is a matter of only a few months.

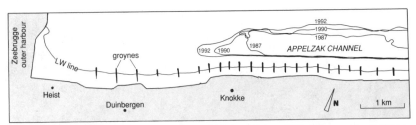

Figure 5:   Sketch map of the Appelzak tidal channel since 1987. The expansion is shown
                using the 1987, 1990 and 1992 -6m depth contour line.

The analysis of this morphological evolution shows that no clear-cut solution to
Knokke's erosion problems is available. As long as no fundamental change can be
imposed on the Appelzak's evolution, new beach nourishments will continue to be
needed.

The Knokke case study clearly shows that the beach evolution is closely linked to the
behaviour of the nearshore. Obviously, the analysis of the morphological evolution
is facilitated by the integrated approach, in which aerial remote sensing height data
are combined with bathymetric data.

Summary and conclusion

Ten years of monitoring of the Belgian coast by remote sensing show a rather
complex alternation of stretches of erosion and accretion. Overall however,
approximately 30 % of the Belgian coast is erosive, a same portion is accretional,
while the remaining 40 % can be classified as stable. With the exception of the
accretional coast immediately west of Zeebrugge, almost all of the accretion can be
related to human intervention such as beach scraping (along 17 % of the coast),
beach nourishments (8 %) and the construction of new, 300-400 m long groynes.
Without this successful interference, over 60 % of the Belgian coast would suffer
erosion.

The combination of the results of two observation techniques, one for the shore and
dune foot area based on aerial remote sensing, and the other for the nearshore using
hovercraft bathymetric soundings, has proved very effective for the monitoring and
understanding of coastal morphodynamic processes.

References

Bruun, P., 1983. Beach scraping - is it damaging to beach stability ? *Coastal
    Engineering*, 7, p. 167-173

De Moor, G., 1979. Recent beach evolution along the Belgian North Sea coast. *Bull.
    Soc. belge de Géologie*, 88, p. 143-157

De Moor, G., 1981. Erosie aan de Belgische kust. *De Aardrijkskunde*, 1981-1/2, p.
    279-294

De Putter, B., De Wolf, P., Van Sieleghem, J. & Claeys, F., 1992. Latest
    developments in hovercraft bathymetry. *Paper presented at Hydro 1992*,
    Copenhague

Helewaut, M. & Malherbe, B., 1993. Design and execution of beach nourishments in Belgium. *Paper present in this volume.*

Kerckaert, P., De Wolf, P., Van Rensbergen, J. & Fransaer, D., 1989. Measurements of Sedimentary Processes from a Fast Moving Air Cushion Platform in Combination with Remote Sensing Techniques. *in :* Coastal Zone '89 (Proceedings of the Sixth Symposium on Coastal and Ocean Management, Charleston, South-Carolina, July 1989, ed. by O.T. Magoon, H. Converse, D. Miner, L.T. Tobin & D. Clark), p. 729-734

Maes, E., 1985. De ontwikkeling van een meetvaartuig voor snelle lodingen in de Noordzee : het 'BEASAC'-project. *Water (Tijdschrift over Waterproblematiek),* 25 (nov.-dec. 1985), p. 174-177

Rijkswaterstaat, 1988. Handboek zandsuppleties (Manual on Artificial Beach Nourishment). *Dienst Getijdewateren,* Middelburg, The Netherlands, 310 p.

Verhagen, H.J., 1989. Sand Waves along the Dutch Coast. *Coastal Engineering,* 13, p. 129-147

Design and Execution of Beach Nourishments in Belgium

M. Helewaut[1], B. Malherbe[2]

Abstract

When a severe coastal erosion problem has to be solved, two kinds of remedial/ coastal protection works are usually taken in view: the first one, most often used in the past, requires the construction of structures such as groynes and breakwaters; the second one aims at providing a coastal protection solution by restoring or modifying the beach profile. Factors to be considered when evaluating the possible solutions are environmental conditions and sediment transport mechanisms.

In fact, engineering design methods and human concern, are evolving, which leads to a new concept of coastal defence works.

This trend is illustrated by case studies of projects carried out along the Belgian coast.

Introduction

The Belgian coastline is characterized by a 65 km long stretch of sandy beach/dune barrier. The coastal barrier is a dynamic system in balance with the prevailing hydrometeorological conditions (sea-level, waves, tides, winds) and is thus subjected to fluctuations in location and height of the beaches and dunes (fig. 1).

Since the early 1900's several constructions along this coastline have been built aimed at providing coastal protection for the polder hinterland or at allowing tourist resorts and urban developments.

These infrastructures for coastal protection or coastline fixation are essentially combinations of sloped sea-walls, groyne-fields and beach profiling/replenishment works.

The growing urban development and touristic exploitation of the coastal area, together with the assessment of eroding coastal sections, create the need for integrated and flexible coastal protection actions. Therefore several beach nourishment projects have been designed and/or executed during the last 15 years.

---

[1]   Ministry of the Flemish Community, Service of Coastal Harbours, Vrijhavenstraat 3, B-8400 Oostende, Belgium

[2]   Harbour and Engineering Consultants, Deinsesteenweg 110, B9810 Gent, Belgium

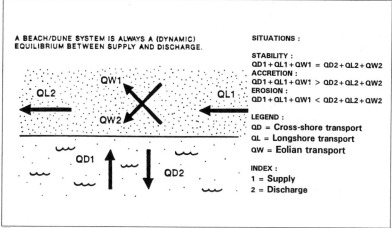

A BEACH/DUNE SYSTEM IS ALWAYS A (DYNAMIC)
EQUILIBRIUM BETWEEN SUPPLY AND DISCHARGE.

QW1
QL2
QL1
QW2
QD1
QD2

SITUATIONS :

STABILITY :
QD1 + QL1 + QW1 = QD2 + QL2 + QW2
ACCRETION :
QD1 + QL1 + QW1 > QD2 + QL2 + QW2
EROSION :
QD1 + QL1 + QW1 < QD2 + QL2 + QW2

LEGEND :
QD = Cross-shore transport
QL = Longshore transport
QW = Eolian transport

INDEX :
1 = Supply
2 = Discharge

Figure 1 :   Equilibrium mechanisms governing the stability of a beach/dune system.

Coastal defence works

Common shore defence systems include hard structures such as groynes, detached breakwaters and sea walls.

Groynes aim to reduce the longshore sediment transport in a given coastal area. Detached breakwaters are build in order to reduce the wave action on the coastline.

Sea walls are build to prevent the coastline from further regression and have sloped masonry structures to achieve a progressive wave breaking.

Beach restoration/replenishment works aim at restoring the deficitent sand balance of the system and at restoring a natural flexible coastal protection. In some cases projects combine sand suppletion with construction of hard defence structures.

In lowlands such as Belgium and the Netherlands, many beach/dune systems act as the ultimate coastal protection. These sandy beach/dune systems locally suffer from erosion. This erosion is a result of a desequilibrium between sand supply and sand-discharge by longshore, cross-shore and/or eolian transport. A deficit in sand-supply causes erosion of the beach/dune system. This deficit may originate from natural causes (e.g. regional, sedimentological variations, sea-level changes) or man-made causes (e.g. coastal structures, sand-excavations, dredging).

Groynes and breakwaters will only affect the wave regime and sediment transport in the immediate area. They do not create solutions for the adjacent coastal areas where the sediment transport capacities are unchanged. On the contrary, they may cause a spectacular increase in sediment transport on the lee-side of the structure causing renewed coastal erosion. In fact, the problem has merely been moved from one area to another.

Knokke-Heist project (1977-1979)

The winter storms of 1976 left a seriously eroded 9 km long beach in front of the seawall and groyne field of the Knokke-Heist municipality. Predicted sand losses and coastline regressions were compared, which corresponded to different alternatives including sand suppletion and groynes. In fact, total capital and maintenance costs of beach nourishment, with or without groynes, were not significantly different.

According to the design, coastal protection works were executed consisting of a beach replenishment scheme, whereby ca. 8,5 million m³ of a selected sand-type was landfilled on the existing eroded beach-profile. A cross-section of the beach replenishment realized by means of hydraulic pumping is shown in figure 2.

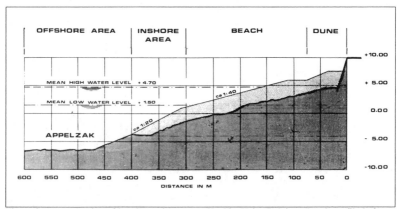

Figure 2:   Design of beach nourishment at Knokke-Heist, Belgium (beach profile before and after nourishment).

The sand was trailed/dredged in sea (access channels, sandbanks) and dumped in a borrow pit in the Zeebrugge harbour, where a cutter suction dredger excavated it to pump it as a hydraulic mixture via a 800 mm beach pipeline with booster stations (max. 5) down to the fill area. The works started in 1977 and ended in 1979 ; a fill density along the 9 km coast of ca. 1000 m³ of sand per running metre was realized.

The monitoring of this beach nourishment project shows clearly that material is eroded by waves, tides and winds from the dry and tidal beach area and displaced to the nearshore area (below the LW-line by cross-shore transport) and the dune area (by eolian transport). The net erosion of this coastal section however seems to be linked to the presence and the evolution of a nearshore flood tidal gulley, the "Appelzak" (Kerckaert et.al., 1985; De Wolf et al., 1993).

Due to the local strong erosion, an additional beach replenishment of appr. 1 million m³ of sand was executed in 1986. This beach replenishment volume has now (1993) almost been completely eroded from the dry beach.

Koksijde project

The beach at Koksijde suffers from regular and very local erosion due to the presence of a tidal gulley "Potje" close to the coastline. This coastline is naturally sheltered from strong wave-attack by shallow sand-banks. Despite a groyne-field the erosion continues.

The area around Koksijde is characterized by an active flood channel, whose shallow North flank (i.e. the Broersbank) plays a large part in influencing the wave height at the beach, and apparently reduces it. At the flood channel head, sand transport occurs from a SW - NNW wave sector.

Due to the orientation of the flood channel, currents are forced to group together between the flood channel and the coast. This narrow band of currents is obviously the reason for the locally severe erosion.

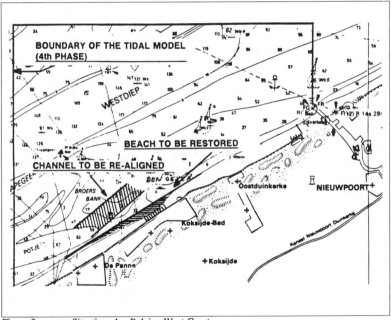

Figure 3:   Situation plan Belgian West Coast
Description of retained compensation dredging.

Five possible solutions were investigated :

A.   Rectification of the flood channel by means of excavation and nourishment techniques (compensation dredging): a.   excavation of the northern slope of the channel head and b. pumping this dredged sand back into the beach whilst maintaining the classic, hydraulic cross-section of the beach.

The head of a flood channel is usually made up of coarser sand so that the solution with combined excavation and nourishment back at the beach is usually the most attractive (fig. 3).

B.   Rectification of the flood channel by excavation and beach nourishment, including disposal of coarser sand (Kwinte Bank) as a protective cover layer.
C.   A connecting channel to be cut from the secondary flood channel north of the Broersbank.
D.   Extension of the existing beach groynes.
E.   Construction or installation of underwater groynes and beach nourishment.

An attempt was made to expose each possible solution to a multi-criteria analysis. The following criteria were used to weigh each solution :
    1.   Efficiency (coastal defence in the long-term).
    2.   Direct or indirect effects on coastal protection.
    3.   Hydraulic influence.
    4.   Technical feasibility.
    5.   Cost of execution.
This multi-criteria analysis isolated alternatives A and C as being most suitable for further investigation. The two alternatives were built into a mathematical tide model and the rate of decrease of the longshore current was calculated.
Solution A gave a sizable reduction in current velocities (up to 20 % at spring tide) near the major erosion areas. The influence of a possible connecting channel (solution C) gave only a slight velocity reduction in the extreme west of the problem area. The design study resulted in a solution of compensation dredging with subsequent beach nourishment and lengthening of existing groynes.

According to the design, the northern bank of the tidal gulley is to be excavated to deviate the strong flood currents in offshore direction. The sand to be excavated by a seagoing cutter suction dredger equiped with a floating discharge pipe (l = 2000m) will be re-used for the beach nourishment in Koksijde. Thus a total length of 3000m of beach could be restored with a fill-density of ca. 450 m³/m. At present, these works have not yet been started.

Ostend project
Since the end of the 19th century the beaches at Ostend, Belgium have been popular holiday resorts. However, steady erosion has occured, especially on the beach section between the Casino-Kursaal and the harbour entrance (fig. 4). Indeed, the beach front promenade was constructed along the line of the 17th century fortifications of the city, which protrude into the sea beyond the natural coastline over ca. 150 m. A system of huge perpendicular groynes, ca. 200 m long could only reduce the erosion. A flood protection improvement programme, realized after the dramatic storm surge of February 1953 and extended after the winter storms of January 1976, resulted in a small storm surge barrier wall on top of the seawall up to level +10.20 m and an associated storm surge sewer system. As this programme could have some results only on the short term, a long-term solution had to be proposed.

The requirements to be taken into account are as follows:
-    reduction of wave overtopping of the existing sea wall;
-    prevention of the existing port access channel from higher sedimentation due to the coastal defence works;

- consideration of future recreational beach activities ;
- limitation of the project within the existing groyne structure extension.

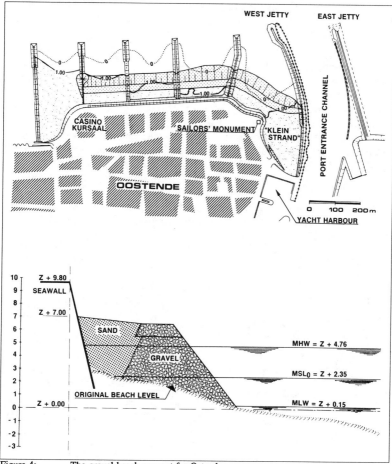

Figure 4:   The gravel beach concept for Ostend

After coastal engineering studies and model testing, a gravel beach supported by a sandfill core has proven to be a feasible and cost-effective beach restoration and flood protection alternative (Simoen *et al.*, 1988). The gravel beach concept is integrated in the existing coastal protection infrastructure, i.e. a sloped seawall with groyne-field. To execute the design illustrated on figure 4, a gravel/sand fill density of ca. 600 m³/m over a length of 850 m is required. At present, these works have not yet been started.

Some additional results could also be achieved with this flexible solution:
- new beach recreational facilities in safe conditions for the public;
- the economic incentive for the much needed re-development of this part of downtown Ostend and the beach front;
- the conservation to a large extent of the present sea-view from the sea promenade.

De Haan project (1990-1992)

The Belgian North Sea Coast at De Haan was seriously hit by the storm of 28.02.90 - 01.03.90 causing an intense erosion of the upper-beach/dune system. Due to the coastal protection function of this coastal barrier for the polder lowlands, the Administration of Waterways decided to execute coastal protection works.

The coastal protection study showed that the major coast-erosion mechanism is an intense cross-shore redistribution of sand in a ridge- and runnel beach system. The L.W.-bar is hereby acting as a buffer in which sand is accumulated in storm periods and from which sand is remobilized in onshore direction during calmer weather periods.

The same study showed that the most appropriate and cost-effective coastal protection system relies on the combination of a profile nourishment with a feeder berm on the L.W.-bar (fig. 5).

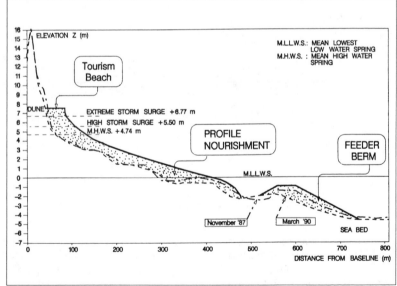

Figure 5:   Typical cross-sectional view of the dune/beach system at De Haan (Belgium). Illustration of a design with profile nourishment and feeder berm at the L.W.-bar.

This selection was done after a careful multi-criteria analysis of different solutions. The original concept of this beach protection/restoration system - first application in Europe - relies on the knowledge of the natural sediment transport and distribution mechanism.

Both structures, profile nourishment and feeder berm, aim at a restoration of the cross-shore sand budget maintaining the original beach slope (Malherbe B., 1989). The coastal protection system involves the nourishment of ca. 1.300.000 m³ of sand (fill density : ca. 750 m³/m) and is designed to use exclusively natural North Sea sediments excavated for the deepening of access channels and by the use of appropriate dredging equipment. The works were executed according to a Quality Assurance Plan. The shoreface nourishment works (1000 - 2000 m³ split hopper dredgers) for the 2,200 km long feeder berm started in February 1991 and ended in March 1992. The profile nourishment works were executed by 2 self-pumping hopper dredgers, a sinker pipeline (1500 m long), a booster with 2 discharges pipelines and bulldozers for beach profiling; the works started in April 1992 and ended in May 1992 (fig. 6).

Figure 6:   Profile nourishment works at De Haan (Belgium)

The results of the monitoring of the project area (by terrestrial levelling and nearshore echosounding) during the first year indicate that the feeder berm progressively feeds the upper beach up till L.W. + 1 m.

Eolian transport is quite significant. Therefore the design has foreseen the building of a wind-screen network on the dry-beach. This seems to be most effective after the first W-NW-storms of November 1992.

After these storms the natural beach evolution indicated a strong cross-shore redistribution of sand in a ridge and runnel system whereby the natural (and predicted) beach slope is obtained.

Despite the particularly windy period in the North Sea in October/November 1992 no significant sand losses could be measured. An exact project-evaluation will be done aftersome years of monitoring.

References

De Wolf P., Fransaer D., Van Sieleghem J., Houthuys R., 1993. Morphological trends of the Belgian coast shown by 10 years of remote sensing based surveying (*this volume*).

Kerckaert P., Wens F., De Wolf P., De Candt P., De Meyer Chr. P., Grobben A., 1985. Beach nourishment: a soft method for coastal protection, *26th AIPCN*, Brussels.

Malherbe B., 1989. Sediment management: a new concept for cost-effective programmation of dredging works, *Leading Ports II*, Buenos Aires.

Simoen R., Verslype H., Vandenbossche D., 1988. The beach rehabilitation project at Ostend, Belgium, *21st. ICCE*, Malaga.

Coastal Erosion Management of Sandy Beaches in Denmark

Ch. Laustrup[1]

Abstract

The Danish North Sea coast is a very exposed coast and has therefore suffered from severe erosion problems during centuries. Since 1982, coastal protection has been carried out in an increasing scale. The means used in coastal protection has shifted from hard structures to nourishment. Also the strategy of nourishment has changed. The reasons for this and the future strategy is discussed.

Introduction

The Danish North Sea coast is partly a sandy coast and partly a wadden sea coast. The Wadden Sea coast is the northern part of the Dutch-German-Danish Wadden Sea system (fig. 1). The morphogenesis is described in the paper 'The Morphology of the Danish North Sea Coast' (Jakobsen & Toxvig Madsen, 1993). The subject of this paper is the coastal erosion management of the sandy coast.

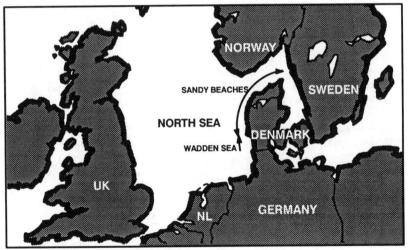

Figure 1:   The North Sea countries

---

[1]   Chief Engineer, Danish Coastal Authority, Højbovej 1, DK7620 Lemvig, Denmark

The sandy coast is probably one of the most exposed coasts in the North Sea area. Figure 2 shows a statistic of extreme wave heights recorded on 20 m water depth off Thyborøn. The length of the sandy coast is approximately 340 km. It is characterized by wide beaches and dunes. The tidal range in the northern part is 0.3 m and in the southern part 1.4 m. Besides there is a considerable wind setup during storm. This relationship is reflected in the statistics of extreme water levels (fig. 2).

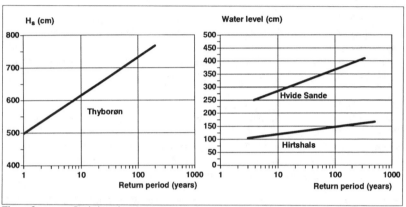

Figure 2:   Statistics of extreme waves and water levels

Erosion and sediment transport

The dominant direction of the wind with respect to sediment transport is WNW. This determines the direction of the sediment transport, (fig. 3) which north of Lodbjerg is largely to the north and south of Lodbjerg largely to the south.

There is however a secondary nodal point at Vrist which implies that the transport on the barriers between the two nodal points is to the south on the northern barrier and to the north on the southern barrier. The sediment from the barriers is carried through the inlet and deposited in the Limfjord to an amount of 1 million m³ per year.

This secondary nodal point was a result of the formation of a gap in the barrier in 1862. Before that time the barrier was unbroken. From 1862 till about 1900 the erosion of the barriers was so strong -10 m per year on average- that the barriers were reoriented and the direction of the sediment transport on the southern barrier was reversed. To reduce this large erosion a group of 54 groins was built from 1875 till 1908. The groins are 250 m long and they reduced the coastline retreat to 1 to 2 m per year.

In the 1930s this groin group was extended south of the secondary nodal point which resulted in severe lee side erosion south of the group with an average of 11 m per year for a number of years.

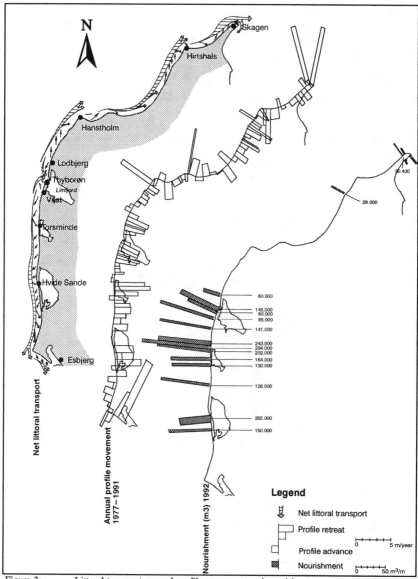

Figure 3:   Littoral transport, annual profile movement and nourishment

In this century the harbours at Hvide Sande and Torsminde were built. Both harbours caused severe lee side erosion south of the harbours.

Problem identification

North of Lodbjerg the land is generally high. South of Lodbjerg the land is generally low. The wide and high dunes in this area form the primary protection against flooding during storms.

As a result of the erosion the annual net loss of land in the period from 1977 till 1992 was 320,000 m². In this period shore protection works have been carried out (see below) but also there has been a higher intensity of storms than average.

In particular the erosion of the dune areas is a problem since the dune besides being an important recreational area also is the protection against flooding. Figure 4 shows the areas which are below a level of 2.5 m.

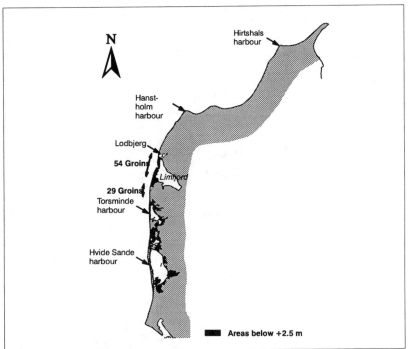

Figure 4:   Harbours, groynes and low areas

The policy of coastal erosion management

Since 1982 the policy of the Danish government has given priority to safety against flooding and to the protection of buildings, roads etc against erosion. This has led to the following guidelines in coastal erosion management:

-    To re-establish a safety level against flooding not below 100 years return period.

- To stop erosion where towns are situated close to the beach
- To reduce erosion on sections of the coast where erosion will reduce the safety against flooding below 100 years in the near future.

The 1992 budget excluding VAT for new structures and nourishment is appr. 62 million Dkr (10 million US $ Dec. 1992). A recent analysis has shown that this budget is not sufficient to stop the ongoing decrease in dune safety. Within the next 10 years this safety will come below 100 years if the budget is not raised. As a consequence of this a budget rise is being considered.

The funding is provided by the government and the local authorities. Coastal protection works carried out before 1982 was funded only by the government. From that year on, the local authorities paid half and the government paid half when new sections of the coast had to be protected.

Coastal erosion management since 1982

To re-establish the safety against flooding to a level of 100 years return period in particular two methods were used:
- Where the dune area still has some "depth", the safety is increased by building artificial sand dunes. This includes a reinforcement of local weak parts and a general reinforcement of the dune so that it has enough volume to resist a 100 year storm. The reinforcement is designed so that the dune still looks as natural as possible.

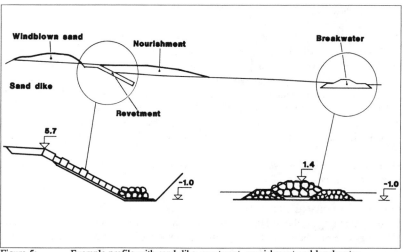

Figure 5:   Example profile with sand dike, revetment, nourishment and breakwater

- On sections where no dune was left after severe erosion, a sand dike was built. Normally there was not enough land available to build the sand dike so wide that erosion during storm could be allowed. Consequently the dike was built with a minimum width to prevent overtopping during the design storm and protected by a revetment to prevent erosion. The most common type of revetment is a concrete block type as shown on figure 5 and 6. Until the end of 1992, a total length of 20 km of this structure had been built and new dunes grew up on the top and the backslope of the artificial sand dike. As it is the policy to place part of the beach nourishment sand on the back beach, the revetment is normally always covered in sand and the structure looks like a natural narrow dune.

To stop or slow down coastal erosion in particular two methods were used:
- For a number of reasons which are mentioned below an important means to slow down coastal erosion on highly exposed sections was the use of detached low breakwaters as shown on figure 5 and 7. The typical breakwaters were built on 1 m water depth, they are 60 m long and built in groups with gaps of 40 m. With this design tombolos will exist under nearly all conditions but the breakwaters will not effect the sediment transport on the bars.
- The purpose of designing the breakwater protection as described was to maintain a relatively high beach without creating severe lee side erosion. It is therefore necessary to nourish the profile seaward of the breakwaters and also for exposed areas to nourish landward of the breakwaters. The need for nourishment landward is typically reduced by 50% or more by the breakwaters.

In the beginning of the 1980s, sand for beach nourishment was normally unloaded in a harbour and distributed along the beach by dumpers. The seaward end of the profile was nourished by dumping sand on the outer bar. Later the sand for beach nourishment was pumped directly on to the beach and the seaward profile was nourished by the dredger pumping over the bow. This method is considered more reliable than dumping on the outer bar since the sand will be placed landward of the inner bar. It is therefore more likely that the sand stays longer in the section where it is needed. This has been confirmed by analyses of field measurements.

The sand is dredged seaward of the 20 m contour except for the sand dredged in harbour entrances.

Trends in coastal protection management
In the beginning of the 1980s nourishment played a much smaller role than today. This was due to a number of reasons of which some of the important are mentioned below.

For historical reasons, especially the local politicians trusted in hard structures like groins and breakwaters. The main reason for that was the successful design and building of the large groins around the turn of the century. However, the argument that building long groins in a lee side erosion area would create more erosion was

Figure 6: Set concrete blocks revetment. Low and high beach

accepted. It was therefore agreed to build breakwaters with the main function to protect the beach.

The principle of beach nourishment was new to politicians. The argument that 'the erosion of the nourishment sand during storm is part of the plan' was hard to sell.

Figure 7:   Aerial photo of beach with breakwaters

The technology used for pumping sand onto a beach as exposed as the Danish North Sea coast was not very developed. Consequently, the price of sand delivered on the beach was relatively high and it was cost efficient to save sand by building breakwaters. The price of a m³ of sand delivered on the beach in 1985 was typically 30 Dkr (5 US $). Today the price is typically 18 Dkr (3 US $).

The environmental aspects of coastal engineering were not given the priority which they have today. Therefore the environmental advantages of using beach nourishment instead of hard structures was not a convincing argument.

The development in the use of nourishment and hard structures is shown on figure 8. In 1992, 80% of the budget was spent on nourishment and 20% on hard structures.

If the funding is increased in the coming years an increasing part of the budget will be spent on nourishment. The nourishment will be designed to serve two purposes which are a) to maintain the beach and the foreshore during normal weather conditions and b) to protect the dune and the dune revetment during storm conditions.

This means that a certain amount of sand should be placed on the beach and on the foreshore and a certain amount should be placed as a stockpile at the dune foot.

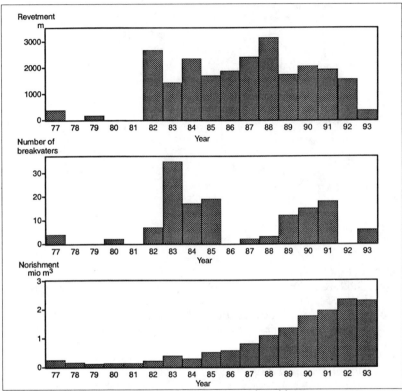

Figure 8:   Development in the use of hard structures and nourishment

According to experience with nourishment in different sections of the profile, the sand will probably be pumped directly ashore or pumped over the bow of the 129 dredger. Over-the-bow pumping can be used with good result where the water is deep enough for the dredger to get close to the beach so that the sand can be placed landward of the inner bar. It may then be combined with beach scraping so that a stockpile at the dune foot can be established. Since no pipeline is needed to pump the sand onto the beach, even small amounts may be delivered in a cost efficient way by this method.

Figure 9:   Beach nourishment over-the-bow

References

Jakobsen P.R. & H. Toxvig Madsen (1993). Morphology of the Danish North Sea coast. *This volume.*

Bezuijen, A., Wouters, J. and Laustrup, C., 1988. Block Revetment Design with Physical and Numerical Models, *21ᵗʰ Int. Conf. on coastal engineering, 1990,* p. 2159 - 2173.

Laustrup, C., 1988. Erosion Control with Breakwaters and Beach Nourishment. *Journal of Coastal Reasearch,* 4, p. 677 -685.

Laustrup, C., 1991. Beach Nourishment on the Danish North Sea Coast. *P.I.A.N.C. Bulletin,* no. 72.

Disposal of Dredged Material from the Estuaries

F.P. Hallie[1], G.A. Beaufort[2], M. van der Doef[2]

Abstract

The silting up of harbours and the lower parts of rivers are serious problems. These problems can be characterized in terms of quantity and quality. Two separated cases are described. One dealing with relatively large amounts of relatively clean dredged material. The other one dealing with smaller quantities, but far more contaminated dredged material.

Coastal Waters
The annual amount of dredged material from the Rotterdam harbour area varies in the period 1980-1990 between 7 and 23 million m³. This paper deals with the latest developed solutions.
A feasibility study including a policy analysis approach, resulted in several possible solutions which are generated and screened on technical, economical and environmental criteria. Indepth study is made for three alternatives: continue the present situation; a alternative deeper dumpsite offshore the present one and the use of created pits for dumping. The disposal area itself as well as the impact on the coastal zone and the adjacent Wadden Sea is taken into account. As a final result the ranking of the alternatives is discussed.

Estuaries and Inland Waters
As pollution particles adsorb strongly to silt and clay, the net result is a widespread pollution problem in all places where sedimentation takes place. Subsequently river mouths and harbours are the most severe pollution problems.
Real cleaning of soils, i.e. removing the pollutants from the silt and clay particles, is being studied. For the majority of the polluted sites, however, this is not an affordable solution and certainly not possible within a reasonable period of time.
For many areas in the Netherlands, Environmental Impact Assessment (EIA) studies are being made in order to develop acceptable, safe and isolated disposal sites.
For two large areas large dumpsites (30 - 40 million m³) are necessary. For these two sites individual EIA's are made.

---

[1] Ministry of Transport, Public Works and Water management, North Sea Directorate, P.O. Box 5807, 2280 HK Rijswijk, The Netherlands

[2] Ministry of Transport, Public Works and Water management, Civil Engineering Division, P.O. Box 30000, 3502 LA Utrecht, The Netherlands

Different types of depot can be designed: a depot on land, a pit or an island/atoll depot. Determination of the size of the depot and the use of local material is discussed.

An important issue for the disposal site is the principle of design; minimization of the emission (insulation), control and monitoring of a disposal site. Different methods of insulation are discussed and the benefit of these techniques is evaluated. Studies on long-term behaviour and construction methods for a number of liner-material continue.

Introduction

The Netherlands are formed by the sea and rivers. Large amounts of sediments were settled down in the delta of the rivers Rhine, Meuse and Scheldt. One can speak indeed of the "low lands". That means many advantages, such as harbour developments. It means also several disadvantages, such as the silting up of the lower rivers and estuaries. Dredging activities are necessary for maintaining the nautical depths of sea port entrances and harbours.

The lower part of the Netherlands is one of the most important economic zones, because of the harbours for very large ships and the related industries.

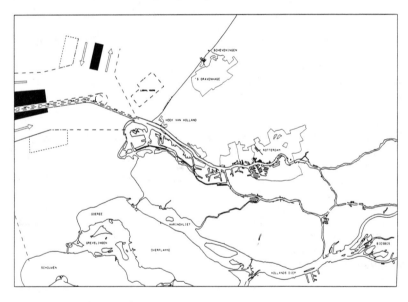

Figure 1:   Access to Europoort

Depending on the definition which is used, this part of the Netherlands can be called "coastal zone". The dependence of the sea, the land and their interaction is obvious. In this paper two problems will be discussed. One is the sedimentation of sediments in the entrance channel and harbours of Rotterdam (fig. 1). The main issue is where the dredged material can be disposed. The second is related with the contaminated aquatic sediments in the lower rivers and also upstream. This as a result of several human activities. For instance, building dams in the estuaries for coastal protection and deepening the lower rivers for nautical reasons.

The industrialization of western Europe has caused many pollution problems. One of these is the pollution of the water and the aquatic sediments. As most pollution elements adsorb strongly to silt and clay particles, the net result is a widespread pollution problem in all places where sedimentation occurs. In the past it was no problem to use the dredged material for landfill or to fertilise the land for agriculture. Nowadays the dredged material is very suspicious and it is hard to store or to (re)use it.

Due to these environmental problems a system is set up to classify the dredged material. This system will be briefly discussed. With the classification system in mind it is more easy to understand that not all the dredged material can be handled in the same way. Only light contaminated material is permitted to be disposed at sea. More severe contaminated material has to be stored in a depot. Depending on the origin of the dredged material and by using the classification, a decision can be made how to handle it.

Material to be dredged

The yearly amount of maintenance dredging is high. The total amount is appr. 50 million m^3. About 40-45 million m^3 is disposed at sea, about 7 million m^3 is stored in depots.

Besides the maintenance dredging the problem exists of the polluted aquatic sediments (Ministry of Transport and Public Works, 1989). If the aquatic sediment is polluted to such extent that it represents a serious danger to public health, the environment, the functioning of the aquatic ecosystems and/or the use of groundwater, it is cleaned up. The total amount of polluted sediment is estimated to reach about 100 million m^3. The locations are all over the Netherlands. For this cleaning up operation large-scale storage depots have to be build. In 1995 at least two of the required depots have to be realized. The part concerning the inland waters will focus on the design aspects.

Quality of the dredged material

Five categories of pollution have been defined for to the dredged material (Ministry of Transport and Public Works, 1989).

A warning value is given for the aquatic sediment cleaning policy. If this value is exceeded then research into necessity of cleaning is urgent. The dredged material must be stored under strict ISM-conditions (ISM = Isolate, Store and Monitor) on land or in deep underwater pits.

Material of a quality between the warning value and lower test value should, when possible and sensible, be stored under ISM-conditions. The strictness of the conditions depends on the degree of pollution of the spoil.

Material of a quality between test value and the general environmental quality can, under certain conditions and depending on the local situation, be disposed into the water. The quality of the aquatic sediment may not be worsened.

Material of a quality equal to or better than the environmental quality can be disposed into the water, as long as no worsening occurs.
Material which complies with the target value can be disposed into the water without problems.

Coastal waters
In the Netherlands the yearly amount of dredged material for maintaining the nautical depths of sea entrances and harbours is about 40-45 million m^3. Nearly half of that is dredged in the port of Rotterdam and its surroundings (PIANC, 1989a). The amounts increased during 1960. Also did the industrial activities in the area itself and upstream, for instance in Germany. The quality of the dredged material changed dramatically. Once a friendly product, now useless and hard to dispose. It became very necessary to find a way to deal with this problem.
The originally designed classification system was to separate the dredged material into geographical categories (PIANC, 1989b). Near the river outflow in the sea, the sediments consist mostly of clean sediments from the sea, only a small percentage is polluted river sediments. Therefore this sediment is only lightly polluted and the dredged material can be disposed at sea (category I).
More upstream there is more mixing between sediments from the river and the sea. Therefore the sediment is polluted and the dredged material must be stored in a depot (category II and III).
In some harbours the sediment is heavily polluted due to local activities and/or discharges. Therefore the dredged material must be stored in a special depot (category IV).

This method was suitable for the first years. Later on, the system had to be changed, because the quality of the sediment improved and the system could not be used outside the Rotterdam area.
A new system was set up to regulate the dumping of the dredged material. The new system is based on quality of the dredged material and the loads of contaminants (SACSA, 1990). Until now this system works, but it is a complicated process to make the decision for the kind of disposal that is necessary.
Since the new regulation system has been introduced, the amount of dredged material which is disposed at sea decreased to about 10-15 million m^3 per year. The disposal site, called dumpsite North (fig. 2) gradually shifted towards its more northern section because of the induced rise of the sea bottom and also of the larger draughts of modern hopper dredgers (Spanhoff et al., 1990).
In the last ten years the disposal of dredged material at sea (PIANC, 1986) and the dumpsite itself has been the subject of many studies. The main purpose was to monitor the development of the surroundings for management of the dumping area. For this reason a 3-D mathematical model was developed for prediction of the transport and fate of the dredged material, especially the fine sediments.
The results of the field studies and model computations suggest a relatively strong return flow of mud dumped at the present dumpsite. This situation is largely due to an observed gyre in the residual current and to density differences in the coastal zone (Spanhoff et al., 1990).

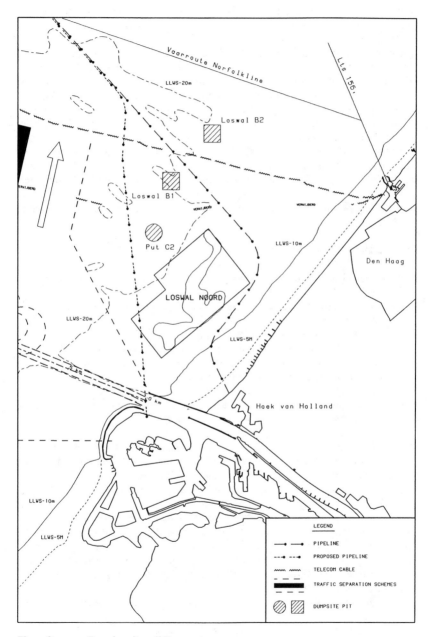

Figure 2: Dumping sites off Europoort

These results give an impulse for the idea that another dumpsite can be more profitable than the present one. This means that an economic evaluation has to be made, including transport and shipping costs. The environmental awareness increased during the last years. The coastal zone along the Dutch coast is vulnerable due to the great ecological importance. The coastal zone receives a special level of protection by implementation of strict regulations on contaminating and other disturbing human uses of the marine environment. This zone is called the "Environmental Zone" and is introduced in the North Sea Water System Management Plan 1991-1995 (Ministry of Transport and Public Works, 1992). The present dumpsite is situated in the environmental zone. The question is: will another dumpsite, outside the environmental zone, benefit to a better functioning of this zone? The economic and environmental questions together gave rise to the start of a project concerning the possibilities and implications of another dumpsite (Rijkswaterstaat, 1992). The project is a cooperation between Rijkswaterstaat and Municipality of Rotterdam. Both are involved in the problems concerning the disposal of dredged materials.

Policy analysis approach
Nowadays, only a multidisciplinary approach can cope with the combination of economic, environmental and spatial problems. The main questions and related problems are:
- where to dispose in relation to channel and harbour sedimentation; especially as a result of the residual current to the harbour;
- where to dispose in relation to the shipping costs;
- where to dispose in relation to the environmental conditions and effects;
- which quality and quantity to dispose in relation to the environmental impact as a result of the polluted dredged material.

The policy analysis approach more or less guaranties that all relevant aspects are taken into account. It is a sound method for dealing with different interests and the several alternative solutions which sometimes contradict.
The following interests are taken into account:
- spacial, the dumpsite in relation to other kinds of use (traffic, pipelines);
- location, in relation to the environmental zone and the transport along the coastal zone and sedimentation in the Wadden Sea;
- maintenance dredging, reducing the return flow of dumped material to the Rotterdam harbour;
- shipping costs, to minimize the costs of the hopper dredgers in use;
- sand mining, is it possible to fill the created pit in the seabed with the dredged material and use the extracted sand for landfill or beach nourishment;
- environment, what is the impact of dumping and the distribution of the dredged material, at the dumpsite and along the coastal zone.

The problem is now defined as follows:
- is it possible to reduce the return flow in order to reduce the dredging costs; and,
- is it possible to create a less disturbed coastal zone?

Alternatives

To tackle the above described problem, several alternative solutions are possible, i.e. another dumpsite at land or sea, changing the method of dumping (more concentrated or more diffusion), changing the dredging technique by dumping on the tide, reducing the amount of dredged material with the help of some kind of infrastructure or re-use the material.

A quick screening resulted in only three alternatives that are worth working out in more detail, namely the use of the present dumpsite, another dumpsite shifted to deeper water and a created pit in the seabed.

These three alternatives are reviewed for distribution of silt including the contaminations, for reducing the amount of dredged material, financial benefits, and environmental impact, in particular for the coastal zone and the Wadden Sea, two vulnerable regions.

Results

The review in more detail resulted in two options, the shifted dumpsite and the pit in the seabed. The created pit including the use of the sand extracted from this pit on the sand market gives the most positive cost-benefit ratio. Also the shifted dumpsite to deeper water has a positive ratio.

During the study it became clear that there is a lack of knowledge for the physical processes, the composition of the dredged material (percentage sand/silt), the kinds of contaminants and its impact on the environment. Especially the impact of dredged material on the environment and biological processes has to be better understood for a more sound decision.

Estuaries and inland waters: the problem

The problems to solve are the large shallow areas with heavily polluted sediment. In the Ketelmeer, a lake at the end of the river IJssel, a branch of the river Rhine, waves and watermovement by shipping resuspend the particles continuously. PAC's, PCB's etc. come into the biological circuit and also leach out and seep into the groundwater. And although it is impossible to prove that people die because of it, abnormalities in insect larvae have been found and studied to investigate the severeness of the problem.

In the Hollandsch Diep (fig. 1), a closed branch of the river Rhine, the pollution problem is the same. A complication occurs however. The water is relatively deep which results in less resuspension. Because siltation continues with cleaner material, the polluted material is buried under it and removing becomes more difficult and expensive.

Most pollution occurred in the 1950-1970's by industry and agriculture. In the meantime legal measures and environmental awareness have resulted in a dramatic decrease of pollution sources.

As a result of the improved situation (cleaner sediments), an ambitious clean-up plan for the sediment for the lower branches of the river Rhine is being developed. Real cleaning of soils, i.e. removing the pollutants from the silt and clay particles is being studied. For the majority of the problem sites however, this process is yet too expensive and it cannot be done within a reasonable time span.

In the third National Policy Document on Watermanagement (Ministry of Transport and Public Works, 1989) suggestions were made to tackle the problems in the Ketelmeer and Hollandsch Diep with large depots for contaminated dredged material, not only for the lakes themselves, but also for the surrounding area. Therefore major efforts are focused on removing of the contaminated sediment and constructing safe disposal sites for the dredged material.

Disposal sites for contaminated dredged material

Environmental Impacts Assessment (EIA) studies are being made in order to develop acceptable disposal sites for Ketelmeer and Hollandsch Diep. Both sites should be able to contain approximately 30-40 million m^3 of material.

In the local EIA studies, many alternative locations and construction alternatives for depots are being developed. Parallel to these EIA's, in-depth studies on chemical and groundwater processes in and around these sites are being made. Possible effects of all thinkable insulation measures are part of it, the main complication being that even if you see dry land instead of water, the groundwater table is only one foot below the surface in this country.

For each alternative mathematical model simulations on leachate of pollutants to the groundwater and the surfacewater are made. Comparison with the present situation gives insight on the positive effect of each alternative.

When comparing alternative solutions the time scale is an important issue. Should a depot solve the problem forever and what is "forever"?

Leachate calculations resulted in the conclusion that after 25,000 to 100,000 years pollutants will have leached from a depot, even if constructed to the highest standards. Dilution is vast of course over such a long period of time.

In fact, constructing for a lifetime of 25,000 years is unrealistic, if you consider the possible changes in habitation, sea levels and geological situation. It has been suggested therefore to value restriction of leachate on "short" term more than for the very far future.

A risk analysis is used as a tool to come to logical designs. "Natural" redundancy and independency of human actions seem to be key elements in the design. For the comparison of dumpsites leachate calculations for 250, 2500 and 25,000 years have been made

In the overall comparison system the following aspects are being used:
-       improvement of environmental situation in % of present situation;
-       location implications on landscape, shipping, recreation and agriculture.

In a comparison matrix all aspects and subaspects are shown. The validation of each aspect against the other is a matter for the steering group, advising policy makers to decide on. All relevant organisations are represented in this steering group. The result of the study for alternatives, including the comparison has to be presented for the independent EIA commission and for public hearing.

Depot design: different types of depot

The following types of depot are distinguished (fig. 3.):

1. depot on land, disposal above the (ground) waterlevel;
2. pit, disposal below waterlevel;
3. island/atoll, a (deep) pit with a ringdike, disposal mainly below waterlevel.

The preferred depot-type for both large disposal sites is the island/atoll-type.

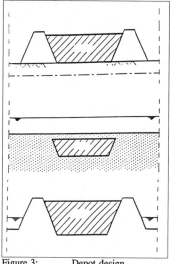

Figure 3:   Depot design

Size (volume) of the depot for the Ketelmeer

First the size of the depot has to be determined. The inventory of the contaminated aquatic sediment shows 10 million in situ m³ in the Ketelmeer and 5 million in situ m³ in the adjoining areas. The density of the in situ material is approximately 1400 kg/m³.

Dredging of contaminated sediments will be done by normal or adapted hopper- or cutter-suction dredgers. During the dredging process, large quantities of water are added in order to be able to dredge and transport the material. Also deeper dredging is necessary to be sure that all contaminated sediment is removed.

Because of these two reasons, much more space is required in the depot than the original 15 million m³ of contaminated dredged material.

However, sedimentation and consolidation during exploitation cause a reduction of space needed. The resulting contents of the depot has to be between 22 and 26 million m³.

Dike/pit construction - use of local materials

In order to create a disposal site of this size (22 to 26 million m³), large amounts of material have to be moved. Therefore, it is profitable to use local materials i.e. the sand and clay from the pit for the construction of the dikes around the pit. However, a surplus of sand remains. By dividing the depot into two compartments and delaying the building of the second compartment, it is possible to sell this surplus of sand on the regular market, without disturbing the price.

Principle of design

The basic principle of design is to minimize the emission of pollutants during the lifetime of the disposal site (insulation), to be able to control the situation, even if insulation should fail, and to monitor the disposal site and its environment.

Because of the fact that (at present) there are no fixed permissible fluxes for long-term disposal, the design of the depot includes a basic design (i.e. no additional insulation) and a set of insulation techniques. The selection of a disposal site and its insulation techniques can then be based upon a cost/"benefit" analysis, discerning best practical means and best available techniques.

Insulation towards groundwater

Basic methods of insulation are chemical (immobilization), (geo-)hydrological and liner technology. In order to minimize the impact on groundwater, transport of pollutants both by advection and diffusion, should be prevented.

In case of a disposal site of the type "island/atoll", the reduction of advection can (relatively) easily be achieved by controlling the hydraulic conditions, i.e. by lowering the watertable within the ringdike according to the watertable of the underlying aquifer.

In the case of the depot of Hollandsch Diep, this is the natural situation. But in the case of the depot Ketelmeer, pumping will be required over a period of centuries and possibly the effluent will need to be purified. This can be an important disadvantage of the depot Ketelmeer.

Transport by diffusion can be reduced effectively only after a very large reduction of the advection is reached.

The reduction of transport by diffusion is, however, far more difficult. Therefore, the mechanism of diffusion was studied with special interest to the effect of imperfections of a lining.

Based upon this research, possible materials have been selected and were studied with respect to their long-term behaviour. With these means, it even seems possible to reduce the amount of pollutants, transported by diffusion. Possible materials include concrete, asphalt, composite plastics and materials leading to precipitation (gypsum). Long-term stability and methods of construction for these materials are being researched and are considered promising.

Benefit of the disposal site

In case of the Ketelmeer, the removal and disposal of the contaminated aquatic sediments into a depot, even without additional insulation techniques, already leads to a reduction of emission of pollutants by a factor of about one hundred compared to the present situation.

By eliminating the advective transport, the impact on groundwater is further reduced, but no more than by an (extra) factor 3 or 4.

Whether techniques to reduce the diffusion will be considered, will depend on both effectiveness and cost. Therefore a reduction by a factor 10 is aimed at, primarily based upon the high cost of this kind of insulation.

Conclusions

- The large amounts of dredged material from entrances to harbours, harbours itself, estuaries and the inland waters, can be dealt with in a sensible way;
- It will cost society a certain price by doing it so, i.e. complex studies and procedures to generate acceptable solutions or building large storage depots;
- In the coastal waters the choice is between concentration in a created pit or distribution from a dumpsite in deeper water;
- In the inland waters the choice is how to design the storage depot, especially the insulation and the lifetime;

- Depots do have an (overall) positive effect on environment;
- After 25,000 years - 100,000 years all relatively mobile pollutants will have leached into the environment, for less mobile pollutants this takes up to several millions of years;
- With additional insulation, even further reduction of emission is possible; Whether future disposal sites actually will be equipped with these linings, largely depends on future regulation for the design of disposal sites;
- If sources of pollution for aquatic sediments are not strongly reduced, more and more disposal sites will be needed and insulation of disposal sites seems a waste of money.

References

Ministry of Transport and Public Works (1989), Third National Policy Document on Watermanagement; Water in The Netherlands: a time for action.

Ministry of Transport and Public Works (1992); North Sea Water System Management Plan 1991-1995.

Ministry of Transport, Public Works and Watermanagement, Directorate General for Environmental Affairs (1992); EIA for disposal of dredged material (in Dutch).

PIANC (1986), Disposal of dredged material at sea; *Report of a working group of the Permanent Technical Committee II, Supplement to PIANC bulletin no 52.*

PIANC (1989a), R. van Veghel, C.W.A.O. van Raalte, R.G.J. van Orden and M. Veltman. Modern dredging and disposal of (contaminated) dredge spoil from maintenance works in fairways to and harbours in the port of Rotterdam and its surroundings. *In: PIANC bulletin no 65*

PIANC (1989b), R.G.J. van Orden. Policy plan for the disposal of dredged material from the port of Rotterdam. *In: PIANC bulletin no 65*

Rijkswaterstaat, North Sea Directorate, Tidal Waters Division and Municipality of Rotterdam (1992). Investigation future location and method for disposal of dredged material from Rijnmond (Rotterdam and surroundings) (in Dutch); Nota NZ-N-92.09

Rijkswaterstaat, Civil Engineering Division (1993). Results of the second phase Reference Depot (in Dutch)

SACSA (1990), Seventeenth meeting of the Standing Advisory Committee for Scientific Advice. Regulating policy for dredged material presented by The Netherlands

Spanhoff R., Tj. van Heuvel and J.M. de Kok (1990). Fate of dredged material dumped off the Dutch shore. *In: Proceedings of the International Conference Coastal Engineering 1990.*

Aquatic Disposal of Dredged Material in the Belgian Part of the North Sea

B. de Putter[1], P. de Wolf[1], B. Malherbe[2]

Abstract

Extensive research and study on the behaviour of dredged material after dumping in sea has been conducted on the disposal grounds "S1", "S2" and "Zb Oost" located in the southern Bight of the North Sea. Investigations of recirculation on alternative disposal grounds is currently ongoing using radio-active tracer-techniques for direct field measurements. The parameters that are relevant during the dumping operation are now better defined and well-known.

Introduction

Annually, about 20,106 tons of dry solids weight are dredged for maintenance and dumping purposes in the Belgian maritime harbours and the access-channels to these harbours and to the Westerscheldt river. These sediments are dumped on allocated dumping grounds.

Since 1984 a lot of effort has been put into the research of the behaviour of this dumped material under natural circumstances. Special focus has been given to the study of the dispersion and recirculation to the dredging areas of the fine grained material because this has major impacts on:

1. economy of the dredging operation;
2. technology of dredging/dumping process;
3. ecology of this part of the North Sea.

Dispersion of fine-grained material under wave, current and wind action is almost impossible to simulate exactly with mathematical models without a thorough understanding of the natural processes. Therefore, and in view of later developments of simulators, an extensive field investigation program on open water disposal in the Belgian North Sea has been executed, including the following items:

1. evaluation of the efficiency of the dredging process;
2. residual sediment transport patterns using field measurements and sediment trend analysis techniques;
3. calculation of the global sediment budget;

---

[1] Ministry of the Flemish Community, Service of Coastal Harbours, Vrijhavenstraat 3, B8400 Oostende, Belgium

[2] Harbour and Engineering Consultants, Deinsesteenweg 110, B9810 Ghent, Belgium

4. soil investigation;
5. evaluation of the efficiency of the dumping process;
6. field-measurement of recirculation of dump losses.

Dumping grounds

Maintenance and capital dredging works in the Belgian maritime harbours and the access channels to the harbours and to the Westerscheldt river all yield huge quantities of dredged material for disposal. With the exception of a minor quantity of dredged material from certain parts of the harbour of Zeebrugge, all the dredged material meets the criteria of the Oslo Convention for dumping in sea. One of the major Belgian dumping grounds is the "S1" north of the shipping lane "Scheur-West", the major access to the Westerscheldt (harbour of Antwerp) and the harbour of Zeebrugge. In total 5 allocated dumping grounds are currently in use in this area (fig. 1).

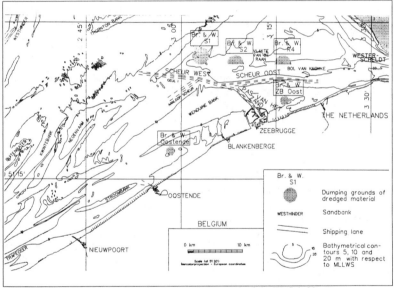

Figure 1:   The Belgian nearshore area and the mouth of the W. Scheldt estuary.
               Indication of major dumping grounds used for dredged material.

The dredged material is of a wide variety: dredging in the harbour of Zeebrugge and the Pas van het Zand concerns muddy sediments (hopper densities between 1.15 and 1.25). In the Scheur shipping lane, the dredged material is sandy mud or fine sand (d50 = 0.180 mm). This wide variety of sediments is dredged mainly with trailing suction hopper dredgers.

A sensitivity analysis was performed taking into account the parameters which could influence the efficiency of the dredging process. Based on the results of this sensitivity analysis the relevant parameters could be identified i.e.:
- the maximum achievable mixture density in the hopper to increase production;
- the dumping efficiency influenced by the site and the dredged material mixture density;
- the recirculation degree of fine-grained material.

Residual sediment transport patterns and the turbidity maximum area

The knowledge of residual sediment transport patterns is essential for the understanding of the recirculation mechanism of the sediments, i.e. the recirculation of fine-grained dump losses back to dredging areas.

An extensive overview of the investigations conducted in Belgium since 1976 to determine residual transport patterns is presented by De Meyer & Malherbe (1987) and Malherbe (1989). Techniques used for the sedimentological investigations were tracers (bed-load, suspension, dumping, overflow and recirculation tracers), differential (morphological) mapping and, more recently, sediment trend analysis (STA).

Sediment Trend Analysis (STA) is a technique in which the statistical comparison of granulometric data, deduced from bottom samples, is interpreted in terms of dominant residual transport directions (Mc Laren & Bowles, 1985). The method has been used on several sedimentological data-bases in order to define the areas in which recirculation is likely to occur (fig. 2).

The interpretation of the field investigations have led to the conclusion that the area in front of the Belgian coast is characterized by a marine turbidity maximum area (TMA) in which mud sediments are hydraulically trapped.

This TMA is characterized by a loose mud deposit layer (thickness: 0.30 - 0.60 m) trapped within the TMA delimitation and with a strong interaction deposit (on seabed) - suspension (in water-column). The volumes of loose mud trapped within this marine TMA are significant quantities and may amount to some tens of millions of tons of dry solids.

Soil and seabed investigations

The different techniques used for the soil investigation of the dumping grounds include sampling, natural radio-activity and vibrocoring.

Parallel to these soil investigations, the morphological evolution of the seabed of the dumping grounds was monitored and assessed by using differential mapping techniques (fig. 3) based on digital terrain modelling (DTM) of sounding data.

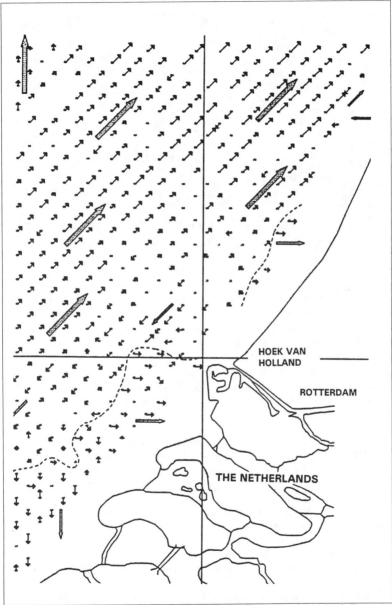

Figure 2:   Example of residual sediment transport directions calculated with Sediment
              Trend Analysis (STA). Scientific evaluation of Sediment Trend Analysis
              techniques (ISS1165.00032).

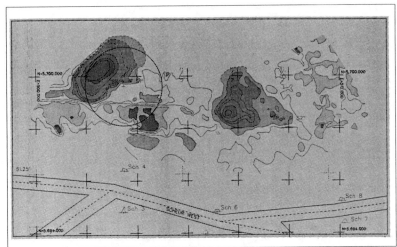

Figure 3:   Detailed differential map of the S1 and S2 dumping grounds (the circle and
half circle delineate the dumping grounds). Differential contourlines of seabed
evolution are also shown.

All these soil investigations have led to the following conclusions (Malherbe, 1989):
- between 0 and 40 % of the total dumped material remains on the dump sites: the
   rest, i.e. the dump-losses, is dispersed by wave and current action;
- the sediment fraction remaining on the dump sites is almost pure sand (d50 =
   0.180 - 0.200 mm): the fine-grained mud is washed out.

Recirculation of dumped material

From the results of the previously mentioned investigations it appears that the
recirculation of dump losses may have a significant impact on the technical,
economical and ecological aspects of the dredging/dumping process.

It is however quite difficult to simulate the recirculation mechanism using
sedimentological mathematical models because the sediment-dynamics of such loose
mud deposits is still poorly understood; hence, the physical equations governing these
dynamics are not yet established to compute nature-like behaviour.

Therefore, direct field measurements were preferred to determine the recirculation
mechanism. The only feasible technique appeared to be the use of long-living, low
activity radioactive tracers.
For these recirculation studies a special tracer technique whereby 2 different
mud-tracers are used simultaneously, was developed (Malherbe, 1989).

The Belgian national and Flemish regional authorities have now started up a study
to develop a mathematical sedimentological model in which the results of the
recirculation tests will be incorporated.

Up till now 4 recirculation tracers were injected in front of the Belgian coast on the following dump sites (see also fig. 1):

1. Br & W S1:   - 1 flood tracer (Tb) and 1 ebb tracer (Hf)
                  - injection: 28.02.1988
                  - monitoring period: 149 days

2. Br & W S2:   - 1 tracer (Tb)
                  - injection: 22.04.1992
                  - monitoring period: 70 days
3. Br & W ZbO: - 1 tracer (Hf)
                  - injection: 22.04.1992
                  - monitoring period: 70 days

The tracing of the mud has been executed onboard the research vessel "Speedy II" by the SAR (Section of Application of Radio-Elements, Saclay, France).

The tracers were detected by taking large samples (approx. 10kg) of superficial sediments (scraper-sampler) or small cores (mud coring device). The samples were put into hermetic polyethene-containers for spectrometric analysis.

The results of the recirculation tracer tests generally give the tracer-activity levels detected within the samples. With the help of the spectral analysis, the activities of both tracers can be determined simultaneously.

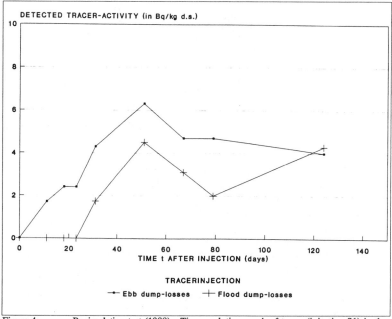

Figure 4:   Recirculation test (1988) - Time evolution graph of tracer (injection S1) in the
                 harbour of Zeebrugge

The detected activities (Bq/g.d.s.) are presented in time evolution graphs and tracer-dispersion maps. Time-evolution graphs show the activity-evolution of both tracers at one single sampling station (fig. 4 and 5). A snap-shot of the geographical dispersion of the tracer is presented in figure 6.

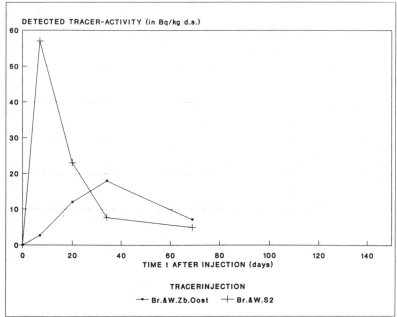

Figure 5:   Recirculation test 1992 - Time evolution graph of tracer (injection Zb Oost + S2) in the harbour of Zeebrugge.

The time-evolution graphs lead to the following conclusions for the 4 recirculation tests:

a. Recirculation of dumped dredged material can be clearly assessed with the tracer technique and seems to be quite significant (intensity and speed);

b. There is no significant difference in recirculation between the ebb-dumped and the flood-dumped material;

c. The recirculation to the nearshore region (TMA-zone) is very rapid, i.e. within approx. 12 days;

d. After approx. 1 month there is no significant time-variation, i.e. the activity of both tracers remains constant. A peak of measured activity occurs however within the first week after injection linked to the rapid recirculation;

e. There is no major difference in the recirculation pattern resulting from dumping on dumping grounds located close to the shore (e.g. Zb Oost) and on dumping grounds more offshore;

f. From figures 4 and 5 it can be deduced that the recirculation of dumped dredged
    material remains fairly constant nonwithstanding continuous maintenance dredging
    in the harbour- and waterways.

Figure 6 shows the situation after 78 days of the tracer which was flood dumped on
distant S1 dumping ground: the tracer is concentrated over a relative small area below
the Wenduine Bank. The tracer is present with very high levels of activity. From the
tracer budget it can be deduced that approx. 98 % of the total injected tracer mass
was present in this area. The tracer was dispersed over an area of ca. 1000 km².

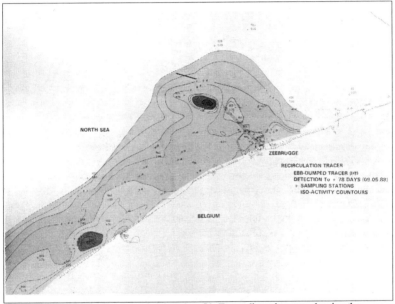

Figure 6:   Recirculation tracer injected on S1. Tracer dispersion map showing the
                   dispersion of the tracered dumped dredged material

Everything seems to happen as if very little sediment is evacuated out of the system
(closed box principle of TMA), as if no significant material is brought into the
system except (traced) dumped material and as if the same material is constantly
moving within a closed circuit.

Environmental impact assessment
Based on these investigations and a special sampling and analysis program, a
preliminary environmental impact assessment (EIA) has been done (De Wolf &
Baeteman, 1991).
Some major conclusions of this EIA can be formulated as follows (Malherbe, 1989):
-   Cohesive, fine-grained sediments act as natural filters for contaminants present in
    water. Consequently they support a wide variety of contaminants due to their

adsorption characteristics (e.g. heavy metals, organics). Moreover, sediments are the life environment for probably the most pollution-sensitive part of the food chain, i.e. the benthos.
Since the seabed lithology of the Belgian nearshore area, which is relevant for benthic life, is affected by the dispersion and recirculation of dumped dredged material over an area of more than 1000 km^2, it is necessary to assess clearly the sedimentological impact of dumping operations before doing any further statements about physical, chemical or even biological impacts.
-   The actual state of the research, allows only to assess in a provisional way the possible environmental impact of dumping dredged material off the Belgian Coast.

It can be presumed that the influence will be felt mainly in the following area's:
*   *Physical:*
-   on the dumping ground, by bottom covering by the sand fraction;
-   on wider areas, permanent change in the bottom composition by dispersion of the fine mud fraction;
-   an increase in turbidity of the water column, which occurs only in the case of dumpings executed in short intervals.
*   *Chemical:*
-   through the mobilisation of pollutants out of the spoil, when the dredging operation brings the spoil in a environment with different physico-chemical conditions (e.g. oxygen, salinity);
-   by a global pollutant transfer when the dredged material is considered as the carrier of a significant "contaminant load".
*   *Biological:*
-   direct impact on the bottom life, especially on the sedentary benthic fauna and on the breeding- and spawning sites;
-   an eco-toxicological impact which is at present difficult to evaluate.

On the basis of the existing Guidelines of the Oslo Commission, it was established that, with the exception of a minor quantity of dredged material from certain areas in the Zeebrugge Harbour, all dredged material meets the criteria for dumping in the sea.

A monitoring and research program has now been started up by the Belgian national and regional authorities to gain a better insight in the effects of dumping of dredged material in the North Sea.

Conclusions
The accumulated knowledge of residual sediment transport mechanisms in the area of the North Sea in front of the Belgian coast indicates the existence of a "turbidity maximum area" (TMA) in open sea. Such a TMA governs the sediment accumulation in channels and harbours and consequently the maintenance dredging process management.

For the particular case of the dump sites used for dredged material in the Belgian North Sea it appears that primary disposal efficiency, i.e. the fraction of total dumped material which remains on the disposal ground, and the recirculation of the dump losses, back to the dredging areas have a major influence on the efficiency of the dredging and dumping operations and on the environmental impact. An area of ca. 1000 km² is affected by the dispersion of dumped material on the sites "Br & W S1", "Br & W S2", and "Br & W Zb Oost".

The field investigations of the dumping process clearly revealed the separation between the less mobile sand, which remains mainly on the disposal ground, and the highly mobile mud fraction, which is dispersed and recirculated.

The measurement of the migration and the recycling of the fine-grained dumped material indicates that it becomes hydraulically-trapped within an interactive system of low-density sedimentation and suspension. This hydraulic "trapping" is a further argument that mud accumulation in this part of the North Sea is due to a TMA.

For the measurement of the migration of dumped dredged material a special new tracer technique was developed and applied; this technique proved very successfull.

The preliminary environmental impact assessment (EIA) indicates that the major environmental impact of dumping of dredged material in open areas in front of the Belgian coast is a sedimentological one. Attention thus was also given to changes in bottom lithology which affect benthic ecology. It was evident that such impact evaluation has to be extended from the disposal ground to the whole area influenced by the migration of dump losses.

References

Bastin A., Caillot A. & Malherbe B., 1983. Zeebrugge Port Extension - Sediment transport measurement on and off the Belgian coast by means of tracers. *8th International Harbour Congress,* Antwerp.

De Meyer Chr. & Malherbe B. 1987. Optimization of maintenance dredging operations in maritime and estuarine areas. *Terra et Aqua* no. 35.

De Wolf P. & Baeteman M., 1991. Ecological impact of dredged material disposal in Belgian coastal waters. *Congres on Characterization and Treatment of Sludge* (CATS), Ghent.

Malherbe B., 1989. A case study of dumping of dredged material in open areas: the S1 dumping ground in the southern bight of the North Sea. *International Seminar on the environmental aspects of dredging activities,* Nantes.

Mc Laren P. & Bowles D., 1985. The effects of sediment transport on grain-size distributions. *Journal of Sedimentary Petrology,* Vol 55 No 4.

Case Studies for Coastal Protection:
Dithmarschen; Eider Estuary; Sylt

W.D. Kamp[1], P. Wieland[2]

Abstract

Three cases along the German North Sea coast, north of the Elbe estuary are discussed.

Land-reclamation by building dikes started in the Dithmarschen-area almost eight hundred years ago. The construction of the closure dam in the Bay of Dithmarschen provided areas for agriculture and recreation, and resulted in lakes which facilitated the drainage of lowlands and offered possibilities for boating.

The first closure of the Eider-river in 1937 led to silting-problems. Thus the next closure by a storm surge barrier (1967/72) had been established near the mouth of the estuary. The length of the dike had been cut down from 60 to 5 km and a better drainage of an intake area of more than 2000 km² has been guaranteed without disadvantages for shipping. The construction changed the morphodynamics of tidal channels and created morphological reactions in the adjacent Wadden Sea that are still obvious.

Since centuries the cliffs of the island of Sylt and its sandy beaches encounter erosion. Groynes had been constructed in the western part of the coastline since 1869 but they were not able to reduce erosion significantly. From 1907 to 1954 a seawall had been established to protect the main city (Westerland). The solid constructions stopped dune erosion, but they were not able to prevent scouring of the beach. Beach restoration by artificial sand nourishment started in 1984 and has been repeated several times up to now. The different choices made on Sylt to solve the problems will be presented.

Dithmarschen

Development of the coast as a result of nature and human activities

The coast of Dithmarschen is formed in front of a cliff-like, 20 m high glacial area. In the last 2500 years marine sediments have silted up the area, which resulted in the present low intertidal area. This area has nowadays a width of 7 to 18 km. The area was already in use by farmers and fishermen in the years B.C. Because of a rising

---

[1] Ministry for Land and Water Management, State of Schleswig-Hollstein, P.O. Box 1440, D 2250 Husum, Germany

[2] Ministry for Land and Watermanagement, State of Schleswig-Holstein, P.O. Box 1640, D2240 Heide, Germany

sea level, those inhabitants first raised their small villages, but later (in the 13th century) the first dikes were built. Due to these dikeworks on the growing salt marshes, cultivated areas (polders) were created, so-called Köge. At present day one can distinguish 54 of them, all with different sizes.

Diking of the intertidal area

The first dike-like structures were built in the end of the 11th century as roads, somewhat higher than the surrounding landscape (Prange, 1986), connecting the living mounds. The first closed dike-circle was created in the 14th century, on the line Brunsbüttel - Marne - Meldorf - Wöhrden - Wesselburen - Lunden (fig. 1). From this dike in a natural way the intertidal land in front silted up. Later this silting process was speeded up by technical measures. When the land had silted up to a level high enough, a new enclosure-dike was constructed. Until 1590 more of these polders (Köge) were constructed in a strip from Brunsbüttel to Wöhrden and near and north of Wesselburen. In the next 200 years more Köge were created in the south, off Marne, in the north, off Wesselburen and north of the Eider-river. The island of Büsum was connected to the mainland by diking the land on both sides of the connection dam (from the year 1600). This area silted up quite fast.

Because of the meeting of salt and fresh water, an increased deposition of silt occurred in the mouth of the Elbe river. With exceptional speed (much faster than in other areas) new land developed northwest of Marne. In 1890, after the diking of this area, a peninsula was formed, protruding far into the sea, creating the 100 km^2 Dithmarscher Bucht. For a short time there was even a polder on the dune island of Trischen. In the leeside of high dunes land silted up. After several storms this polder had to be abandoned in 1942. The last diking of area silted up sufficiently high, took place in 1934/35, south of the peninsula of Friedrichskoog.

Works on the intertidal lands

In front of the western point of the Friedrichskoog peninsula land silted up. This accretion was removed by a tidal channel. This channel moved towards the sea dike, while it was deepening itself. This caused a substantial danger to the stability of the sea dike (fig. 2). To avert this danger in 1930/31 a revetment with pitched stone (basalt) was constructed and several steel sheetpiles where placed. However, these works were not able to stop the protruding channel (Fisher, 1957). After detailed hydraulic and morphological studies, the following (successful) solution was found and executed in 1934/36 (Wieland, 1992). The channel had deepened in the mean time from MSL - 4 to MSL -7 m. It was closed by a dam and diverted in a western direction by a 2.2 km long dam with a pitched stone (basaltic columns) revetment. The crest hight of this dam was 1.5 m above Mean High Water (fig. 2).

The very heavy floods of February 1962 and January 1976 lead to a dike breach in the inner side of the Dithmarscher Bucht, and made clear that the existing sea dike was not sufficient. Also the landward, swampy floodplain was often affected by river-floods. To protect and to improve the inhabited and cultivated land, a master plan for coastal protection was made. In the period 1970/78 this plan was executed.

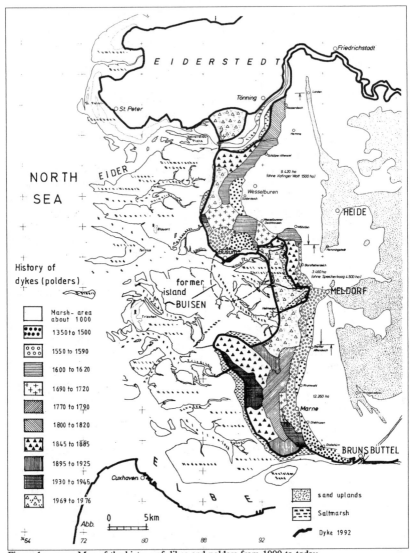

Figure 1:   Map of the history of dikes and polders from 1000 to today

By placing the sea-dike in a more seaward direction, an area of 54 km² was transformed into land protected from the floods of the sea. This work had the following aims:

*   Protection against storm surges (the total sea-defence line was shortened from 30.6 km to 14.8 km, 96 km² was protected directly by the new dikes);

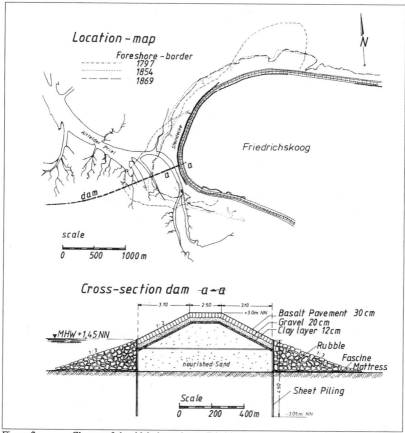

Figure 2:   Closure of the tidal channel near Friedrichskoog

*   Creation of a second defence line for the 424 km² polder area behind the
     Dithmarscher Bucht;
*   A storage area for water to improve the fresh-water management in the area;
*   Construction of a lock to allow shipping to the existing port and rehabilitation
     of this port for fishery and recreation;
*   The creation of 13 km² agricultural area;
*   Create a nature reserve.

In the first period (1970/73) the southern part (16 km²) was diked. The crest-height
of the new dike became 8.8 m above mean sea level. A first discharge sluice was
constructed for the drainage of totally 424 km². The cross section of the sluice is 36
m². In the next five years in the northern part 38 km² was diked, drained with a
second discharge sluice with a cross section of 27 m². Behind this sluice a 2 km²
storage basin was constructed. A lock (9.5 m wide and 55 m long) was constructed
for a connection from the rehabilitated port to the sea (fig. 3).

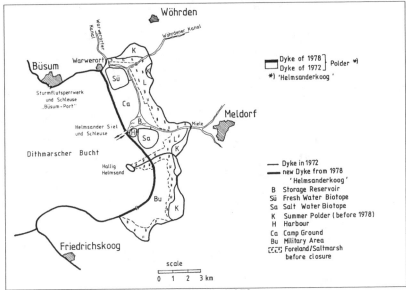

Figure 3:   Closure of the Dithmarscher Bucht 1970/78

In a research programme, the hydraulic and morphological changes were monitored.

The closure of the Eider-estuary

History of the first Eider barrage

The marsh area in the mouth of the Eider suffered for centuries from both high river floods and storm surges (fig. 4). Because of the intensive use of the land, more and more of the tidal marsh was reclaimed by the construction of dikes. Because these dikes were not built on a stable subsoil, they suffered heavily from settlement, and consequently were overtopped regularly by high floods and high tides. Also the wet cross-section was decreased, causing more resistance to the discharge towards the sea, and inducing in this way an extra increase in the waterlevel. All this caused an increase in the frequency of flooding of the area (Wieland, 1992).

Already in 1570 a tributary, the Treene, was dammed at the mouth. This dam caused water management problems in the area. After a sequence of 8 damaging floods in the years 1906 - 1929 it was decided to reach a complete protection against storm surges, but without damaging the interests of navigation. This goal was reached by the construction of a dam in the Eider (1934 - 1936), approx. 30 km upstream of the mouth near Nordfeld. In the dam a discharge sluice and a navigation lock were constructed. These works resulted in an improvement of the situation for a short period (Rohde, 1965), because soon the increased tidal current velocities caused damaging erosion in the delta area:
*   decrease of the tidal prism;

Figure 4:   The Eider-mouth and the Dithmarschen area

*   change of the shape of the tidal wave (flood velocity was increased and ebb-velocity decreased);
*   increased sedimentation (until 1965 approx 65 million m³);
*   rise of the low water level (until 1965 with approx 1 m).

The Eider-barrage between Hundeknöll and Vollerwiek
Unbearable discharge problems and the bad experience of the storm surge of February 1962 lead to the conclusion that the dike along the 30 km long estuary of the Eider river was too low and too weak. Therefore it was decided to close the mouth of the estuary (Rohde, 1965; Petersen, 1967; Cordes, 1972). The works were executed in the period from 1967 to 1972 with the aim:
*   to protect the area against storm surges;
*   to optimize the water management in the drainage area of the Eider;
*   to continue shipping in the Eider estuary.

In order to disturb the natural balance as little as possible, the construction had to fulfil the following requirements (Petersen, 1972; Cordes, 1972):
*   Construction of a seadike, able to withstand a design storm surge, according to the latest knowledge in dike-engineering, with a minimization of the construction and maintenance costs by shortening the total length of the sea-defence;
*   The creation of an optimal storage basin downstream of Nordfeld for storage of the drainage water during high tides at sea;
*   Maintain the minimum waterdepth for navigation along Nordfeld for commercial traffic and fishery by flushing the channels;
*   Maintain the free tidal movement in the estuary till Nordfeld.

Technical solution
The technical solution (Cordes, 1972) was as follows:
*   Construction of a 4.8 km long seadike with a crest height of 8.5 m above Mean Sea Level in the line from Hundeknöll (Dith-marschen) and Vollerwiek (Eiderstedt). Because of this the total sea defence length was shortened by approx. 55 km;
*   Construction of a barrage, including discharge sluice, with 5 gates of 40 m with segment doors on both sides;
*   Construction of a navigation lock of 75 m long and 14 m wide, with a heaving bridge over it;
*   Diking of a 1250 ha area of intertidal salt marsh with a 6 km long dam with a height of 3 m above Mean Sea Level.

The sluicing programme is based upon the following (often contradictory) interests:
*   protection against storm surges;
*   prevention against river floods;
*   shipping;
*   fishery;
*   ecology.
Priorities have been determined resulting from weighing all these aspects. Operation is guided from on-line hydrographic data of a special section of the managing service.

Although this most recent damming of the Eider was done as careful as possible with regard to the powers of nature, a number of adaptions of the system were required:
*   moving the tidal channel of the Eider for 500 m in a northern direction, through the sluicing work;
*   reduction of the cross-section and fixation of the channel near the sluice;
*   reduction of the flood-area upstream of the Eider dam by diking 125 km² of intertidal marshland;
*   timely stopping the tidal movement by using the barrage.

Impact of the Eider-barrage on morphology and hydrography
  Due to the construction several changes occurred in the morphology and hydro-graphy of the area, which have to be compensated with regulations (Wieland, Kraatz, 1991). Before this damming, the Eider channel could freely meander on the location

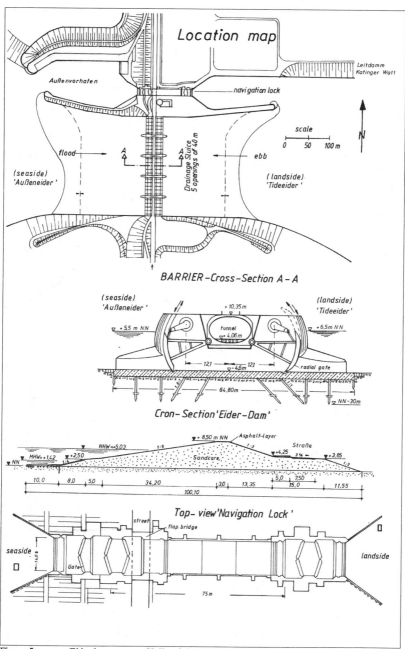

Figure 5: Eider barrage near Vollerwiek

of the present dam over a width of appr. 4 km. After the bed was fixated by the sluice, the downstream reaction was a more severe meandering behaviour of the channel (shorter curves, fig. 6). The northern channel moved in a northward direction with a speed of nearly 1 m/day, deepening itself to 12m below MSL. At that moment the channel was approx. 60 m from the Vollerwieker dike, and was endangering this dike. For the protection of this dike the following measures were taken, using the forces of nature:

*   moving the northern channel into the southern channel by dredging a connection 2 km west of the dangerous bend; this reduced the discharge and velocities; the southern channel increased its cross-section by a natural scouring process;

*   damming the bend in the channel at the point most near to the land; this was done while the barrage was completely closed by hydraulic fill of sand.

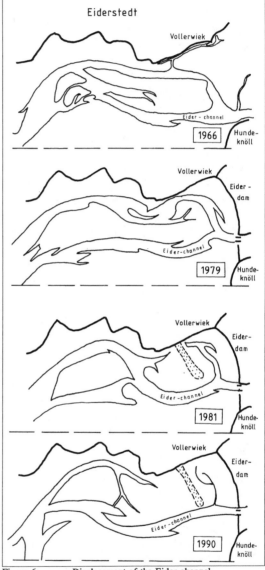

Figure 6:   Displacement of the Eider channel

The separated channel sections decreased to minor channels due to the decreased tidal flow. The sand dam is now the separation between the higher and lower section of the tidal flats.

The deposition of sand in the channels was initially increased by the damming of the Eider. Previously in the outer and inner Eider approx. 0.4 million m³/year was transported inward. The low-water level increased with approx. 0.3 m. In order to decrease the import of sand during every tide the discharge of water into the estuary was controlled by a stepwise opening of the gates. This is done since 1980, it resulted in a decrease of the sediment load to 0.3 million m³/year. The low water level has not increased since 1980 (Wieland, 1991).

Research programmes accompany the developments, until a new dynamic equilibrium is formed.

Protection of the island of Sylt

Development of the island of Sylt
The island of Sylt is one of the German North Sea islands that takes a special position in shape and location. The island's west coast (with a length of 40 km) expands deeply into the North Sea and is therefore affected by storm and wave forces.

The core of the island is the remainder of former mainland. About 8000 years ago the island of Sylt had the shape of a round oval moraine, which was located several kilometres further west than today (Hoffmann, 1981).

With the rising of the sea level after the last ice-age, the moraine of Sylt got increasingly affected by the North Sea waves and tidal currents eroded material from the west side of the moraine and transported the sand to the north and the south, where in the course of the millennium the nowadays existing dunes have been blown up. In this way the long shape of the island with the old core of the moraine and the adjoining dunes in the north and the south developed (fig. 7). The dunes are also affected by recession due to steadily progressing erosion from west to east.

With the rising sea level, the present shape of the island's base also emerged. The cross-section of fig. 8 demonstrates, how the island's base rises gradually from the deeper sea ground which is not reached by wave forces to the cliff or the bottom of the dunes. In the foreshore zone, sandy reef is located on the island's base. Shore and reef together function as a transportation mean. While waves and tides continuously erode and move away material from the sandy shore

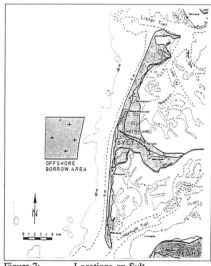

Figure 7:   Locations on Sylt

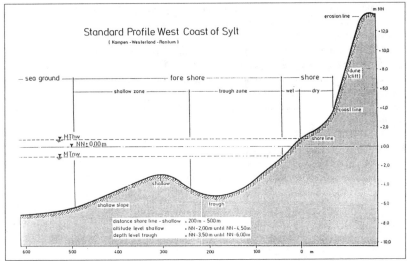

Figure 8:   Standard profile of the westcoast of Sylt

and foreshore area, dunes cliffs are only affected in irregular intervals during storm surges. The eroded material is moved in cross-shore direction and into foreshore areas and is steadily transported to the north and the south by surf waves and tidal induced currents parallel to the coast; then it is deposited in shallow areas in front of the "Lister Tief" and in the "Hörnum Tief" or is moved ahead the inlet, lost for the sediment balance of the island (fig. 7).

The average annual retreat at the west coast of Sylt from 1870 to 1950 has been 0.9 m. The frequency of storm surges with longer lasting high water levels has increased considerably during the past 35 years. Thus the impact of wave energy at the west coast has grown. The annual average coastal recession corresponds to the increase in storm surges. The annual losses of material from the southern part averaged over the last 35 years has been calculated to 1.4 million m³/year.

While the whole western coastline, except the middle section in front of the city of Westerland -which is protected by a seawall- has steadily moved to the east by erosion, the losses at the southern and northern ends are more extensive. The reason is that cross-shore sediment transport in the middle coastal zone causes a lateral sediment exchange, whereas the sand from the ends moves into the tidal channels and therefore is lost. As a result of dune erosion during storm surges and sediment transport into the Hörnum tidal inlet, the west coast of southern part -"Hörnum Odde"- has shifted a few hundred meters to the East. The southern end of the island has moved about 500 meters to the North during the past 15 years.

For the protection of the front dune and its buildings, in areas likely to be flooded during storm tides, dikes of a total length of 3 km along the coast with integrated

seawall and revetments were constructed between 1907 and 1954 (Dette, Gärtner 1987).
All these solid constructions show the same side effects that are also known from other armoured coastal areas. The reflection of breaking waves at the revetments leads to increased turbulence and thus to erosion in front of the construction. Lee-erosion appears at the edges of the fortified sections, which increase the damages at the neighbouring coastal areas. As the west coast of Sylt is an erosive coast and as the coastal protection works in its area prevents from supply material by erosion from the island's substance above Mean High Water into the coastal longshore transport, the foreshore and the beach are increasingly deepened and the stability of the toe protection is more and more endangered.

As a result of the storm surges of the past years, the managing state agency in Husum has worked out a special project "Coastal protection on Sylt" in order to obtain a general concept for the future coastal protection measures on Sylt (intermediate planning period: 10 years). The project has been based on comprehensive investigations (Führböter et al. 1972, 1976). In this special project the solution offered by the latest developments and technologies (e.g. solid constructions: every kind of bank revetment, groynes) and sand nourishments were examined. It turned out that continuously repeated sand nourishments are an appropriate solution from the technical, economical and environmental point of view (Kamp, 1991).

Construction work
The sand nourishments consists of a backfill material and a stockpile of sand. In front of the erosion line an artificial dune is established that functions as a stockpile of sand. In the case of extremely severe storm surges, the artificial dune (recession depot) in front of the cliff will be affected and eroded but not the core of the island

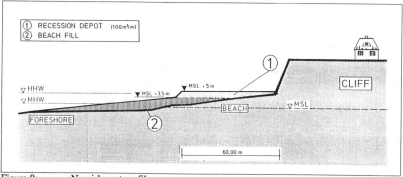

Figure 9: Nourishment profile

itself (fig. 9). The 60 m wide artificial dune works as a wearing structure and has to be restored in certain time periods or after a severe storm flood.
The beach nourishment, which means also the filling of sand into the underwater profile, creates a wide and high beach and keeps away moderate storm surges from

the stockpile of sand. The material which is eroded from the nourished beach area to the north and the south due to the daily impact of waves and tides has to be replaced by beach nourishments carried out in certain periods of time so far as the beach is not filled by material from the erosion of the artificial dune (sand depot) during severe storm tides.

Sand sources suitable for the sand nourishment have been tapped west of the island of Sylt (fig. 7). The borrow area for the nourishment of sand, i.e. according to the project, during the following 10 years is located 7-10 km in front of the islands's west coast. The average water depth in this area is 14 m. According to geological and

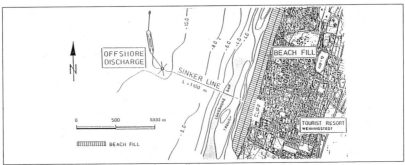

Figure 10:   Offshore discharge form trailer

hydrological knowledge no impact on the island is expected from the planned sand borrowing. An observation of the sand borrowing area is planned in order to recognize unfavourable developments in time.

The excavation of sand from the sea bottom in the borrow area is effected by automatic trailing suction dredgers (hopper dredgers) following either the trailing heads or production dredgers (fig. 10).

Analysis of the natural forces and development numerical models.
The sand nourishments are accompanied by a research program with the aim to improve knowledge about the development at the west coast of Sylt and to optimize the method of sand nourishment.
Therefore the of State Schleswig-Holstein, with support of the German Ministry for Research and Technology (BMFT), has launched a large research program "Aims of the optimisation of Coastal Protection on the Island of Sylt", in which several institutions participate.

Above all the research program is aimed at:
-        the hydrodynamical pressure on the west coast of Sylt and both edges for the island by wave and tidal forces;
-        their effects on coast erosion and sediment budget.

Therefore it is necessary to analyze the sediment transport in front of the coast of Sylt and how it depends on waves and tides, including artificial interferences. Subsequently it shall be examined how the sediment transport can be influenced positively with regard to the protection of the island, how the sand nourishment can be optimized, and how additionally economic measures may be able to reduce the impact of the energy on the west coast of Sylt and its edges.

The evaluation of the interdisciplinary research program, which consists of a natural monitoring program and model testing, include the following main points:

-   recording of the sea motion (tides, waves) in deep sea areas in front of Sylt, analysis of the wave climate and its influence on the sand transport, survey of tidal conditions;
-   hydrological measurements in front of the coast and in distinguished test fields for the recording of water levels, waves (height, frequency) and current characteristics as well as suspension measurements in the surf zone;
-   regular survey of the beach and the dune areas along the west coast of Sylt as well as the survey of the sea test fields;
-   sedimentological analysis of chosen testfields with the aim to record the periodical season-caused sediment along the coast;
-   development of hydrodynamical models in order to:
    a.   evaluate the tidal- and wave-caused littoral sediment transport;
    b.   investigate infringements in the beach-trough-reef area;
    c.   estimate measures to be carried out at the northern and southern ends of the island;
-   development of a computer-aided data management system in order to optimize and transfer the measured field and numerical simulation data for the purpose of engineering planning and evaluation of structural measures.

Analysis of structural measures

The primary aim of the project is to investigate the structural measures which may enable an optimization of the beach nourishments. Included in these investigations are measures which protect the ends of the island. Therefore it is necessary to consider the problem of coastal erosion, sediment transport at the reef-trough-system as well as the erosion and transport processes in the channel systems and the adjacent shallow areas.

The coastal protection measures are investigated on the basis of the present field data and applicable models which have been developed from the participating institutions.

To meet the appropriate measures it was determined to find solution concepts and possible general conditions. Based on the solution concepts and above all to preserve the present-day coastline it was agreed to define the following supporting requirements:

-   preservation of the natural beach area;
-   preservation of the surf zone (possibly in a weakened manner);
-   prevention of negative influence not only upon the adjoining areas but also on the sands and islands south of Sylt;
-   reduction of influence on the environment.

To meet the above mentioned requirements satisfactory solutions concentrate upon:

1. Beach nourishment and an additionally optimization of position, shape, and number of periodically intervals;

2. Construction of membrane systems or similar structural elements to be built into the artificial dune body which are able to diminish damages and protect the endangered areas as well as to prevent dune erosion during severe storm surges;

3. Structural measures (underwater groyne systems) along the beach-trough-reef area in order to increase the residence time of the replenished sand material;

4. Measures along the reef area, which could influence the energy system onto the beach and through which the influence on the system and questions about the stability have to be investigated;

5. Measures in front of the ends of Sylt which may protect against erosion, stop the development of troughs in the foreshore area, and cease the formation of shallow areas.

The examinations to be carried out have the aim to maintain the existing island without giving up the typical characteristics like cliff, shore, dunes and breakers.

References

Cordes, F: Eiderdamm, Natur und Technik, Hans Cristians Verlag, Hamburg, 1972.

Dette, H.H., J. Gärtner (1987) : The history of a seawall on the island of Sylt; *Coastal Sediments 1987, Proceed.*, Vol. I ASCE.

Fisher, O.: Das Wasserwesen an der schleswig-holstinischen Nordseeküste, Bd. 5, Dithmarschen. Verlag Reimer, Berlin, 1957

Führböter, A., R. Köster, J. Kramer, J. Schwitters, J. Sindern (1976) : Beurteilung der Sandvorspulung 1972 und Empfehlungen für die künftige stranderhaltung am Weststrand der Insel Sylt, *Die Küste*, vol. 29.

Führböter, A., R. Köster, J. Kramer, J. Schwitters, J. Sindern (1972) : Sandbuhne vor Sylt zur Stranderhaltung, *Die Küste*, vol. 21.

Hoffmann, D. (1981): The Geological Development of the North Frisian Islands; Wadden Sea Working Group, Sect. "Geomorphology", Final Report, Leiden.

Kamp, (1991): Küstenschutzmassnahmen des landes Schleswig-Holstein an der West-küste Sylt, in *Optimirung des Küstenschutzes auf Sylt, proc. Workshop Federal ministry of Research and Technology, Bonn* (1991), p.125-138.

Petersen, M: Der Eiderdamm Hundeknöll-Vollerwiek als Folge künstlicher Eingriffe in den Wasserhaushalt eines Tideflusses; *Materialsammlung der Agrarsozialen Ges. e.V.*, nr. 62, Kiel, 1967

Prange, W: Die bedeichungsgeschichte de Marschen in Schleswig-Holstein. In *Pro bleme der Küstenforschung in südlichen Nordseegebiet*, *bd.16*, Verlag Lax, Hildesheim, 1986

Rohde, H: Die Veränderungen der hydrographischen Verhältnisse des Eidergebietes durch künstliche Eingriffe; *Sonderheft der Deutschen Gewässerkundlichem Mitteilungen*, Koblenz, 1965.

Wieland, P. and Kraatz, D: Dynamisches System Eider. *Büsumer Gewässerkundliche Berichten des ALW Heide*, H41/1981

Wieland, P.: Rinnenmorphologische und hydrologische Veränderungen im Eidergebied nach ter Abdämmung bei Hundeknöll - Vollerwiek. *Büsumer Gewässerkundliche Berichten des ALW Heide*, H59/1991

Wieland, P: Küstenschutz und Binnenetwässerung in Dithmarschen. In *Historischer Küstenschutz - Deichbau, Inselschutz und Binnenetwässerung an Nord- und Ostzee* Verlag Konrad Wittwer, Stuttgart, 1992.

Coastal Protection in the Past
Coastal Zone Management in the Future?
Case Study for the Ems - Weser Area, Germany
H. Kunz[1]

Abstract

The mainland along the German North Sea coast is low lying land; parts are sheltered by islands. History of coastal protection started about one thousand years ago; in the 13th century the embanking of the whole coast was largely completed; in the second half of the 19th century all exposed dikes were fixed. Eroding parts of the islands have been stabilized by solid constructions since the middle of the last century. After the severe storm floods of 1953 and 1962 the coastal protection line has been intensely strengthend. The basis for the actual coastal protection works are legal plans. They promise security against flooding by a coastal defence line according to fixed design criteria. Coastal protection, stabilization of islands, and land reclamation created static boundary conditions and limited substantially the ability of the Wadden Sea to react dynamically and develop naturally.

Throughout the last millenium the coastal community agreed on the absolute necessity of land reclamation and coastal defence. Since some decades the targets of nature conservation have become increasingly relevant. They interfere with the established coastal protection strategy: the latter favours the existing defence line, while nature conservation promotes substantiel retreat. However, sidespecific compromises can be achieved, especially since land reclamation is no goal any longer.

Since some years global warming has become a topic. Although we can't detect by the analysis of tide gauge-data any changes that significantly relate to global climate change, we have to face an accelerated rise of MSL and higher tidal ranges as a realistic expectation. If we relate to the IPCC - best estimate scenario, the security amount implemented in the German design criteria would be worn out in 20 to 40 years. The technique is available to deal with this rate of SLR and even more by adapting the defence line without substantiel retreat. Nevertheless, the concepts for the future protection of the German coast must be adjusted, if society really wants to meet the goals of nature conservation, especially in the Wadden Sea National Parcs.

The morphological reaction of the tidal flats on SLR, on higher tidal range and on changes of wind climate will be of crucial importance, as well for the unprotected parts of the saltmarshes as for the islands. If society wants to preserve those areas

[1] Niedersächsisches Landesamt für Ökologie - Forschungsstelle Küste (Coastal Research Station), An der Muhle 5, D2982 Norderney, Germany

in a natural stage, we have to review the actual coastal defence line strategy. Numerous demands need to be taken into account besides environmental aspects. This leads to concepts of Coastal Zone Management. But on a long term basis they can only be effectively applied, when the coastal community agrees on the necessity to provide parts of the mainland and of islands for natural development processes, which means losses by erosion. With respect to history and to the established infrastructures and uses it is not likely that a decission on substantial retreat will come up within the next decade. Hence, according to the legal plans, the coastal defence system needs to be completed at first. Since environmental aspects are more appreciated nowadays they will have an important influence on the design of the completion. This allows to implement compromises towards more nature conservation, even if it would require more expensive solutions. An intermediate next step could be a "coastline management plan" providing more flexible responses. If we want to meet the multi-functional demands stated by society in an appropriate way, we need to promote further towards "coastal zone management plans".

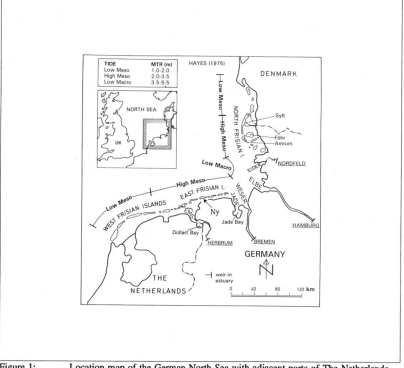

Figure 1: Location map of the German North Sea with adjacent parts of The Netherlands and Denmark. Classifications related to tidal range.

Introduction

The present shape of the southern North Sea coast is the latest transitional stage of a changing and in no way completed geological process (Cameron *et al.*, 1993). The German part of the North Sea coast and adjacent parts of The Netherlands (west) and Denmark (north) are drawn on fig. 1. The displayed classification based on tidal range (Hayes, 1975) describes the German part as high mesotidal and low macrotidal for the area between Jade- and Eider-estuary. The East Frisian islands are sandy barrier islands (dune islands) formed by the coincidence of tides, currents, surf and wind-born accretion (Lüders, 1953; Streif, 1990). The North Frisian islands are the remainder of former mainland: the three large islands in the northern part (Sylt, Föhr, Amrum) have Pleistocene and Tertiary cores and outcrops of moraine (Streif & Köster, 1978); the islands in the more southern part totally consist of marshland.

In the low macrotidal area there are open tidal flats with high sand banks as well as bays of estuarine flats. Sheltered tidal flats extend behind the islands and salt marshes spread behind them as a more or less extended transitional zone.

The mainland along the entire German North Sea coast is low lying marshland with intercalated peat layers and areas of peat bogs (high- and lowmoor). Only in very small parts the Pleistocene hinterland (higher than NN +5.0 m; NN is datum: app. MSL) reaches to the coast. Especially along the estuaries the lowlands (below MHW-level) extend far into the mainland (distances from the seaward limit: 130 km along the Weser to Bremen; 170 km along the Elbe to Hamburg).

History of flood protection

The recent history of the people living in the coastal areas of the southern North Sea is a history of retreat: since ice started melting about 15 000 years ago, the sea level has risen; reliable data for this area reach back to about 9,000 years B.P.; at this time, the coast line had been located about 250 to 300 km further seawards (Behre, 1987). Since that time the relative sea level has risen about 50 m (Cameron *et al.*, 1993; Streif, 1990). With respect to coastal protection we have to focus on the last 1,500 years. Fig. 2 highlights, that the knowledge about sea level variation is still poor: the drawn curves and single information published by different authors vary substantially. Reliable information on tidal range can only be gained from tide gauge data (Amsterdam/Netherlands since 1700; Cuxhaven/Germany since

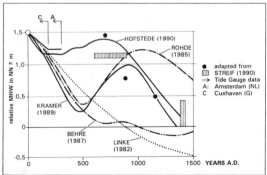

Figure 2:   Rise of relative mean high water level (relative MHW "secular rise") along the German North Sea coast during the last 1500 years

1788 /1855 (Jensen *et al.*, 1993)); however, if tidal range has increased throughout the last 1500 years, only a moderate rate is likely.

From numerous investigations we learned, that about 1800 years ago people incrisingly started to build "Warfen" (Warf or Wurt: a mound of hillock constructed out of earth or stable manure) and continiously raised them higher (fig. 3). Parts of "Warfs" addionally had been protected by embankments build of clay; by this we can date back the beginning of coastal protection in Germany to the 2nd century A.D. (Haarnagel, 1979).

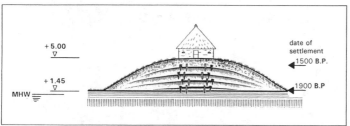

Figure 3:   Schematic cross section through a "Wharf" (adapted from Bantelmann, 1966)

The marshland along the German North Sea coast in the undiked stage must be pictured as flat wetlands, intersected by numerous creeks (fig. 4, upper part). In the first centuries after the turn of the millennium everywhere along the German North Sea coast embankments (dykes) were built. At first they protected polders (Koog: deep lying marshland enclosed by dikes). In the course of the 13th century the embanking of the whole German North Sea coast was largely completed (Rhode, 1978). The poldered areas had been drained and tide gates (sluices) had been placed in the embankments to improve the agricultural use of the area. Additionally the unfertile peat layers had been removed (salt fabrication, burning, mixing with sandy soil) and converted into farmland (fig. 4, lower part). Dyking necessitated drainage and the history of coastal protection has to include the impacts of drainage and agriculture, such as subsidence or topsoil-erosion (Kramer, 1992).

The sudden flooding of a polder during storm flood as a result of a breach in the dike is more dangerous than the gradual inundation of unembanked areas. This is presumeably the reason, why reports of exceptionally high storm surges and numerous losses of life, cattle, property are first reported in the period when the embanking of the coast was already far advanced. Man-induced subsidence and lowering of raised bog-areas aggravated this situation.

From about the end of the tenth century we are supplied with information about storm surges. This is thanks to literate Christian priests, who kept a record, because they interpreted such great misfortunes as punishments from God. The first particularly severe flood on record is the "Julian flood" of the 17th February 1164 (Rohde, 1978). The last historic storm flood catastrophe on the German North Sea coast was the "Christmas" flood of 24th December 1717 with 11,300 victims (fig. 5).

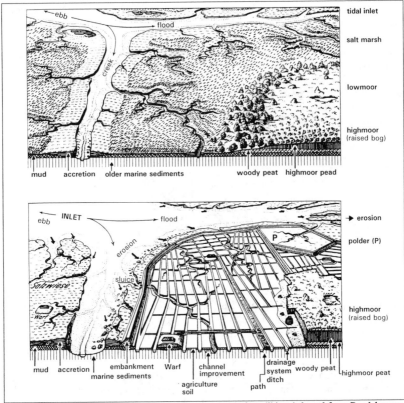

Figure 4:   Landscape of the German Wadden Sea before diking (adapted from Bandelmann, 1966)

The present day outline of the German North Sea coast was largely shaped by the severe floods of the 14th century, especially by the "Second Marcellus flood" of 1362. The coastal inhabitants tried to reclaim parts of these land losses. An example is given by fig. 6, which shows the reclamation of the Ley Bay since the 15th century (Homeier, 1969). Since the 15th to 16th century, land reclamation has become important along the entire German coast. To accelerate the raise of higher tidal flat areas, reclamation fields were provided with a network of groyne-like dams ("Lahnungen") build from brush-wood. By digging small ditches and spreading out the excavated deposits on the area between an increase of 10 cm a year can be achieved. This technique is still used nowadays, not for land reclamation but for the protection of the foreland-face, it is called "active coastal protection" (Erchinger, 1970)

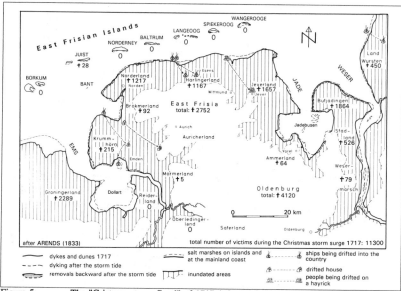

Figure 5:   The "Cristmas storm flood" of 1717 (adapted from Homeier, 1970, 1980)

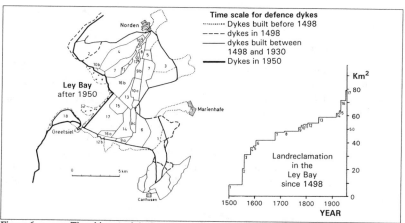

Figure 6:   Time history of the Ley-Bay reclamation since the 15th century; location of dikes
                 (left); area of polders (right) (adapted from Homeier, 1969)

Development of techniques for coastal protection

The technique of coastal protection work changed with time. This has been described
by many authors. The special volume dedicated to the ICCE-Conference, Hamburg
in 1978 contents a report on "coast protection" (Kramer, 1978) which provides an
overview. Recently the report has been complemented and updated (Kramer, 1989).
A few technical aspects related to dikes shall be addressed in this paper.

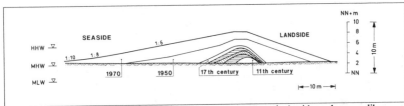

Figure 7:   Comparison of historical embankments (cross-section) with modern sea-dikes
                 (adapted from Kramer, 1978)

For many centuries there had been only the technique of embankments/dikes
available. The first dikes in the 11th century had a very small cross section; they
were low earth walls, steeply sloping and narrow; with time they became higher and
got flatter slopes on the seaside (fig. 7). The dikes developed site-specifically;
however, a general idea how height, width and cross section altered over time can be
roughly drawn (fig. 8).

At first embankments were
built at a larger distance from
the shoreline. If the foreshore
had been eroded, people tried
to protect the exposed lower
parts of the dike against
damage by vertical walls of
wood ("wooden dike"), later
by fixing the flatter designed
slopes with turf- or straw
band-layers ("straw dikes").
By the 18th century the first
solid stone constructions had
been implemented and in the
second half of the 19th century all exposed dikes were
fixed by bricks or stones.

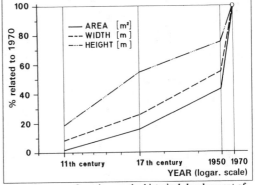

Figure 8:   Overview on the historical development of
                 dike-parameters (cross-section, area, width,
                 height) over time (data from Kramer, 1989)

This coastal defence line has become the basis of the German coastal protection. We
need to be aware, that all of this tremendous work had been done by man with spade
and wheel-barrow (only 40 m³ clay placed in the dike by one man in an eighty hour
week!).

Fig. 9 demonstrates schematically the modern technique of dike construction. The old
clay dike is transformed into two embankments (1), sand dredged out of the tidal flat
is pumped into the slashed dike (2) and profiled as a core (3). Afterwards the clay
out of the deposits or addionally dug out of the foreland or polder is placed on the
sandy core as a cover layer (4) to (6). If necessary the dike will be provided with
revetments and solid toe armoring (e.g. Kramer, 1989).

The progress of available techniqual means enabled man to increasingly implement
solid coastal protection work (revetments, groynes, seawalls) since the middle of the

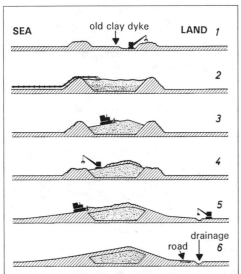

Figure 9:   Schematic display of the construction of modern dikes with a sandy core and a clay cover layer (adapted from Kramer, 1989)

last century. There was a growing demand for those works because of: industrialisation including the construction of shipping channels in the estuaries and of harbors, settlements, agriculture, tourism. This is demonstrated by some examples of the East Frisian islands.

Stabilization of barrier islands

The islands along the German North Sea coast are part of the Wadden Sea; they shelter the mainland against the impact of wave action (Ehlers & Kunz), 1993). The barrier islands of East Frisia have supposedly existed for several thousand years. They permanently changed their shape, predominently as reaction on the dynamic migration of the inlets.

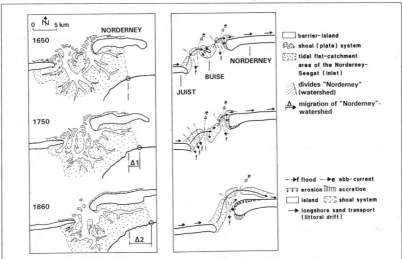

Figure 10:   Reconstruction of the development of the tidal inlet between Juist-island (J) and Norderney (N) from 1650 - 1860 (adapted from Luck, 1976).

From charts and written historical documents we know these processes quite well. Fig. 10 gives an example for the tidal inlet ("Norderneyer Seegat") between Juist and Norderney (Kunz, 1991a) for different stages between 1650 and 1860: initially there were two hydraulical equivalent inlets separated by a small middle island. Within two centuries the middle island disappeared and the single shaped inlet profile shifted to the East, eroding the west spit of Norderney (left). The development can be explained by the connection between the tidal inlet and the tidal prism, taking into account the migration of the affiliated catchment area of the tidal flat (see Δ 1 and Δ 2 on fig. 10, left; O'Brien, 1931; Walther, 1934; Lüders, 1953; Bruun, 1978; Fitzgerald et al., 1984). A model which explains the dominant coastal processes as a state of dynamic equilibrium, affected by the active processes in the tidal inlet followed by a passive response of the island itself (fig. 10, right) has been developed by Luck (1976). In the 1860 situation the sandbars of the reef bow (ebb delta shoals) are forced seawards and don't approach the west end of Norderney any longer. This status of negative sand balance has been fixed by sea walls and groynes, producing static conditions which otherwise only would have existed temporarily (Kunz, 1987, 1991a).

History of the East Frisian Islands recently has been newly reviewed (Streif, 1990). The example given by fig. 11 for Wangerooge island demonstrates how the location and shape of barrier islands changed with time: as reaction on the shifting and turning of the inlet the western part had been eroded and accretion on the eastern spit occured (Krüger, 1911; Sindowski, 1973; Streif, 1990). The turning of the Harle inlet in a clockwise direction can be seen from fig. 12. In order to stop this turning and the deepening of the developing "Dove Harle"-rim the groyne "H" had been strengthend and extended to a length of almost 1.5 km between 1938 and 1940 (fig. 12 middle; Lüders, 1952). By this, the Harle inlet had been forced to turn counter-clockwise back into the older direction (fig. 12, right). The enormous effort to stabilize Wangerooge island had been mainly justified by the fact, that the harbor of Wilhelmshaven had become important since the middle of the last century and the shipping channel "Jade" had to be maintained (Krüger, 1921; Witte, 1970).

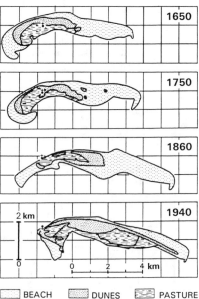

Figure 11: Morphological changes of the east Frisian islands Wangerooge since the last 300 years (data from Homeier, 1962, 1974).

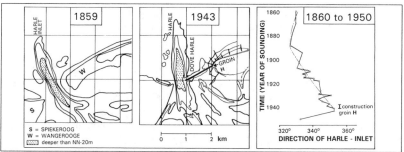

Figure 12:   Location of Harle inlet and Wangerooge island in 1859 and 1943 after the
               construction of groyne H. Change of direction of the Harle inlet over time (1860 -
               1950)

The history of the protection works on the East Frisian islands is well recorded
(e.g.) Witte, 1970; Kramer, 1989). By construction of solid structures since 1850 the
natural migration of the inlets was almost stopped and the islands itself became
stable: man reduced the ability of dynamic responses and forced the inlet-island-units
in an almost static stage. This had a substantial impact of stabilization on the tidal
flats and the mainland shoreline. The coastal community at that time considerd this
benificial (Krüger, 1921). The loss of morphological dynamics is demonstrated in fig.
13 by the migration of the tidal flat watersheds over time: some became almost
immediately static after the first seawalls and groynes had been constructed in the
middle of the last century; nowadays over the whole stretch between Juist and
Wangerooge (fig. 16), the watersheds are in an almost fixed position (Luck, 1975).

Actual coastline protection in Germany

After the severe storm floods in 1953 and
especially in 1962 the rising and strengthening of the whole flood defence system
along the German North Sea coast begun
again. Based on the recommendations of a
special task group (Küstenausschuss Nord-
und Ostsee, 1955, 1962) design criteria had
been fixed for dikes that shall withstand a
storm flood of at least 100 year period
without severe damage. The crest level of
dikes, embankments, walls etc. were
determined by semi-emperical methods that
take into account: mean and highest known
waterlevels, sea level (MHW)-rise (25 to 30
cm), wave run up, subsidence by compaction
(e.g. Kramer, 1978). In earlier centuries, the
construction of dikes purely was done by

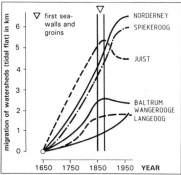

Figure 13:   East-migration of
               watershed over time before
               and after the construction
               of solid coastal protection
               structures (adapted from
               Luck, 1975).

empirical experiences (try, failure, reaction). The change towards coastal engineering in Germany started with Brahms (1767/73), whose work was substantially based on evaluated measurements.

According to legal plans ("Generalplan") nowadays a fixed main flood defence-line covers the entire German North Sea coast. It consists of dikes/embankments, walls (in developed areas such as settlements, industrial regions, harbors), storm surge barriers (mainly in tributaries of the estuaries), weir (artificial end of the estuary) and it contents road-gates (access), tide gates (sluices), pumps (drainage, irrigation), navigation locks (shipping). Case examples are given by Kamp & Wielans (1993). Since 1962 the length of the main flood defence line has been cut down from 1200 km to 700 km. The costs had been about 7 billion DM; approximately another 3 billion DM needs to be spend to meet the legally fixed flood protection goals (Kramer, 1989). The money comes from Federal (70 %) and State (30 %) governments; a symbolic part is raised locally by dike boards.

Development of boards
Coast protection has been a communal activity from the beginning. One can trace agreements and duties back to the 11th century.
Land owners had the duty to build and to maintain embankments/dikes. The first known law that regulated this duty in Germany (Friesland) is the "Sachsenspiegel",

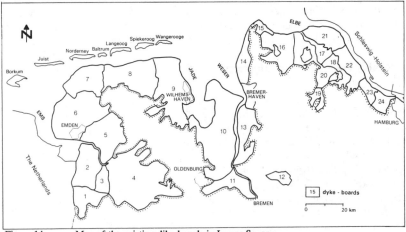

Figure 14:   Map of the existing dike-boards in Lower Saxony.

from early 14th century. The development into the modern dike- and water-boards of today is well recorded (Peters, 1992). These boards cover nowadays much larger areas than they did in former times. They extend from the coast to the contour-line of NN +5,0 m (fig. 14).The dike-boards are part of the democratic decision making procedure for the coastal protection work; in Lower Saxony they are responsible for the maintenance of almost the entire protection system located on the mainland (different regulations in the four German States/Länder along the North Sea coast).

New aspects for the coastal protection in Germany

The addressed situation of the German North Sea coastline can be summerised as follows: the coastal protection of the mainland (Wadden Sea coast and estuaries) is constructed as a defence line that runs near the shore and the river banks; the islands are partly armoured and the tidal inlets are widely fixed. Hence the naturally dynamic behavior of the coastal zone has been converted by technical means into an almost static one. Addionally there is in extended parts of the coastline none or only few foreland that could act as a transition zone for natural development (fig. 15, left).

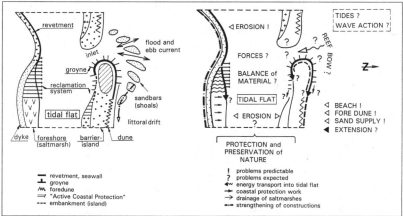

Figure 15:   Schematic scetch of the German barrier islands with coastal protection works (left) and concerning the impacts on the Wadden sea area by sea level rise (right).

The existing legal plans for the coastal protection guarantee the strengthening of the fixed protection line in such a way, that the severe failure during storm flood events is no realistic prospective. This "coastline defence strategy" is nowadays discussed, mainly with respect to two different aspects:

1. Impacts of global warming, especially concerning accelerated sea level rise (SLR)
2. Nature-preservation, especially in the Wadden Sea (National Parcs) but also in the upper parts of estuaries.

Problemes that may arise from accelerated SLR, higher storm surges and waves are presented in fig. 15, right (Kunz, 1991b). According to the actual coastal protection strategy we have to prove, if it is possible to strengthen the existing coast protection line. This leads to the question: can we predict the impacts of expected global climate changes? With respect to the paper of Jensen *et al.*, (1993) in this volume, the discussion is restricted to some remarks.

Extensive statistical investigations based on tide gauge data (relative levels) confirm, that up to now we can't significantly detect an accelerated SLR that relates to global climate change. The same is true for mean sea level (MSL). The averaged linear trend for the relative SLR on the German North Sea coast is about: 25 to 30 cm/100 years for MHW and 10 to 15 cm/ 100 years for MLW (Jensen *et al.*, 1991), Siefert (1989). A natural land subsidence of about 5 cm/100 years for the East Frisian coast is likely. Comparable to other gauges along the German coast, the tidal range has increased during the last decades (Jensen *et al.*, 1993).

The highest storm surges on record for Dangast (southern Jade Bay) and Wilhelms-haven (fig. 1, fig. 14) lead to linear trends of about 30 cm/100 years; no significant increase that might be related to anthropogenic climate change is obvious (Rohde, 1985). This corresponds to the results gained from statistical investigations with storm surge data (Niemeyer, 1987; Halcrow /NRA, 1991). Periodic pattern with return periods from about 7 years (Siefert, 1989) to 60 years (Lüders, 1936) have been detected (fig. 16).

The duration times of tide water levels above a fixed horizontal datum-line show substantial increases for the entire German North Sea coast (fig. 17). This is the effect of the addressed linear sea level rise trend. If we eleminate it, we only have periodic trends along a horizontal water-level-line (Niemeyer, 1987; Jensen *et al.*, 1993).

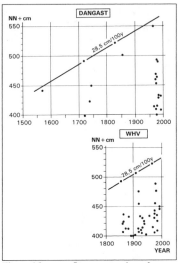

Figure 16:   Storm surge 4 m above MSL at Dangast (south Jade Bay) and Wilhelmshaven (data by Rohde, 1977).

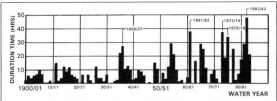

Figure 17:   Duration time per year for floods above MSL + 2 m in hours; time series 1900/01 to 1983/84 at List, Sylt (adapted from Führböter, 1988)

There is a growing consensus on the view that tide gauge data are not yet able to give a reliable answer whether there is an accelerated sea level rise or not (e.g. Jensen *et al.*, 1993), but also on the believing, that an increasing SLR, tidal range and more frequent storm events is a realistic expectation for the next decades (e.g. Mehta & Cushman, 1989); Ipcc/Unep/Wmo, 1990, 1992). At this moment it seems to be reasonable to base considerations on the IPCC - best estimate scenario, as Rijkswaterstaat (1991) did; this would lead to a prediction for mean sea level rise of 60 cm within the next century. That is much more, than the rise of about 30 cm for MHW-level (corresponds to about 15 cm MSL-rise) which the valid regulations in Germany have fixed as a security amount.

Fig. 18 demonstrates, that a MSL-rise of 60 cm/100 years would cut down the security reserve time (estimate had been 100 years) to 25 years (linear) or 40 years (exponential approximation). The result would be: the effort must increase to complete the strengthening of the existing coastal protection line according to the legal plans. After completion the security against storm floods, as guaranteed by special laws (dike law), would be achieved.

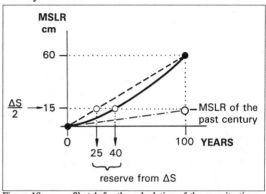

Figure 18:   Sketch for the calculation of the security time reserve.

Parallel to this completion it is necessary to come up with a democratic decision on the guidelines for a future coastal protection-strategy/management at least within the next 10 to 15 years to meet the requirements deriving from security aspects for people and infrastructure as well as those with respect to nature conservation and dynamic development of the Wadden Sea - National Parcs.

Concerning coastal protection techniques we can state: appropriate techniques are available to adapt the established defence line along the German North Sea coast to a sea level rise of 1.0 m or even more. The special problems for the harbours and low lying settlements of Hamburg and Bremen, as well as problems arising from compaction (peat and clay layers) for the heightening of embankments could be dealt with by constructing storm surge barriers in the Elbe-, Weser- and Ems-estuaries, combined with flood basins (temporarily storage of fresh water runoff); technique is also provided to replenish sandy beaches and to protect the parts of the islands that had been armoured by solid groyne- and revetment-systems against failure (e.g. Kramer, 1989; Kamp & Wieland, 1993). Hence it would be possible to protect the German North Sea coastal areas, without substantial retreat throughout the next century, even if the impact of global warming would cause an acceleration of sea level rise in the order as shown by fig. 18. Sea level rise is likely to be accompagnied by changing of atmospheric conditions that might lead to increasing number, height, duration of storm surges and to stronger tide- and wave-induced currents. Additionally SLR may affect the tidal amphidromical system in the North Sea accompagnied by an increase of tidal range. The consequences for morphodynamic processes as well as for the design of coastal structures have to be taken into account.

Concerning nature conservation we have to realize, that the Wadden Sea and the forelands along the estuaries are widely shaped by mans activities throughout the last centuries, as described before. The naturally highly dynamic reactions of the Wadden

Sea units are nowadays substantially governed by man made solid (static) features. The impact of morphodynamic processes, steered by the sea (tides, waves etc.) on beaches, tidal flats, foreland, are reduced by these artificially fixed boundaries: dynamic migration of tidal inlets and estuaries as well as the natural shaping-processes of shorelines by erosion and accretion are not possible any longer or have been substantially restricted. The targets of nature conservation must interfere with the described mandatory protection strategy fixed by law and legal plans. This is especially the case in areas with predominantly erosive trends and small distances between the coastal protection line and the sea.

The contradictory aims can be adapted to each other by compromises in a certain range. An example is the solution for the Ley Bay, where plans of a total closure (poldering as done since the middle age; fig. 6) had been given up. This compromise was possible since land reclamation is no supported target any longer. A balance had been achieved between the protection against inundation due to storm surge/rain discharge, the access to the offshore waterways for fishing/shipping on the one hand and environmental aims/nature conservation on the other hand (Niemeyer, 1991).

The natural reaction of the German Wadden Sea to a moderate sea level rise is well known from the past: shift of barrier islands in a landward direction; erosion of both the islands that have been part of the mainland and the shoreline of the mainland itself. By technical means this has been extensively stopped. But erosion in front and in the vicinity of these structures as well as at the unprotected parts, such as sandy beaches, tidal flats, saltmarshes has increased. This general trend for the Wadden Sea is not only related to SLR, but is a reaction on the decrease of sand supply as an side effect of armouring and additionally of river-regulations and training means. The more and more used technique of artificial nourishment will only help to postpone the erosion of sandy beaches and cliffs. The erosion rate of saltmarshes and of the transition zone to the tidal flats, as the most vulnerable parts of the system, will further accelerate.

The intertidal flats exist in an unstable quasi equilibrium stage. According to relationships between the tidal prism and geometrical parameters of the system, we expect for the relatively large inter- and supratidal areas also a tendency towards more erosion if SLR or (and) tidal range would increase. A mean sea level rise of 60 cm (IPCC - best estimate) means a four times larger rate of SLR than in the last century. The drowning effect will be counteracted by sedimentation of eroded sand from tidal channels and from beaches in the vicinity of the inlet. However, if the sedimentation cannot keep pace with the rate of sea level rise, the inlet and channels will become too large in relation to the tidal volume and erosion of the flats will take place (Vellinga, 1987; Misdorp et al., 1990). Along with this, the waves will become higher and by this the rate of erosion of the salt marshes would further increase.

For the shoreline (islands, salt marshes, Pleistocene cliffs) two response options are available in general:
  - stabilize further, either by armouring and (or) through artificial beach nourishment;

- give up parts of the salt marshes and islands; allow erosion and managed retreat from the shoreline.

The decision making procedure concerning the two options has to consider the aspects of coastal protection and those of nature conservation; additionally the aims of recreation, settlements, industry, shipping etc. need to be taken into account.

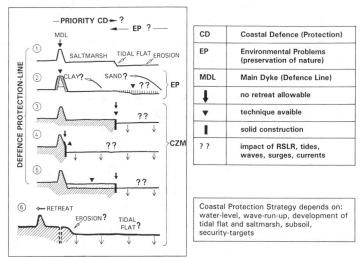

CD	Coastal Defence (Protection)
EP	Environmental Problems (preservation of nature)
MDL	Main Dyke (Defence Line)
↓	no retreat allowable
▼	technique avaible
▌	solid construction
? ?	impact of RSLR, tides, waves, surges, currents

Coastal Protection Strategy depends on: water-level, wave-run-up, development of tidal flat and saltmarsh, subsoil, security-targets

Figure 19:   Sketch on problems (actual and future) concerning coastal protection of the Wadden Sea coast

Fig. 19 addresses schematically the problems for a coastline with an eroding foreland (saltmarsh). The strengthening of the flood defence line (main dike) in the traditional way demands clay from of the saltmarsh and sand dredged out of the tidal flat (2); as the width of the foreland is getting too small, either "active coastal protection" by groyne-systems (2) or armouring of the edge (3) would be implemented. According to the aims of nature conservation, none of the technical procedures is desirable. If no decision (compromise) becomes possible and consequently nothing would be done, the foreland would be totally eroded by natural processes; latest at this moment the legal coastal defence-plans would require the protection of the dike by solid structures (4). At that stage a foreland could only be re-established by artificial nourishment with a stablized edge (5): this will not be a salt marsh, and "active coastal protection" (2) is no alternative option any longer in this stage. In the cases (3) to (5) the ongoing deepening of the tidal flat is likely to occur; it would necessitate further edge-protection-work with time. Retreat (6) is the only way to really meet the targets of nature conservation without an agreement to any kind of compromises. From an engineering point of view the flood protection in the case of retreat is to ensure by simple dike-techniques, if space is provided to re-establish a new coastal defence line.

However, a substantial conversion of the existing coastline-protection strategy into a strategy (management) of retreat will nowadays not be accepted by the coastal community. The technique of "active coastal protection" (2) can postpone the armouring, if certain natural conditions are in place (height of tidal flat, rate of suspended matter, currents and waves). A compromise would than prevent the unfavourable development according to (3) to (5) for a limited time. It would meet the demands of the actual "coastline defence strategy" on the one hand and would postpone the basic decision between the contradictionary options: (3), (4) and (6).

The demands on the design criteria for the "main dike" (MDL) could be lowered, if a "second dike"-line exists or would be constructed (Führböther, 1987). This would further extend the space for compromises.

Compromises on a site-specific level (no clay- and sand-exploitation in the foreland and the tidal flats, dike-reinforcement only in landward direction etc.) are in discussion; those compromises are relevant, but they don't really deal with the basic conflicts addressed before.

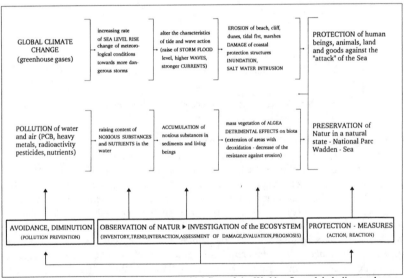

Figure 20:   Synergism of impact on the tidal flats of the Wadden Sea; global climate change and pollution (Kunz, 1990)

The schematically displayed development of the German Wadden Sea coast by fig. 20 is a realistic expectation, even if an acceleration of SLR will not occur. Comparable prospects are true for the islands (Kunz, 1991b). The basic question that needs to be answered is: what kind of Wadden Sea coast do we want to preserve in Germany as a heritage for the generations to come? If we want to preserve the peculiarity of the landscape, at least in certain areas, we definitely need to prove if and where agreements on managed retreat are necessary and can be achieved.

An appropriate discussion of the raised topic needs reliable predictions about the future development of Wadden Sea morphology. With fig. 20 attention shall be drawn to synergetic effects that may affect the stability of the intertidal flats, presumably the most important unit for the determination of future Wadden Sea development: pollution may destabilize the flats by detrimental effects on the benthic biota (sediments); global warming may enlarge the tidal- und wave-induced forces; both effects together would accelerate the erosive processes.

Conclusions concerning the future long-term development of the German Wadden Sea

During the past centuries a sophisticated flood-protection line has been established along the entire German North Sea coast. The recent construction of the protection line is based on legal plans that consider a security amount for future sea level rise. This would be worn out in 20 to 40 years, if we assume an accelerated mean SLR of about 60 cm for the next century (IPCC - best estimate). The existing German coastal protection line could be adapted to this SLR by proven technical means and would presumably include storm surge barriers in all of the estuaries on a long-term basis. This would be applied, if society wants to go on with the valid coastal protection strategy in future. However, the site-specific decisions on those technical solutions would take into account the environmental aspects, as far as possible and additionally, other demands of the coastal community as well as cost-benefit considerations to a certain extend. The legal plans have not been totally fullfilled up to now. Since obligations for the protection of the low lying lands against flooding are promised by law, the strengthening according to sea level rise and modified climatic conditions has to be implemented in time. It is foreseen to complete the work; an adaption of the design to accelerated SLR is not planned yet (but would be taken into account for special constructions like barriers). It is discussed, if it is appropriate to distinguish different security standards and along with this to re-establish or to construct so called "second dike lines".

Concerns and objections against the existing coastal protection strategy arise from nature conservation, especially because this strategy does not include options for substantial retreat. There is a realistic prospective, that without retreat-options we will loose within the next century most of the saltmarshes (foreland) and need to armour further stretches of the coast, especially on the mainland (marshes) where artificial nourishment with sand can not be applied. The scope of the following problem is the basic question: do we want to give up parts of the islands and of the mainland to ensure natural development in these areas, and to preserve the characteristics of the Wadden Sea? This basic question must be answered by the coastal community as soon as possible. The decisions need to be prepared on a broad basis. It is necessary to compare the whole variety of solutions and provide people with reliable and realistic prospectives. Taking a moderate impact of global warming into account, the decission-making-procedure should be finished within the next 10 to 15 years.

Meanwhile, the existing coastal protection needs to be completed according to the legal plans to ensure the promised security standard. As far as consensus is possible at this time, the addressed concerns with respect to the future development of the Wadden Sea will be considered. Since environmental aspects are highly appreciated nowadays, it has become possible to implement substantial compromises towards nature conservation, even if they are more expensive. Additionally those coastal areas which are suitable for retreat or alternatively for the construction (re-establishment) of a "second dike line" should not be developed (no further infrastructures and uses) until the described basic decisions have been achieved. The addressed problem requires new concepts: the different demands of society need to be united in integrated plans; an intermediate next step could be a "coastline management plan" providing more flexible responses. To meet the multifunctional demands stated by society, we finally need to promote "coastal zone management plans".

References

ARENDS, F. (1833): Physische Geschichte der Nordsee-Küste und deren Veränderungen durch Sturmfluten; Woortman jun.-Press, Emden.

BANTELMANN, A. (1966): Die Landschaftsentwicklung im nordfriesischen Küstengebiet, eine Funktionschronik durch fünf Jahrtausende; *Die Küste*, 14, vol. 2.

BEHRE, K.-E. (1987): Meeresspiegelbewegungen und Siedlungsgeschichte in den Nordseemarschen; *in: Vorträge der Oldenburg. Landschaft*, vol. 17.

BRAHMS, A. (1767/73): Anfangsgründe der Deich- und Wasser-Baukunst; facsimile-reprint (1989); edit.: Marschenrat, Schuster-Press, Leer.

BRUUN, P. (1978): Stability of tidal inlets - Theory and engineering; Developments in Geotechnical Engineering, vol. 23, Elsvier Sc. Publ., Amsterdam.

CAMERON, D., D. VAN DOORN, C. LABAN, H. STREIF (1993): Geology of Southern North Sea Basin; CZ '93-Coastlines of the Southern North Sea; *This volume*

EHLERS, J., H. KUNZ (1993): Morphology of the Wadden Sea; natural processes and human interference; CZ '93 - Coastlines of the Southern North Sea; *This volume*

ERCHINGER, H.-F. (1970): Küstenschutz durch Vorlandgewinnung, Deichbau und Deicherhaltung in Ostfriesland; *Die Küste*, vol. 19.

FITZGERALD, D.M., S. PENLAND, D. NUMMEDAL (1984): Changes in Tidal Inlet Geometry due to Backbarrier Filling - East Frisian Islands, West Germany; *Shore and Beach*, vol. 52, no. 4.

FORSCHUNGSSTELLE NORDERNEY (1963): Niedersächsische Küste, Historical Map, 1:50 000, No. 5, Coastal Research Station, Norderney.

FÜHRBÖTER, A. (1987): Über den Sicherheitszuwachs im Küstenschutz durch eine zweite Deichlinie; *Die Küste*, vol. 45.

FÜHRBÖTER, A. (1988): Changes of the tidal water levels at the German North Sea coast; *proc. 6th International Wadden Sea Symposium List/Sylt*, Helgoländer Meeresuntersuchungen.

HAARNAGEL, W. (1979): Die Grabung Feddersen Wierde. Methode, Hausbau, Siedlungs- und Wirtschaftsform sowie Sozialstruktur; Feddersen Wierde, vol. 2, Wiesbaden.

HALCROW/NRA (1991): Shoreline Management: The Anglian Perspective, in: The future of shoreline management - *conference papers*: report of the National River Authority and Sir William Halcrow Partners, GB.

HAYES, M.O. (1975): Morphology of sand accumulations in estuaries; in: L.E. Cronin (ed.): *Estuarine Research*, vol. 2.

HOFSTEDE, L.A. (1991): Sea Level Rise in the inner German Bight since AD 600 and its implications upon tidal flats geomorphology; in: Brückner, Radtke (*eds.*): Von der Nordsee bis zum Indischen Ozean; Franz Steiner Press, Stuttgart.

HOMEIER, H. (1962, 1964, 1974): Beihefte zur Historischen Karte der Niedersächsischen Küste, Forschungsstelle f. Insel- u. Küstenschutz (Coastal Research Station), Norderney.

HOMEIER, H. (1969): Der Gestaltwandel der ostfriesischen Küste im Laufe der Jahrhunderte - Ein Jahrtausend ostfriesischer Deichgeschichte; J. Ohling (publ.): Ostfriesland im Schutz des Deiches, Bd. II, Pewsum.

HOMEIER, H. (1970/80): Reisefibel (Guidebook), Forschungsstelle für Insel- und Küstenschutz (Coastal Research Station), Norderney.

IPCC/UNEP/WMO (1990): Strategies for adaptions to Sea Level Rise; Intergovernmental Panel on Climate Change, Response Strategies Working Group, report Coastal Zone Management Subgroup, edit.: R. Misdorp, J. Dronkers, J.R. Spradley, Rijkswaterstaat, Den Haag.

IPCC/UNEP/WMO (1992): Global Climate Change and the Rising Challenge of the Sea; report IPCC-Response Strategies Working Group and IPCC-Coastal Zone Management Group, Rijkswaterstaat/Tidal Waters Division, The Hague.

JENSEN, J., H.-E. MÜGGE, W. SCHÖNFELD (1991): Development of water level changes in the German Bight - an analysis based on single value time series; *proc. 22nd ICCE, Delft*, 1990, ASCE, New York.

JENSEN, J., L.A. HOFSTEDE, H. KUNZ, J. DE RONDE, &. HEINEN, W. SIEFERT (1993): Long term water level observations and variations; *This volume*;

KAMP, W.D., P. WIELAND (1993): Case studies for coastal protection problems between Elbe and Sylt: Dithmarschen; closure of the Eider estuary; the island of Sylt; CZ '93 - Coastlines of the Southern North Sea; *This volume*.

KRAMER, J. (1978): Coast protection works on the German North Sea Coast; *Die Küste*, vol. 32.

KRAMER, J. (1989): Kein Deich, kein Land, kein Leben - Geschichte des Küstenschutzes an der Nordsee; Gerhard Rautenberg Press, Leer.

KRAMER, J. (1992): Binnenentwässerung und Sielbau im Küstengebiet der Nordsee; Historischer Küstenschutz, Konrad-Wittwer Press, Stuttgart.

KRÜGER, W. (1911): Meer und Küste bei Wangerooge und die Kräfte die auf ihre Gestaltung einwirken; *Zeitschr. f. Bauwesen*, LXI.

KRÜGER. W. (1921): Die Jade, das Fahrwasser Wilhelmshavens, ihre Entstehung und ihr Zustand; *Jahrbuch der Hafenbautechnischen Gesellschaft*, vol. 4.

KUNZ, H. (1987): History of Seawalls and Revetments on the Island of Norderney; *Coastal Sediments '87*, Proc. vol. 1, ASCE, New York.

KUNZ, H. (1990): The impact of an increased sealevel rise on the German Wadden Sea and how the global climate change may affect coastal zone management; *proc. Littoral 1990*, Marseille.

KUNZ, H. (1991a): Protection of the island of Norderney by beach nourishment, alongshore structures and groynes; *Proc. 3rd COPEDEC-conf*, vol I. Mombasa.

KUNZ, H. (1991b): Klimaänderungen, Meeresspiegelanstieg, Auswirkungen auf die niedersächsische Küste; *Mitteilungen d. Franzius-Instituts*, University of Hannover, vol. 72.

KÜSTENAUSSCHUSS (1955): Allgemeine Empfehlungen für den deutschen Küstenschutz; *Die Küste*, Jg. 4.

KÜSTENAUSSCHUSS (1962): Empfehlungen für den Deichschutz nach der Februar-Sturmflut 1962; *Die Küste*, Jg. 10, vol. 1.

LINKE, O. (1982): Der Ablauf der holozänen Transgression aufgrund von Ergebnissen aus dem Gebiet Neuwerk/Scharhörn; *Probl. d. Küstenforschung im südl. Nordseegebiet*, vol. 14.

LUCK, G. (1975): Der Einfluß der Schutzwerke der Ostfriesischen Inseln auf die morphologischen Vorgänge im Bereich der Seegaten und ihrer Einzugsgebiete; *Mitt. d. Leichtweiß-Inst. f. Wasserbau*, University of Braunschweig, vol. 47.

LUCK, G. (1976): Inlet changes of the East Frisian Islands; *Proc. 15th ICCE*, vol. 2, ASCE, New York.

LÜDERS, K. (1936): Das Ansteigen der Wasserstände an der deutschen Nordseeküste; *Zentralblatt der Bauverwaltung*, vol. 50, Berlin.

LÜDERS, K. (1952): Die Wirkung der Buhne H in Wangerooge-West auf das Seegat "Harle"; *Die Küste*, vol. 1.

LÜDERS, K. (1953): Die Entstehung der Ostfriesischen Inseln und der Einfluß auf den geologischen Aufbau der ostfriesischen Küste; *Probleme der Küstenforschung im südlichen Nordseegebiet*, vol. 5.

MEHTA, A.J., R.M. CUSHMAN et al. (1989): Workshop on Sea Level Rise and Coastal Processes; Palm Coast/Florida, National Technical Information Service, U.S. Department of Commerce, Springfield, Virginia.

MISDORP, R., F. STEYAERT, F. HALLIE, J. DE RONDE (1990): Climate Change, Sea Level Rise and morphological developments in the Dutch Wadden Sea, a marine wetland; *in*: J.J. Beukema et al. (eds.): Expected effects of climate change on marine coastal ecosystems; Kluwer Academics Publishers, Dordrecht.

NIEMEYER, H.D. (1987): Zur Klassifikation von Sturmtiden; *Jahresbericht der Forschungsstelle Küste* (Coastal Research Station), vol. 38, Norderney.

NIEMEYER, H.D. (1991): Case study Ley Bay: an alternative to traditional enclosure; *proc. 3. COPEDEC-conference*, vol. I, Mombasa.

O'BRIEN, M.P. (1932): Equilibrium Flow Areas of Inlets on Sandy Coasts; *Journal of the Waterways and Harbors Division*, vol. 95.

PETERS, K.-H. (1992): Die Entwicklung des Deich- und Wasserrechts im Nordseeküstengebiet; Historischer Küstenschutz, Konrad Wittwer-Press, Stuttgart.

RIJKSWATERSTAAT (1990): A new coastal defence policy for the Netherlands; report, IG's-Gravenhage.

ROHDE, H. (1978): The history of the German Coastal Area; *Die Küste*, vol. 32.

ROHDE, H. (1985): New aspects concerning the increase of sea level on the German North Sea coast; *Proc. 19th ICCE*, 1984, vol. 1, ASCE, New York.

SIEFERT, W. (1989): Mean Sea Level changes and storm surge probability; *proc. XXIII IAHR/AIRH-Congress*, Ottawa.

STREIF, H. (1990): Das ostfriesische Küstengebiet - Nordsee, Inseln, Watten und Marschen; Sammlung Geologischer Führer 57, 2nd edition, Gebr. Borntraeger-Press, Berlin/Stuttgart.

STREIF, H., R. KÖSTER (1978): The Geology of the German North Sea Coast; *Die Küste*, vol. 32.

VELLINGA, P. (1987): Sea level rise, consequences and policies; report H 472 delft hydraulics, Delft.

WALTHER, F. (1934): Die Gezeiten und Meeresströmungen im Norderneyer See gat; *Die Bautechnik*, vol. 13.

WITTE, H.-H. (1970): Die Schutzarbeiten auf den Ostfriesischen Inseln; *Die Küste*, vol. 19.

The French and Belgian Coast from Dunkirk to De Panne : a Case Study of Transborder Cooperation in the Framework of the Interreg Initiative of the European Community

A. Bryche[1], B. de Putter[2], P. De Wolf[3]

Abstract

The North Sea coast between the French town of Dunkirk and the Belgian sea resort of De Panne consists of sandy beaches bordered by dunes. A number of smaller sea resorts lie between these two towns. Problems of coastal protection are similar on both sides of the border; particularly the coastline is subject to beach depletion and erosion of the dunes.

The European Community (EC) has recently taken an initiative called "Interreg" which, among other things, promotes the joint execution of transborder programmes by the member states. The initiative permits a substantial financial contribution of the Community to those programmes which meet its objectives. In the framework of this "Interreg" initiative, the French and Belgian authorities have proposed to the EC the joint execution of detailed studies of coastal protection works in the border region. The proposal has recently been accepted by the EC.

The paper first briefly describes the similarity of the coastal morphology on both sides of the border and the solutions that have been applied in the field of coastal protection in both countries. Some difficulties also had to be solved before the joint study programme could be started up: e.g. the difference in chart levels and geodetic systems used on both sides of the border. The paper finally gives a summary of the studies which will be executed in the framework of the Interreg initiative.

Introduction

Until recently, contacts between French and Belgian authorities on aspects of the struggle against coastal erosion in their border region have been rather scarce.

Scientists and technicians have been experimenting and looking for solutions separately on each side of the border, although the southern North Sea coasts,

---

[1] Ingénieur en Chef, Services Techniques Municipaux de la Ville de Dunkerque, 181 rue Militaire, F-59140 Dunkerque, France

[2] Chief-Engineer-Director, Ministry of the Flemish Community, Service of Coastal Harbours, Vrijhavenstraat 3, B-8400 Oostende, Belgium

[3] Senior Engineer, Ministry of the Flemish Community, Service of Coastal Harbours, Vrijhavenstraat 3, B-8400 Oostende, Belgium

whether Belgian or French, have a similar geomorphological structure and face the same kind of problems.

In 1990, two almost simultaneous events brought people, ideas and means together: the violent North Sea storms of 27-28 February 1990 eroded the beaches on both sides of the border (fig. 1); at about the same time, the EC was developing its incentive programme for Interregional Transborder Cooperation, the incentive consisting of financial support to joint operations on either side of the border having the same objective.

In 1990, links were established between a group of several French coastal towns (the Syndicat Intercommunal du Littoral Est de la Région de

Figure 1:   General situation of the study area
              Dunkirk - De Panne

Dunkerque [SILE]) and the Belgian Service of Coastal Harbours (Ministry of the Flemish Community) in order to discuss the common erosion problems and to initiatiate a joint study into the morphological processes and the coastal defence measures necessary to protect the coastal area. Since the mutual interest was evident, it was decided to submit a joint proposal to the EC in 1990.

The "Interreg" initiative of the European Community (EC)

The Interreg initiative was decided by the European Commission in 1990. The objective was to prepare the frontier regions of the EC countries for the opening of the inner boundaries. The initiative distinguished between three types of actions:
-    set-up of joint transborder programmes;
-    measures favouring the relation between public institutions, private associations and voluntary organisations of the frontier zones;
-    setting up joint institutional structures.

Operational programmes meeting the criteria of the Interreg initiative would be co-financed by the member states participating in the programmes and the European Community.

The French-Belgian project proposal consists of detailed studies of coastal protection works in the border region. Although each participating country is responsible for the execution of the necessary studies on its own territory, the preparation of these

studies has been carried out in joint consultation. The cooperation will be increased during the course of the studies.
The proposed project has been accepted by the guidance group of the Interreg programme. The administrative dossier of the project is now (end 1992) going through its last phases.

Coastal morphology of the French-Belgian border area

The general character of the French Flemish coastal area, i.e. the seashore between Calais and Dunkirk, is very similar to that of the Belgian Flemish coastal area. The coastline of the Flemish plain is oriented WSW-ENE; in France it stretches for 54 km from Calais onwards, and in Belgium for another 65 km (fig. 2). A morphological description of this sandy coast is amongst others given by Houthuys et al., (this volume). The coastal zone of the border area, often located some two metres below mean high water level, is protected from inundation by the sea by natural dunes and man-made constructions, usually seawalls: 46 km out of 65 km in Belgium, 18 km out of 54 km in France.

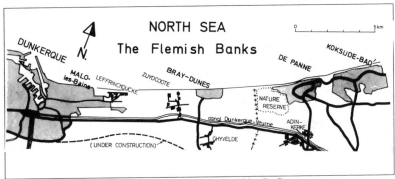

Figure 2:   Generalized map of the border area Dunkirk - De Panne.

French territory

In the French border territory there is a trend towards beach accretion in the area stretching from Calais to Dunkirk and a trend towards erosion on the 14 km stretch between Dunkirk and Belgium (Lepetit et al., 1980; Corbeau, 1991). Some of the heavy military defence works erected on the dune tops by the Germans in 1939-45 are now found on the beach, at 25 or 30 m from the foot of the eroded dune.

Throughout history, the coastal area is heavily affected by storms. On February 1st, 1953, a violent storm broke the dike of a drainage canal. At high tide, sea water entered the canal and flooded large residential districts of Dunkirk. In 1973/74, 1983 and 1987, the beaches and seawalls were damaged by severe storms. At several occasions, the seawalls were repaired at great costs.
During the violent nocturnal storms of February 27th and 28th, 1990, the whole seashore between Dunkirk and the Belgian border suffered damages of a magnitude unknown so far. The NNW winds had generated frontal waves which lowered the

backshore sand level by more than 3 m at several locations, thus altering the beach profile. Dikes collapsed and several dunes were undermined by the sea over a width of 30 m. Twenty million French francs were necessary to undertake the urgently needed protection works and the subsequent repairs.

The damage caused by the sea to the shores to the east of Dunkirk has never been a serious threat to the inhabited areas until so far. However, the beaches are strongly affected: at high tide, there is hardly any space left for the tourist. For this reason, the French local authorities have been searching for solutions since 1974.

Belgian territory

The Belgian part of the study principally aims at defining coastal protection works at the nature reserve between the French border and the outskirts of De Panne, and at De Panne (a seaside resort of 10,000 inhabitants) itself. It is however necessary to widen the study area to the east, and to include the coastal resort of Koksijde (15,500 inhabitants) so as to evaluate the mutual impact of planned coastal protection works.

Regression of the coastline west of De Panne was illustrated in 19970/75 by the fact that, during the removal of the Atlantik Wall (a military defence line built at the foredune during the second World War) several parts were found some 30 to 40 metres seaward of the 1970/75 dunefoot.
The Koksijde beach has also been suffering continuing erosion. The first seawalls were erected here in 1949 and 1951 over a total length of 188 m. After the extremely severe 1953 storm, the seawalls were extended with another 924 m between 1953 and 1957. Further extensions with 189 m of seawalls were added in 1966.

During the 1976 storm the sea breached the first dune ridge west of De Panne in four places and entered the nature reserve. Shortly after this breaching a dune foot protection was constructed, running over 1295 m all the way to the French-Belgian border. This dune foot protection was destroyed in the January 1978 storm. It was rebuilt in 1978-1979 following a firmer design and extended to a length of 2095 m. The early 1990 storms once again caused heavy damage to the dune foot protection: through a 10 m wide gap the waves could reach and erode the dunes, and immediately to the east of the dune foot protection wall a dune foot regression of 10 m occurred. As a consequence, the dune foot protection once again had to be extended further eastward so that its present total length is 2370 m.

During the recent early 1990 storms the beaches between the French border and the seaside resort of Koksijde suffered a heavy loss of approximately 270,000 m³ of sand. However, as no important storms have occurred since, the beaches have been able to recover more or less, and the remaining loss is now in the order of 50,000 m³ of sand.
The Koksijde beach has to be renourished annually in order to allow exploitation for tourism; since 1982, a total of approximately 138,000 m³ of sand has had to be supplied.

From this short overview it is clear that continuous efforts are needed to maintain the western stretch of the Belgian coastline. It is also clear that solutions for the coastline between Dunkirk and Koksijde can only endure if the solutions adopted at both sides of the border are adjusted to each other.

Solutions used so far

French territory

In the 1970s it was decided to build six breakwaters in front of Malo-les-Bains. Two of these submerged breakwaters were built in 1978. At the same time 250,000 m^3 of sand was added to the beach through nourishment. The third breakwater was constructed in 1988, accompanied by a beach nourishment of 165,000 m^3 of sand. Although the breakwaters appear to be effective for protection, they resulted in an increase of erosion at the lee-side (east). The three breakwaters planned in addition to the ones already existing have therefore not been built.

Following the storms of 27 and 28 February 1990, the "soft" solution of mechanic beach scraping has been adopted for the stretch of coast from the easternmost French breakwater to the Belgian border. During these storms, the backshore was lowered by 0.8 m near Dunkirk to 3.5 m near the Belgian border. Beach scraping resulted in the redistribution of sand on the beach: sand was replaced from a 50 m wide strip at the low waterline to a similar strip near the high waterline. In the absence of storms since early 1990, the beaches have almost attained their original profile.

Belgian territory

For an overview of coastal protection works on the Belgian coast, see De Moor and Blomme, 1988. On the particular stretch of coastline east of the French border, the main protection works included :
-     the construction of a seawall at Koksijde, from 1949 to 1966, with a total length of approximately 1,300 m. In 1966 and 1977 additional toe protections were provided;
-     the reprofiling of the top of the beach at Koksijde in 1977 and 1979 over a length of 835 m immediately west of the seawall;
-     the construction of a dune toe protection starting from the French border over a length of 2370 m in 1976, 1978, 1979 and 1992;
-     the construction of six groynes on Koksijde beach in 1956. In the same area four additional groynes were built in 1969. In 1986-88 the length of two of these groynes was increased from 120 to 400 m;
-     beach scraping combined with the supply of 138,000 m^3 of sand at Koksijde beach over the last 10 years;
-     the placing of osier hedges above the high-water mark to trap wind blown sand. They have proved to be very effective.

It is noteworthy that no seawall or dune foot protection has been built over a coastlength of approximately 2.5 km at De Panne. The beach in this area is very wide (350 m to the low-water mark) and has provided adequate protection until now.

Description of the Interreg project

The Interreg project defines morphological studies and evaluations in the French as well as the Belgian border area. Both countries are responsible for the studies in their own territories.

The French and Belgian studies will receive a 50% funding from the European Community. In France the remaining part is financed jointly by the national government and the by the local authorities through the "Communauté Urbaine de Dunkerque" (CUD) and SILE. The Belgian study is financially backed by the Ministry of the Flemish Community.

French study programme

The French study, with an estimated cost of 7,000,000 French francs, started in september 1991 and will last for 4 years. It will cover the coastal stretch between Dunkirk and the Belgian border and will include the dune, beach and shallow shoreface areas (to a water depth of -15 to -20 m).

The study consists of four main phases:

- Phase 1 (1991-92) includes the collection, representation and analysis of already existing data (information on morphology, hydro-meteorology, sedimentology), as well as the establishment of a numerical wave refraction model and of a sediment budget for the Dunkirk coast;
- Phase 2 (1992-1994) involves field surveys relating to the waves and currents situation, sediment, coastal retreat, fore- and backshore morphology;
- Phase 3 (1994) consists of an integrated mathematical model study, following the analysis, synthesis and evaluation of the existing and measured data;
- Phase 4 (1994-95) will include, if this is deemed necessary after Phase 3, a physical model study of selected areas of 2 to 3 km wide. The simulations will allow to refine the design of the works and to examine local impacts.

Belgian study programme

In 1987 a study was commissioned by the Belgian authorities to define adequate measures to fight erosion at Koksijde beach. This study led to the following proposal (Verslype et al., 1990): (1) to dredge the northern flank of a tidal channel situated close to the shore and deviate the strong flood currents responsible for the offshore erosion; (2) to re-use the dredged sand for a beach nourishment at Koksijde; (3) to lengthen existing groynes.

The results of this particular study will of course have to be taken into account in the more integrated study executed in the framework of the EC Interreg project.

Since 1983, a detailed monitoring programme of the beach and dune evolution of the westernmost coast has been commissioned by the Belgian authorities (De Wolf et al., this volume). The monitoring programme includes:

- a twice yearly photogrammetric survey of foreshore, backshore and dune foot area;
- an airborne remote sensing based survey of dune topography and vegetation every three years;

- since 1986, a yearly hovercraft-based bathymetric survey of nearshore and adjoining sea floor, carried out using the BEASAC® (Belfotop Eurosense Acoustic Sounding Air Cushion) platform.

The programme further includes offshore hydro-meteorological measurements from a number of measuring piles and wavebuoys; moreover, since several years a great number of current measurements have been performed along the Belgian coast.

A lot of information is thus already available at the start of the Interreg study. The cost of the Belgian part of the study is estimated at approximately 9 million Belgian francs. The study will consist mainly of three parts:

- coordination between the Belgian and French studies;
- cooperation with the French studies and exchange of relevant measurement data;
- design of coastal protection works at De Panne and the nature reserve west of De Panne.

The problems of geodetic and altimetric reference systems

It is obviously very important that the bathymetric soundings and the topographic surveys planned in the Belgian and French studies can be compared; actually the Belgian and French geodetic and altimetric reference systems are totally different.

A preliminary matching has been performed during the first months of 1992 by a team of Belgian and French specialists. By convenience, it has been decided to choose an international projection system, i.e. the UTM system (Universal Transverse Mercator).

For altimetry, a direct high-precision levelling has been performed between Belgian and French benchmarks. This allowed to determine exactly the offsets between the different datums, and the Belgian levelling system has been used for the sea bottom levels during the bathymetric soundings.

Among other things, it was necessary to establish an initial sea bottom map from De Panne to Dunkirk. This task has been performed by the BEASAC® platform, which is currently used to survey the Belgian nearshore.

Conclusion

The Interreg initiative of the European Community has provided a powerful impulse in bringing together the coastal regions situated at both sides of the French-Belgian border to coordinate efforts in their fight against the sea. The studies executed by the Belgian and French partners will permit in the short and middle long run to define an efficient strategy for coastal protection in the border region. In the long run these studies will be the basis for an ongoing cooperation under the form of a permanent exchange of information and scientific data, and of the coordination of future works and investigations.

References

Augris, C., Clabaut, P., Vicaire, O., 1990. Le domaine marin du Nord - Pas de Calais. Nature, morphologie, mobilité des fonds. *Ed. by IFREMER,* Département Géosciences marines (BP 70, F-29280 Plouzané) and Université des Sciences et Techniques de Lille, UFR Sciences de la Terre (F-59655 Villeneuve d'Ascq Codex).

Corbeau, C., 1991. Bilan sédimentaire pluri-décennal du littoral dunkerquois. Mémoire présenté à l'Université de Paris Sud (Orsay).

De Moor, G., and Blomme, E., 1988. Shoreline and Artificial Structures of the Belgian Coast. In: Walker, J., (ed.). *Artificial structures and coastlines.* Dordrecht, Kluwer, Geo-journal Library, 1988, pp. 115-126, 1 fig.

De Wolf, P., Fransaer, D., Van Sieleghem, J. & Houthuys, R., 1993. Morphological Trends of the Belgian Coast shown by 10 Years of Remote-Sensing based Surveying *(This volume).*

Houthuys, R., De Moor, G. & Somme, J., 1993. The shaping of the French-Belgian North Sea Coast throughout recent Geology and History *(This volume)*

Kerckaert, P., Maes, E., Fransaer, D., Van Rensbergen, J., 1990. Monitoring Shore and Dune Morphology using Bathymetric Soundings and Remote Sensing Techniques. Osaka, *27th International Navigation Congress,* 1990, Section II, Subject 2, pp. 89-95.

Lepetit, J.P., Clique, P.M., Thellier, P., Tranchant, M., 1980. Catalogue sédimentologique des côtes françaises. Tome 1: De la baie de Somme (exclue) à la frontière belge. *Ministère des transports, Electricité de France, Direction des Etudes et Recherches* (6 Quai Watier, F-78400 Chatou).

Verslype, H., Blomme, E., Decroo, D., 1990. Sediment Transport, Beach Nourishment and the Effect of Marine Works on the West Coast of Belgium. Osaka, *27th International Navigation Congress,* 1990, Section II, Subject 2, pp. 7-10.

Engineering Tools and Techniques
for Coastal Zone Management

J. van der Weide[1], F.M.J. Hoozemans[1]

Abstract

In coastal zone development schemes, coastal engineers are confronted with a changing attitude towards human interventions in the coastal zone. Traditionally civil engineering has been applied to support socio-economic activities of man and to protects its life and property against natural hazards. With the increased risk of flooding this task will be important also for the decades to come. However, gradually it became apparent that the effect of human interventions may cause irreversible damage to the natural coastal system. A more flexible response is needed for the coastal zone manager, especially within the concept of a sustainable development. As the coast is a common resource there is a growing awareness that we need to apply standards of sustainable resource allocation and use of this area. This can only be achieved through well-balanced coastal zone management. Coastal zone management aims at solving present and future problems in the coastal zone, by finding an acceptable balance between economic welfare and environmental well being, using a careful analysis of the natural processes and socio-economic developments. Coastal engineering should therefore not only protect man against natural hazards but also support coastal planners in the socio-economic development of the coastal zone and safeguard the natural environment through flexible response to changes.

Current coastal engineering technology should reflect these changing needs;
- *Coastal engineering interventions should be based upon a sound understanding of the processes, active in the coastal zone. The increasing scale of the projects and the complexity of the processes call for a closer cooperation between earth scientists, such as geologists, geomorphologists and hydraulic engineers. Such a cooperation has been established in The Netherlands, within the framework of the Coastal Genesis project;*
- *Coastal engineering interventions should be designed, using new and flexible techniques which can satisfy the needs of both man and nature. Building 'with' and even 'for' nature and dynamic preservation of coastlines by beach nourishment are typical examples of this approach;*
- *Coastal engineering interventions should be implemented as part of coastal zone management, using a thorough understanding of this impact on the existing socio-economic system and natural environment.*

---

[1] Delft Hydraulics, P.O. Box 152, NL8300 Emmeloord, The Netherlands

History of coastal zone management in the Netherlands

The coastal zone, the river delta's and estuaries have been inhabited by man, from the early days of its existence. The sea was a major source of food, the coastal plain offered space for human settlement and related economic activities and the estuaries and rivers could be used for transport and communications with the hinterland. For that reason, many civilisations found their origin in deltaic areas and it is expected that by the year 2000 some 80% of the largest human settlements will be found in the coastal zone. Obviously, these developments have a significant impact on the autonomous natural developments in the coastal zone, which calls for an integrated coastal zone management.

Coastal zone management aims at solving present and future problems in the coastal zone, by finding an acceptable balance between economic welfare and environmental well-being, using a careful analysis of the natural processes and socio-economic developments.

The Netherlands, one of the most densely populated countries in the world, is an example of such a development. Located in the delta of the rivers Rhine, Meuse and Scheldt, it got its present shape through the interaction between river sediments and coastal processes, the wind, waves, tides and the resultant littoral drift.

The social and economic activities in the Netherlands were strongly hampered by the fact that most of the coastal plains were liable to flooding during high tides. Naturally, even nowadays, about 50% of the Netherlands is lying well below Mean Sea Level.

Presently some 250 km of the Dutch coast consists of sandy beaches and dunes. The remaining 100 km is artificially protected by dikes (fig. 1).

From the earliest time the Dutch have tried to defend their land against the onslaught of the sea. Initially, in a passive way by building artificial hills, known as terps, but later-on in a more active way by building dikes to safeguard life and property. Actually, the first treaty on dike construction was written in Holland in 1579, by Andries Vierlingh. He gave valuable advise how to plan and construct

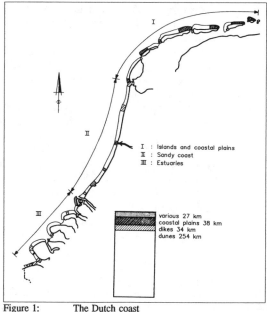

I : Islands and coastal plains
II : Sandy coast
III : Estuaries

various 27 km
coastal plains 38 km
dikes 34 km
dunes 254 km

Figure 1:   The Dutch coast

dikes, using his long-life experience as a practical coastal engineer.

But often the Dutch coastal system appeared inadequate to resist the forces of nature. Storms and high tides in the sea and floods from the rivers caused numerous breakthroughs, resulting in loss of life and land. In order to improve their coastal defences, the Dutch approached the sea in a more offensive way. Supported by research and development and with the help of modern civil engineering technology, they reduced the length of their coastline. Starting with the closure of the Zuiderzee in 1927 and finalizing with the Eastern-Scheldt storm-surge-barrier in 1988, a series of estuary-closures was completed. The remaining coastal defences and dunes were reinforced to be able to withstand a 1/10,000 year design condition. Holland's coastal infrastructure is now ready to cope with the challenges of the 21st century.

And obviously, these challenges are numerous. Due to the effect of climate changes, the sea level in the North Sea may rise some 0.6 m in the next century. This will cause an intensified attack on the Dutch coast, which now already suffers a net sediment loss of more than 5 million m³ per year.

Also the effect of the construction of the enclosure dams and the impact of port expansions at Rotterdam and IJmuiden are now being felt. The natural sediment transports has been affected, creating areas with erosions and areas where new coastal plains and offshore bars are being formed. Combined with the growing concern with respect to the quality of the environment, these problems have stimulated the development of an integrated coastal management policy with the objective:

- to develop an integrated approach to the coastal zone and its hinterland, including old/new land, taking into account the various social and economic functions with their respective physical infrastructure;
- to use this approach when maintaining existing or creating new land and/or new infrastructural elements, by making use of materials and processes present in nature.

The above developments have strongly influenced coastal science and technology in the Netherlands. Starting as a design orientated effort, it has now been developed into a tool for integrated coastal management, aiming at a balance between economic welfare and social well being.

It is now widely accepted that such a controlled development of the coastal zone should be based upon a proper understanding of the processes in the coastal zone, supported by a sound engineering technology and socio-economic skills to obtain an acceptable balance between short term benefits and long term assets.

In summary, Coastal Engineering of the nineties should:

- support socio-economic development;
- protect life and property against natural hazards;
- safeguard the natural environment.

The coastal environment

Boundaries

There is no generally accepted definition for the coastal zone. The 1982 United Nations Conference on the Law of the Sea has delineated various juridical and legislative zones as shown in figure 2. In this context coastal science and engineering is confined to the territorial waters. For practical application, a further refinement is

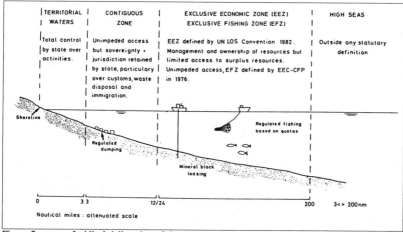

Figure 2: Juridical delineation of the coastal zone

needed, which is often based upon uniformity of the physical and ecological conditions. Boundaries of the study area are selected on a case by case basis, as studies can be performed at a variety of spatial and time scales. Boundaries of the system under consideration and the accuracy of the description of the relevant processes should be selected accordingly.

The longshore dimension of the coastal system is dictated by the requirement that disturbances generated by activities within the system should not propagate beyond the system boundaries for the time-span under consideration.

This may result in system dimensions in the order of 1000 km for long term processes. Coastal engineers normally involved in smaller scale systems, with dimensions in the order of a few to some hundred kilometres. The present paper will be restricted to these applications. A schematical picture of such a system is shown in figure 3.

If the impact of the system on water quality and ecology has to be studied, a larger scale coastal system should be used for the analysis. In that case the study often has to be extended even beyond the territorial waters, which calls

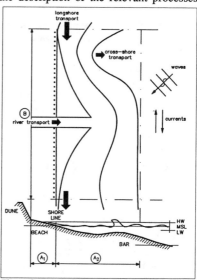

Figure 3: Schematization of the coastal system

for an international approach. This paper is limited to nearshore coastal engineering aspects, however.

The seaward boundary of the system is often located at a depth, where the contribution to the coastal sediment budget is low. This means areas where both cross-shore and longshore transports are small. This is normally the case for depths in the order of 25 to 30 metres. If deep offshore canyons are present their effect should be included as a sediment sink. Obviously, appropriate hydraulic boundary conditions have to be defined at the seaward boundary, to account for the effect of waves and currents, active in the larger coastal system.

The landward boundary is generally selected as the high-water mark. In areas where coastal plains and dunes contribute to the sediment budget, their effect should be introduced as a boundary condition. The same applies for sediment transport from rivers or estuaries, which should be introduced as a sediment source. Again in those cases where water quality is at stake, the total catchment areas of rivers, should be fully included.

Once the system boundaries have been defined, the physical features of the area have to be described by means of topographic maps of the terrestrial zone and bathymetric charts of the marine area. Remote Sensing and aerial photographs offer convenient possibilities to collect the required data. Modern GIS techniques and related DBM systems have provided new possibilities to facilitate the storage, processing and retrieval of such data.

The elements of the marine part of the coastal zone

This paper will focus on the processes in the marine part of the coastal zone. The basic elements of this area are:
- The air, including dissolved gasses and pollutants;
- The water, including dissolved matter, which may be described by it chemical, physical and resultant biological properties;
- The sediments, characterized by their physical, mineralogical and chemical properties and the related hydrodynamic and geotechnical parameters, such as fall-velocity, critical shear stress etc.;
- The marine life, characterized by the type and quantity of the various species.

In general, the properties of basic elements as such are used to monitor the ecological conditions in the coastal area. They are further used as an input for the description of the processes active in the coastal zone.

The processes in the coastal waters

The processes in the coastal waters are complex, often interactions between two or more processes have to be taken into account. In general following types of processes can be identified:
- aerodynamic processes, such as air-sea interaction, and aeolean (wind) transport of sediments;
- hydrodynamic processes, such as waves, tides and resultant water levels and currents;

- morphodynamic processes, such as sediment transport and related changes in the bathymetry and shoreline geometry;
- geodynamic processes, induced by geotechnical instabilities such as subsidence, earthquakes, liquefaction, sliding etc.;
- ecodynamic processes, describing the resultant changes in the ecosystem due to any or all of the foregoing processes and/or elements.

The various processes and their interactions have been described schematically in figure 4.

In this paper, no further detailed descriptions will be given of these processes. Reference is made to relevant text books, such as for instance Carter (1988).

The ecology of the coastal zone

The importance of the coast in ecological terms is well described in Carter, 1988: Coral reefs, salt marshes and wetlands are sensitive to change in water quality. For these coastal formations further insight in the ecology of the area is a must.

Additional functions of coastal ecosystems include the critical role of buttering the joining of land to sea by moderating coastal erosion and large salinity changes.

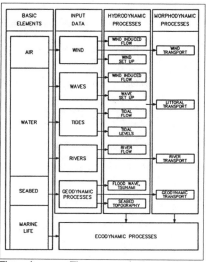

Figure 4:   The processes in the coastal system

Coral reefs, for example, act as barriers dissipating wave and current energy, and mangroves, by occluding and absorbing terrestrial freshwater run off, buffer salinity changes in coastal waters. Degradation of these ecosystems may, therefore, have a serious impact on the coast.

In particular reefs and coastal wetlands are of international importance. The IUCN/UNEP has made a directory of wetlands, in which some 730 thousand square kilometres of valuable areas are listed. The geographical distribution is shown in the followting table, emphasizing the importance of the tropical countries.

Nations who have accepted and signed the Ramsar convention have committed themselves to protect and safeguard these areas.

	Area of Wetlands in km²	Wetland as % of total country area
1. North America	32330	1.639
2. Central America	25319	0.882
3. Caribbean Islands	24452	9.431
4. South America Atlantic Ocean Coast	158260	1.132
5. South America Pacific Ocean Coast	12413	0.534
6. Atlantic Ocean Small Islands	400	3.287
7. North and West Europe	31515	0.713
8. Baltic Sea Coast	2123	0.176
9. Northern Mediterranean	6497	0.609
10. Southern mediterranean	3941	0.095
11. Africa Atlantic Ocean Coast	44369	0.559
12. Africa Indian Ocean Coast	11755	0.161
13. Gulf States	1657	0.079
14. Asia Indian Ocean Coast	59530	1.196
15. Indian Ocean Small Islands		
16. South-east Asia	122595	3.424
17. East Asia	102074	0.999
18. Pacific Ocean Large Islands	89500	19.385
19. Pacific Ocean Small Islands		
20. USSR	4191	0.019
TOTALS	732921	0.846

Source: "A global survey of coastal wetlands, their functions and threats in relation to adaptive responses to sea level rise", paper by Dutch Delegation to IPCC-CZM Workshop, Perth, Australia, February 1990. Directories of Wetlands, issued by IUCN/UNEP (1989-90), 120 countries, excluding amonst others Australia, Canada, New Zealand, USA.

Interaction between man and nature

Effects of man-made interventions
Coastal resources -space, living and non-living resources and energy- are used by man, for a variety of socio-economic functions. Most of these activities are supported by a physical infrastructure and controlled by governmental bodies.

Most natural systems are in a transient state of erosion or accretion. Historic records of coastline evolution and sediment budget analysis can be used to assess the evolution stage of the area under consideration. Few coastal areas are stable, and even these stable coastal systems may change, due to changing boundary conditions, induced by nature or man. Tectonic motions and vulcanism, changing rainfall patterns and related erosion and sediment transport, sea level fluctuations are well known in this respect. But also man-made interventions can have far reaching effects. Deforestation and urbanization in river catchment areas, river dredging, dam construction and/or river regulation, all are examples of human interventions inland which will also affect the coastal zone in the long term.

Human activities in the coastal zone proper, will have a more direct impact. Well known examples are mangrove deforestation, mining of sand and coral, coastal engineering works, urban development and related sewage disposal, fish-farming etc. The impact of these activities on the coastal system is shown schematically in figure 5.

TRIGGER	IMPACT ON	MORPHODYNAMICS	
		HYDRODYNAMICS	ECODYNAMICS
Natural			
• climate change		●	○	●
• tectonic activity		○	●	○
• sea-level change		●	○	●
Human intervention upland			
• deforestation		○	●	○
• river regulation		●	●	○
• dams		●	●	●
• oil, groundwater extraction		○	●	○
Human intervention coastal zone			
• mangrove deforestation		○	●	●
• sand and coral mining		○	●	●
• marine works		○	●	○
• fish farming		○	○	●
• cooling water out fall		○	○	●
• sewage out fall		○	○	●

● primary impact   ○ secondary impact

Figure 5: Triggers for coastal problems

In order to quantify these impacts, the interaction with the coastal processes should be known in more detail.

This knowledge is also essential when the effect of nature on human intervention has to be quantified in order to define,for example, down-time for marine operations workability and design loads for marine structures.

Most common problems associated with the description of these coastal processes and the interaction with man-made interventions are shown in figure 6.

The coastal engineer should be able to identify these problems, to evaluate and quantify its significance and to develop solutions which are technically sound, economically feasible and socially acceptable.

Tools and techniques for problem solving

In-situ measurements and analysis of historical records of hydrodynamic and morphodynamic data can give valuable insight in the governing autonomous natural processes and trends. If no major human interference is foreseen, these trends may be extrapolated to characterize the natural processes for the years to come.

	HYDRODYNAMICS	MORFO DYNAMICS	ECO DYNAMICS
NATURAL SYSTEM	• tides • waves	• sediment transport • coastal morphology	• energy flux • nutrient cycle • biomass production
IMPACT OF STRUCTURES	• wave reflection	• shoreline dynamics • dredge material deposits	• pollution, turbidity • water quality • bio diversity
DESING OF STRUCTURES	• flow pattern • wave penetration • navigation • structural design	• accretion, erosion • siltation in dredged channels	• organic deposids • marine growth • marine corrosion

Figure 6:   Coastal problems and relevant processes

However, in the event that conditions will change due to human interference or changing natural conditions, these effects cannot be predicted from trend analysis. In those cases models are used, which are capable of describing the natural processes and their interaction with human interferences.

In cases where governing equations are known, but that are difficult to solve mathematically, physical models are used to simulate the processes. When appropriate scale-laws are used, model results can then be converted into full-scale values.

If large sets of experimental results are available empirical relations can often be obtained, which can be applied more universally. Stability formulae for breakwater design are good examples of this type of solution technique. If the governing equations are such that they can be solved numerically, mathematical models can be applied.

Selection of appropriate methods

The selection of the solution technique is determined by technical and economic considerations. First of all, the method should be technically feasible, which means that the problem can be formulated mathematically and that numerical techniques are available to solve that class of mathematical equation. The decision whether or not a method is acceptable depends largely on the required degree of accuracy of the solution, which is dictated by the project. Large-scale, complicated projects in high risk areas will need a higher degree of accuracy, than small and easy jobs in relatively unexposed areas.

However, improvement of the quality will be reflected in the cost of the solution technique. It will often pay off to use a sophisticated technique for costly projects but such a technique may not be economically justified for small projects.

Using these two parameters as input -the complexity of the problem and the cost of the project- the range of applicability of various tools is indicated in figure 7.

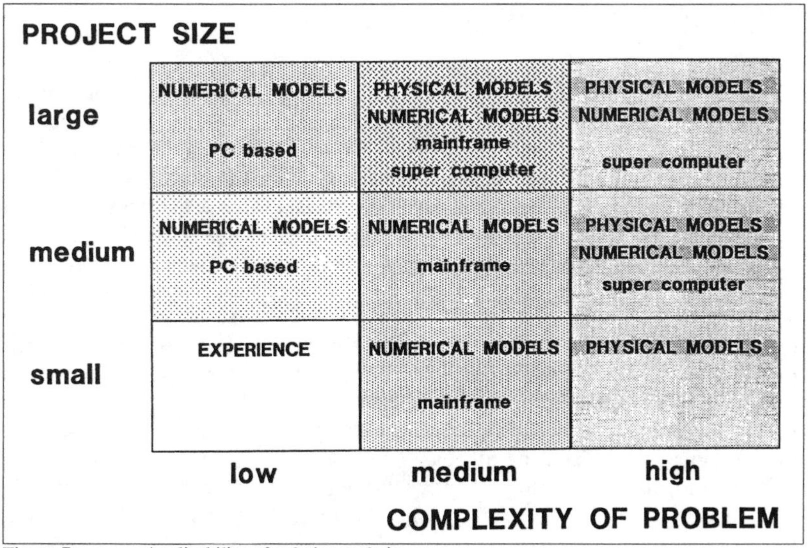

PROJECT SIZE			
large	NUMERICAL MODELS PC based	PHYSICAL MODELS NUMERICAL MODELS mainframe super computer	PHYSICAL MODELS NUMERICAL MODELS super computer
medium	NUMERICAL MODELS PC based	NUMERICAL MODELS mainframe	PHYSICAL MODELS NUMERICAL MODELS super computer
small	EXPERIENCE	NUMERICAL MODELS mainframe	PHYSICAL MODELS
	low	**medium**	**high**
			COMPLEXITY OF PROBLEM

Figure 7:   Applicability of solution techniques

Research in physics and ecology, and developments in computer hardware and software will strongly expand the range of applicability of numerical models, in particularly P.C. based options.

A state of the art review of coastal modelling was given by O'Connor (1991), in a key note address to the Conference "Coastal Engineering in National Development". He concludes:

> *A better understanding of the physics and biology of the coast will reduce the negative impact of human interference with nature.*
>
> *Natural processes can even be used to the benefit of mankind, through building with nature. Models can be a usefull expedient to optimize this concept, as a response to the needs of the nineties.*
>
> *Various types of computer models have been developed over the last decade to describe tide, wave and sediment movement in coastal regions. Both tide and wave models have advanced to the point where they can be used routinely to solve engineering problems. The next decade will see further advances, particularly in the development of 2DP wave models and 3D wave-induced current models. Morphological models have also developed rapidly but they are less reliable than tide and wave models due to the complexity of sediment transport processes. 1D (profile and one-line), 2DP and 3D models exist and have been used in engineering design for both sandy and muddy coastlines. However, acceptable answers can only be obtained, if use is made of considerable amounts of field data. The next decade is likely to see the emergence of more reliable sediment models as experience with existing models and further basic research at laboratory and field scale improves understanding of coastal processes.*

The design of marine structures

General

There is a large difference in the design-methodology for marine and coastal structures, compared to the procedures applied for the design of land-based structures. For land-based structures, operational loads and gravity-forces determine to a large extent the design-loads. Environmental forces such as wind-thrust and hydraulic forces are in general less important. Since operational loads and gravity-forces can be computed accurately, a deterministic design concept can be applied. This means that the structure is designed for a series of well defined loading-conditions.

Marine and coastal structures, however, are located in an environment which determine to a large extent both the operational loads and the design-loads. Since these loads cannot be computed accurately, they are defined in statistical terms. Design-conditions are therefore specified in terms of probabilities, and a probabilistic design concept is applied.

In this paper the principles of coastal engineering will be treated in a global way, using information available in textbooks. As a base-line for the presentation the Shore Protection Manual is used. Suplementary information on the various topics, may be found in more specialized textbooks, such as Pilarczyk (1990), Short and Blair (1986) and Van der Weide (1991).

When designing coastal structures, the following aspects have to be considered:
-    *The function of the structure*
     Coastal structures are used for many purposes, such as high-water protection, coastal defence, wave-attenuation, flow-guidance etc. For each of these functions different functional requirements have to be met and consequently different design-specifications have to be formulated;
-    *The physical environment*
     As stated above, the design-loads on coastal structures are mainly determined by environmental forces such as wave loads and drag forces. Moreover, hydrodynamic processes may induce morfological changes which may affect the performance and the integrity of structures. Finally, the geotechnical characteristics of the site are of importance as they determine the bearing-capacity of the subsoil, an important parameter for the overall-stability of the structures;
-    *The construction-method envisaged*
     The availability of material, equipment and labour determines to a large extent the construction procedure. This procedure in its turn determines the possibilities of the designer. In those areas where there are severe constraints in construction-procedures the designer should be aware of this and should revise his design accordingly;
-    *Operation and maintenance*
     Once the structure is completed, maintenance will be required to secure its performance. The designer should be aware, therefore, of the maintenance-procedures and should take care that maintenance is possible.

A framework for the design

These various aspects, however, can not be studied independently. In order to make a well balanced design, all aspects have to be integrated in a framework, which describes the various aspects and the mutual interaction between these aspects (fig.8).

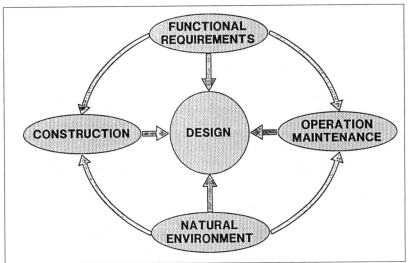

Figure 8:   The framework for design

The natural environment is first described in terms of hydrodynamics and morphodynamics. Subsequently, the various functions of coastal structures are defined and the corresponding functional requirements are formulated in terms of objectives and performance criteria. Finally, procedures are outlined for the conceptual design and the detailed engineering, taking into account the elements of the design-framework. This approach is further explained in the next sections.

The design process

Ideally, the above aspects should be considered during all the stages of the design-process. During the design, following stages are identified (fig. 9):

- *Conceptual design*

    In this stage a number of different alternatives are generated, which all meet the functional requirements. Designs are general and only main dimensions are given. At this stage the relevant aspects should be identified, which determine the technical and economical feasibility. Their relative importance should be evaluated and rough figures should be obtained to quantify them;

- *Preliminary design*

    In this stage, a limited number of alternatives is selected after a screening procedure which focusses on the technical feasibility. If required the conceptual stage should be repeated when the first set of alternatives does not meet these criteria. At this stage, structural dimensions are quantified in some detail and a check is made on the economic feasibility. Again the design process may have to be repeated when the project becomes economically unfeasible;

- *Detailed engineering*
At this stage the structural details are designed and detailed design drawings are made. Parallel to this the possibilities for financing are explored, and environmental and socio-political aspects are considered in greater detail. Technical, economical and socio-political criteria are used in this stage to screen the feasibility of the proposed design;

- *Construction*
When the design stage is completed, tender documents are prepared and the tendering procedure is started, resulting in the construction.

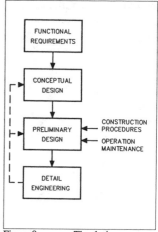

Figure 9:   The design process

Although this stage is not part of the actual design-process, the designer should be aware of the constructional constraints imposed by this stage.

Availability of material, accessibility of the area and the corresponding limitation in the use of equipment, feasibility of intermediate construction stages are some of the aspects to be considered. Especially in the detailed engineering stage these aspects should be evaluated;

- *Operation and maintenance*
Again this stage is beyond the actual design-process, but should be considered by the designer. Especially maintenance can have an impact on the selection of the design. When possibilities for maintenance are poor, the initial design should be such that the structure can operate without regular maintenance. If local labour-cost is cheap and capital investments are difficult, a cheap structure may be more appropriate, when regular maintenance can be guaranteed.

Conceptual design

General
As stated in the previous chapter, the design should be based upon the functional requirements taking into account the environmental conditions in the project area and giving due regard to constructional aspects, operation and maintenance.

The functional requirements
The primary function of coastal structures is to protect the hinterland against the adverse effect of high-water and waves. If high-water protection is required the structure should have a height well above the maximum level of wave-uprush during storm-surges. This normally calls for high crest elevations.

If, however, some overtopping is allowed in view of the character of the hinterland, the design requirement is formulated in terms of the allowable amount of overtopping. In the netherlands a value of 2 liters per second per running metre of dike is accepted for instance. Obviously crest-elevations can be reduced considerably in this case.

Another function of the structure is to improve operational conditions.

For structures, such as breakwaters, where wave-reduction is the main objective, a further reduction in crest-height can be applied. Wave-heights due to transmission and overtopping should be negligible during operational conditions but may reach va-lues of 0.5 metres in extreme design conditions. If flow guidance is the main objec-tive, training walls are mainly used to direct flow. The crest-elevation is mainly determined by constructional aspects, which implies that a minimum level of 1 metre above mean high water should be applied to guarantee an uninterrupted progress of work. Wave overtopping during operational and extreme conditions are of less con-cern in this case.

Selection of the structural concept

The selection of the structural concept depends on the function, the local environmental conditions and the constructional constraints. The governing criteria are the technical and economic feasibility for the project under consideration. A review of possible concepts is shown in figure 10.

Basically, a simple sheet-pile wall will be sufficient to provide the required height. Such a concept is feasible only in small water-depths with moderate wave-action. In deeper areas the coffer-dam concept has to be applied, which is more complicated to build, especially in areas with frequent wave-agitation. Another method to stabilize the sheet-pile is the use of anchors. All methods are particularly usefull for slope and bank-protection in waters which are well protected against waves. Also for flow guidance the method is feasible. Vertical-face structures can also be constructed by means of gabions, block type dams, or caissons. In this case the stability is derived from the weight of the structure, which are therefor called gravity-type structures. Often, the performance of such structures in terms of overtopping, is improved by using a parapet or crown-wall. Gravity-type structures may be used in moderate to large water-depths, provided that no breaking waves occur. Due to the potential foundation problems as a result of dynamic wave-loading, such conditions should be avoided. This implies that the water-depth under design conditions should be at least twice the design wave-height. Obviously, the construction, the transport and the positioning of caissons requires knowledge and experience, which makes the method often impractical. Especially in areas where the weather-windows for construction are small, the caisson concept could be a good solution.

In many areas in the world wave-heights and foundation conditions are such that no gravity-structures or sheet-piles can be used. Sloping structures are a solution then, as wave-loads on these structures are more easily accounted for. Moreover, foundation-loads are more evenly distributed and differential settlements can, to a certain extent, be accepted. The slope-type structure is most widely used since it is a versatile concept which can be constructed also by less-experienced contractors.

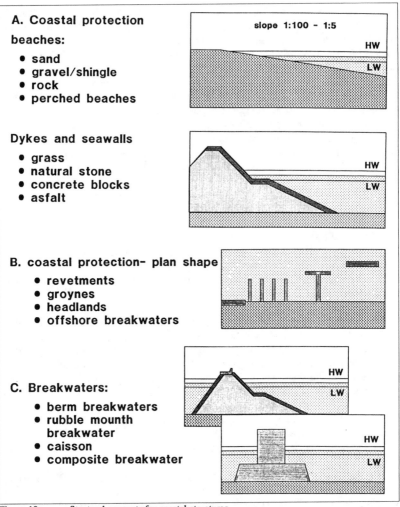

Figure 10:   Stuctural concepts for coastal structures

The structure can be used in moderate to deep water, and can be designed to withstand severe waves. With the increasing size of the structures, however, the limits of the application are more or less reached. Stability of the armour layer and geotechnical stability of the core are points of concern, particularly for larger breakwaters.

Tools and techniques for problem solving

In the conceptual design stage, the performance of the structure is characterized by the amount of water which passes the crest or by the wave height transmitted through the structure. Generalized design graphs are used to quantify these parameters. Typical examples are given in Carter (1988), CUR (1991) and CUR (1987).

References

Carter R.W.G., 1988, Coastal environments. An introduction to the Physical, Ecological and Cultural Systems of Coastlines. London Academic Press.

Manual on the use of rock in coastal and shoreline engineering, 1991. CIRIA-CUR, Gouda.

Manual on artificial beach nourishment, 1987. CUR-Rijkswaterstaat-Delft Hydraulics, Gouda.

O'Connor, B.A., Coastal Modelling. Key Note address. *Conference Coastal Engineering in National Development. Kuala Lumpur.*

Pilarczyk, K.W., 1990, Coastal Protection. A.A. Balkema - Rotterdam.

Shore Protection Manual, 1984. Coastal Engineering Research Center. Dept. of the Army. Waterways Experiment Station Corps of Engineers, P.O. Box 631, Vicksburg (Miss).

Short, N.M. & Blair Jr., R.W. (editors), Geomorphology from Space. Chapter 6. Coastal landforms. NASA Scientific and Technical Information Branch, Washington DC, 1986.

Van der Weide, J., 1991, Tools and techniques to solve coastal engineering problems of the nineties. Delft Hydraulics, Delft, November, 1991.

SUBJECT INDEX
Page number refers to first page of paper

Beach nourishment, 41, 65, 188, 258
Beaches, 267, 336
Belgium, 245, 258, 288

Case reports, 277, 298
Coastal engineering, 131, 298, 344
Coastal management, 1, 27, 52, 145,
    178, 188, 202, 214, 227, 233, 314,
    344
Coastal processes, 27, 41, 65, 85, 96,
    245
Coastal structures, 267
Comparative studies, 162
Contaminants, 214
Cooperation, 1, 202

Data collection, 110
Denmark, 227, 267
Design criteria, 258
Disposal, 288
Dredging, 277, 288

Ecosystems, 178, 344
Environmental impacts, 52
Environmental surveys, 162
Equilibrium state, 96
Erosion control, 233, 267, 298, 336
Estuaries, 277
Evolution, development, 14, 52, 178,
    245
Examination, 233

Fisheries, 214
Flooding, 314

Geology, 14, 27
Germany, 314
Great Britain, 233

Harbors, 277
History, 14, 27, 85
Human factors, 65, 85, 178, 214, 344

Input, 162
International compacts, 1
International development, 202, 336

Maps, 14
Mathematical models, 96
Modeling, 96
Models, 52
Monitoring, 41, 245
Morphology, 1, 41, 52, 65, 85, 96, 245,
    336

Netherlands, 188
North Sea, 1, 14, 27, 41, 85, 96, 110,
    131, 145, 162, 178, 202, 214, 227,
    298
Numerical models, 131

Oceanography, 131

Planning, 344
Policies, 145, 188, 227
Pollution control, 145
Project evaluation, 258

Radioactive tracers, 288
Recreation, 227
Research and development, 288
Rivers, 233

Sedimentation, 27
Shore protection, 65, 188, 202, 245, 258,
    298, 314, 336
Silts, 277
Statistical analysis, 110

Trends, 110, 162, 267, 314

Water levels, 110
Water quality control, 145
Wave measurement, 131

AUTHOR INDEX
Page number refers to first page of paper